Wood Protection and Preservation

Wood Protection and Preservation

Special Issue Editor

Christian Brischke

MDPI • Basel • Beijing • Wuhan • Barcelona • Belgrade • Manchester • Tokyo • Cluj • Tianjin

Special Issue Editor
Christian Brischke
University of Goettingen
Germany

Editorial Office
MDPI
St. Alban-Anlage 66
4052 Basel, Switzerland

This is a reprint of articles from the Special Issue published online in the open access journal *Forests* (ISSN 1999-4907) (available at: hhttps://www.mdpi.com/journal/forests/special_issues/wood_protection_preservation).

For citation purposes, cite each article independently as indicated on the article page online and as indicated below:

LastName, A.A.; LastName, B.B.; LastName, C.C. Article Title. *Journal Name* **Year**, *Article Number*, Page Range.

ISBN 978-3-03936-332-2 (Hbk)
ISBN 978-3-03936-333-9 (PDF)

Cover image courtesy of Christian Brischke.

Contents

About the Special Issue Editor

Christian Brischke studied Wood Science at the University of Hamburg. After several years of research at the Center of Wood Science and at the Federal Research Centre for Forestry and Forest Products, he finished his doctoral studies in 2007, with a thesis entitled "Investigation of decay influencing factors for service life prediction of exposed wooden components". Between 2007 and 2017, Dr. Brischke worked at the Institute of Vocational Sciences in the Building Trade at Leibniz University Hannover, and was granted the Venia legendi in 'Wood Technology' in 2016. Since April 2017, he has been working at the Department of Wood Biology and Wood Products at the University of Goettingen. The focus of his research is on wood protection, wood modification, durability and moisture performance of wood and wood-based products.

Editorial

Wood Protection and Preservation

Christian Brischke

Wood Biology and Wood Products, University of Goettingen, Buesgenweg 4, D-37077 Goettingen, Germany; christian.brischke@uni-goettingen.de

Received: 9 May 2020; Accepted: 12 May 2020; Published: 13 May 2020

Abstract: Wood is an advantageous building material in many respects, but it is biodegradable and therefore requires protection when used in highly hazardous applications. This Special Issue on 'Wood Protection and Preservation' comprises 19 papers representing a wide range of aspects related to the field and gives timely examples of research activities that can be observed around the globe.

Keywords: biological durability; decay fungi; wood borers; wood protection by design; wood preservatives; chemical wood modification; thermal wood modification; water repellants; test methods; service life planning; performance specification

---

Globally, the use of non-renewable resources needs to be reduced. In this respect, wood and wood-based products can play a key role as they are generally low in embodied CO_2 and can be gained from sustainable forest resources. Wood has numerous advantages compared to other building materials, such as a high strength–weight ratio, good thermal insulation, easy machinability and appealing aesthetics. However, its durability against different biological agents is limited and requires consideration when wood is exposed to moisture, and thus to favorable conditions for decay.

In highly hazardous applications, the natural durability of wood can be insufficient, and wooden elements need to be protected by design. Alternatively, wood durability can be enhanced through wood preservatives or modification systems. In recent years, several highly effective wood preservatives have been banned in different countries as they harm human health and the environment. Innovative approaches for improving wood durability are being sought.

We encouraged studies from all fields, including method development, experimental studies, monitoring approaches and models, to contribute to this Special Issue, to promote knowledge about wood durability mechanisms and strategies for the protection and preservation of wooden structures and wood-based building materials.

The Special Issue comprises 19 papers by authors from 14 countries in Asia, North America and Europe. They represent a wide range of aspects related to wood protection and wood preservation and give timely examples of research activities that can be observed around the globe. Several authors reported on processes of thermal modification [1–6] and different chemical wood modification techniques [2,3,5,7–10], which are among the latest alternative wood protection methods without the use of biocides. New preservatives and assessment methods of preservative-treated wood products are presented [10,11], as well as studies on the natural durability of wood [12], fire-retardant treated wood [13,14], the effect of concrete on wood durability [15] and different novel surface modification techniques using plasma [13,14,16,17]. Besides biological durability [3,6,10,12,15,18,19], the mechanical properties [3,8,11], moisture performance [1,3,5,12,14,18], bonding properties [6,14] weathering stability [4] and the corrosiveness [7] of differently treated wood were investigated and reported within this Special Issue. Examples of research on fungal biology [9], service life planning with wood [18] and test methodology [12] were also included and complete the spectrum.

Conflicts of Interest: The author declares no conflicts of interest.

References

1. Zelinka, S.L.; Kirker, G.T.; Bishell, A.B.; Glass, S.V. Effects of wood moisture content and the level of acetylation on brown rot decay. *Forests* **2020**, *11*, 299. [CrossRef]
2. Kamperidou, V. The biological durability of thermally- and chemically-modified black pine and poplar wood against basidiomycetes and mold action. *Forests* **2019**, *10*, 1111. [CrossRef]
3. Reinprecht, L.; Repák, M. The impact f paraffin-thermal modification of beech wood on its biological, physical and mechanical properties. *Forests* **2019**, *10*, 1102. [CrossRef]
4. Cui, X.; Matsumura, J. Wood surface changes of heat-treated cunninghamia lanceolate following natural weathering. *Forests* **2019**, *10*, 791. [CrossRef]
5. Lovaglio, T.; Gindl-Altmutter, W.; Meints, T.; Moretti, N.; Todaro, L. Wetting behavior of alder (*Alnus cordata* (Loisel) Duby) wood surface: Effect of thermo-treatment and Alkyl Ketene Dimer (AKD). *Forests* **2019**, *10*, 770. [CrossRef]
6. Chang, C.-W.; Kuo, W.-L.; Lu, K.-T. On the effect of heat treatments on the adhesion, finishing and decay resistance of Japanese cedar (*Cryptomeria japonica* D. Don) and Formosa acacia (*Acacia confuse* Merr.(Leguminosae)). *Forests* **2019**, *10*, 586. [CrossRef]
7. Zelinka, S.L.; Passarini, L.; Matt, F.J.; Kirker, G.T. Corrosiveness of thermally modified wood. *Forests* **2020**, *11*, 50. [CrossRef]
8. Bollmus, S.; Beeretz, C.; Militz, H. Tensile and impact bending properties of chemically modified scots pine. *Forests* **2020**, *11*, 84. [CrossRef]
9. Kölle, M.; Ringman, R.; Pilgård, A. Initial *Rhodonia placenta* gene expression in acetylated wood: Group-wise upregulation of non-enzymatic oxidative wood degradation genes depending on the treatment level. *Forests* **2019**, *10*, 1117. [CrossRef]
10. Casado-Sanz, M.M.; Silva-Castro, I.; Ponce-Herrero, L.; Martín-Ramos, P.; Martín-Gil, J.; Acuña-Rello, L. White-rot fungi control on *Populus* spp. wood by pressure treatments with silver nanoparticles, chitosan oligomers and propolis. *Forests* **2019**, *10*, 885. [CrossRef]
11. Sharapov, E.; Brischke, C.; Militz, H. Assessment of preservative-treated wooden poles using drilling-resistance measurements. *Forests* **2020**, *11*, 20. [CrossRef]
12. Brischke, C.; Wegener, F.L. Impact of water holding capacity and moisture content of soil substrates on the moisture content of wood in terrestrial microcosms. *Forests* **2019**, *10*, 485. [CrossRef]
13. Chu, D.; Mu, J.; Avramidis, S.; Rahimi, S.; Liu, S.; Lai, Z. Functionalized surface layer on poplar wood fabricated by fire retardant and thermal densification. Part 1: Compression recovery and flammability. *Forests* **2019**, *10*, 955. [CrossRef]
14. Chu, D.; Mu, J.; Avramidis, S.; Rahimi, S.; Liu, S.; Lai, Z. Functionalized surface layer on poplar wood fabricated by fire retardant and thermal densification. Part 2: Dynamic wettability and bonding strength. *Forests* **2019**, *10*, 982. [CrossRef]
15. Nicholas, D.; Rowlen, A.; Milsted, D. Effect of concrete on the pH and susceptibility of treated pine to decay by brown-rot fungi. *Forests* **2020**, *11*, 41. [CrossRef]
16. Li, B.; Li, J.; Zhou, X.; Zhang, J.; Li, T.; Du, G. Study of gliding arc plasma treatment for bamboo-culm surface modification. *Forests* **2019**, *10*, 1086. [CrossRef]
17. Köhler, R.; Sauerbier, P.; Ohms, G.; Viöl, W.; Militz, H. Wood protection through plasma powder deposition—An alternative coating process. *Forests* **2019**, *10*, 898. [CrossRef]
18. Humar, M.; Kržišnik, D.; Lesar, B.; Brischke, C. The performance of wood decking after five years of exposure: Verification of the combined effect of wetting ability and durability. *Forests* **2019**, *10*, 903. [CrossRef]
19. Brischke, C.; Emmerich, L.; Nienaber, D.G.; Bollmus, S. Biological durability of sapling-wood products used for gardening and outdoor decoration. *Forests* **2019**, *10*, 1152. [CrossRef]

Article

Assessment of Preservative-Treated Wooden Poles Using Drilling-Resistance Measurements

Evgenii Sharapov [1,2], Christian Brischke [3,*] and Holger Militz [3]

[1] Volga State University of Technology, Lenin sq. 3, 424000 Yoshkar-Ola, Mari El Republic, Russia; SharapovES@volgatech.net

[2] Northern (Arctic) Federal University named after M.V. Lomonosov, Severnaya Dvina emb. 17, 163002 Arkhangelsk, Russia

[3] University of Goettingen, Wood Biology and Wood Products, Buesgenweg 4, 37077 Goettingen, Germany; hmilitz@gwdg.de

* Correspondence: christian.brischke@uni-goettingen.de; Tel.: +49-(0)-551-392-29514

Received: 27 November 2019; Accepted: 18 December 2019; Published: 21 December 2019

Abstract: An IML-Resi PD-400 drilling tool with two types of spade drill bits (IML System GmbH, Wiesloch, Germany) was used to evaluate the internal conditions of 3 m wooden poles made from Scots pine (*Pinus sylvestris* L.). Drilling tests were performed on poles that were industrially vacuum-pressure-impregnated with a copper-based preservative (Korasit KS-M) and untreated reference poles. Both types of poles were subject to 10.5 years of in-ground exposure. Wood moisture content (MC) was measured using a resistance-type moisture meter. MC varied between 15% and 60% in the radial and axial directions in both treated and untreated poles. A higher MC was detected in the underground, top, and outer (sapwood) parts of the poles. Typical drilling-resistance (DR) profiles of poles with internal defects were analyzed. Preservative treatment had a significant influence on wood durability in the underground part of the poles. Based on DR measurements, we found that untreated wood that was in contact with soil was severely degraded by insects and wood-destroying fungi. Conversely, treated wood generally showed no reduction in DR or feeding resistance (FR). DR profiling is a potential method for the in-situ or in vitro assessment and quality monitoring of preservative treatments and wood durability. The technological benefits of using drill bits with one major cutting edge, instead of standard drill bits with center-spiked tips and two major cutting edges, were not evident. A new graphical method was applied to present DR data and their spatial distribution in the poles. Future studies should focus on the impact of preservative treatments, thermal modification, and chemical modification on the DR and FR of wood. This may further elucidate the predictive value of DR and FR for wood properties.

Keywords: decay; drilling-resistance measurements; internal defects; nondestructive wood testing; preservative treatment; wooden poles

---

1. Introduction

Roundwood is still globally used in utility poles, piles, and structural elements in wooden constructions. Wood is a natural and organic material, and is therefore susceptible to biological degradation and destruction due to internal stresses. Thus, one of the main problems associated with the safe use of wooden poles is the evaluation of internal defects that may lead to structural failures. Different techniques for the evaluation of the internal condition of wooden poles have been developed. These include drilling-resistance (DR) measurements. From the first prototype of DR measurement (penetration resistance) [1] to basic tool design [2] and advanced drilling tools, the in-situ assessment of timber- and utility-pole structures has always been the main application of this method.

DR measurements are a nondestructive way of indirectly evaluating wood properties [3–5]. DR is based on the use of thin-boring drill bits (e.g., 3 mm diameter) to drill into wood while continuously monitoring energy consumption. Energy consumption is correlated with the physical, mechanical, and technological properties of wood. The main advantages of the DR method are that the drilling tool is portable, measurements are made in situ, it is quick and minimally invasive, and it has high sensitivity for fungal decay and other wood defects [6–9]. Furthermore, DR can be used to predict the density and other elastomechanical properties of wood [3,10–24].

The structural condition of wooden bridges was assessed by Brashaw et al. using a DR device (IML RESI F300-S, IML System GmbH, Wiesloch, Germany) [25]. The authors reported on the potential for DR measurements to detect internal decay, and the need to combine different test methods, including visual inspection and acoustic methods. Kappel and Mattheck [26] reported that DR is a good tool to detect internal defects in timber structures, such as cracks and decay. They recommended transversal drillings in a radial direction in the wood. In cases where longitudinal drilling in early wood or in cracked areas of wood cannot be avoided, additional transversal drilling is helpful to better determine the extent of wood damage.

The efficiency of DR measurements in determining the damage and residual cross-section of decayed wood in timber structures has been presented by different authors [27,28]. Imposa et al. [29] evaluated the extent of decay in ancient wooden trusses and concluded that DR measurements allowed for the quantification of material loss and microvoids in wood.

More detailed analysis of the suitability of DR measurements for the in-situ assessment of structural timber was presented by Nowak et al. [30]. They concluded that DR allows for the detection of internal defects in wooden structures, but that it is influenced by many factors such as moisture content (MC), drill-bit sharpness, and drilling direction relative to grain direction. An IML-RESI F-300 drilling tool was used by Gezer et al. to inspect wooden utility poles [31]. DR measurements were made at breast height and underground with a 45° angle between the direction of penetration and the pole axis; approximately 90% of wood deterioration occurred underground. However, the authors pointed out that, although internal defects in poles could be accurately detected, information from DR measurements was limited to the site of drill-bit penetration.

A comparative assessment of utility poles using three inspection techniques was presented by Reinprecht and Šupina [32]. DR in poles made from Norway spruce (*Picea abies*) and preserved with creosote was measured in radial directions using an IML-Resi F-400 drilling tool. Strong linear correlation ($R^2 = 0.96$) was found between mean DR measurements obtained from two orthogonal drillings in a radial direction in the same plane.

To enhance the durability and prolong the service life of wooden poles, wood is often treated with preservatives such as pentachlorophenol (63%), creosote (16%), and copper chrome arsenate (16%), as reported for the United States [32]. The mechanical properties of treated wood can be affected by the impregnation method, type of preservative, and uniformity of the treatment [33]. Precipitation and soil moisture led to the wetting of poles. Wood MC greater than 20% increases the chance of decay, and MC above 25%–30% indicates a high likelihood of extensive decay [34,35]. Furthermore, wood MC can have significant impact on DR measurements [36–39].

The aim of this study was to determine the influence of preservative treatment and wood MC on the condition of wooden poles above and below ground using DR measurements.

2. Materials and Methods

2.1. Specimen Preparation

Ten logs of Scots pine (*Pinus sylvestris* L.) with a length of 9 m were crosscut into 3 equal sections. Half of the 3-m sections (poles) were industrially vacuum-pressure-impregnated (3 kPa, 180 min, and 0.8 MPa, 180 min) in an autoclave with Korasit KS-M (Kurt Obermeier GmbH and Co. KG, Bad Berleburg, Germany). Korasit KS-M is a waterborne copper-based preservative. Impregnation was conducted at industrial impregnation plant Carl Scholl GmbH (Cologne, Germany). The mean preservative retention of the examined poles was 25.5 kg/m^3. Treated and untreated wooden poles were vertically buried in the ground at a depth of approximately 0.5 m at a field test site in Goettingen, Germany (51.6° N, 9.9° E). In total, 6 samples of treated (n = 3) and untreated (n = 3) poles were removed from the soil in January 2019 after 10.5 years of in-ground exposure. They were then evaluated.

2.2. Drilling-Resistance Measurements

An IML-RESI PD-400 tool and 2 types of spade drill bits were used (IML System GmbH, Wiesloch, Germany) for DR measurements. Both types of drill bits were almost 400 mm long, and had a thin shaft of 1.5 mm diameter and a 3-mm triangular cutting part with a hard chrome coating. The first type of drill bit (Type 1) consisted of 2 flattened and symmetrical major cutting edges that were perpendicular to the rotating axis (axis of the cylindrical drill-bit shaft). The center-spiked tip of drill bit Type 1, which was designed to stabilize linear penetration of the drill bit during the drilling process, was about 400 μm in height from the level of the major cutting edges. The second type of drill bit (Type 2) did not have a tip and only had 1 major cutting edge (Figure 1).

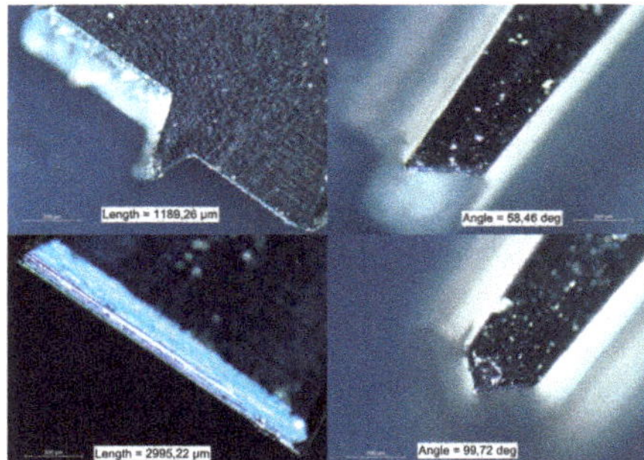

Figure 1. Drill-bit types and their main geometrical parameters. Note: length, length of major cutting edges, angle, angle sharpness for major cutting edges.

DR measurements were taken radially, in 1 plane, and in a north–south direction (Figure 2). Drilling was first done in the ground-level parts (level 0 in Figure 3), followed by the above-ground parts in 300-mm intervals (levels 1–7 in Figure 3), and then the underground parts in 100-mm intervals (levels −1 to −4 in Figure 3).

Figure 2. Drilling-resistance measurements in a preservative-treated pole.

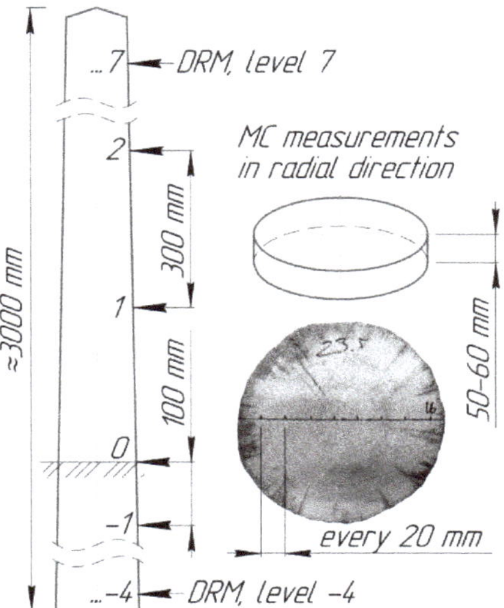

Figure 3. Schematic diagram of pole-analysis method. Positions of drilling-resistance measurements (DRMs, arrows in figure) and sites of moisture-content (MC) measurements along pole radius.

Upon completion of DR measurements, poles were dissected, and a 50–60 mm thick disc was obtained from each drilling position. MC measurements were then made using a resistance-type moisture meter. Measurements were radially taken from the end-grain surface of the discs every 20 mm along regions of drill-bit penetration (Figure 3).

Due to different amplitudes in DR and feeding resistance (FR) for the types of used drill bits, drill bit Type 1 was set to a feed rate of 1.5 m/min and a rotational frequency of 1500 min^{-1}, whereas drill bit Type 2 was set to feed rate of 1 m/min and a rotational frequency of 2500 min^{-1}. DR and FR were measured and digitally recorded for every 0.1 mm of drilling depth. DR data were saved and processed using PD-Tools PRO software (IML System GmbH, Wiesloch, Germany), Microsoft Excel® (Microsoft, Redmond, WA, USA), and SigmaPlot 14 (Systat Software Inc., San Jose, CA, USA). DR profiles obtained from each measurement included a relative DR curve reflecting the torsion force on the drill bit and an FR curve reflecting the force needed to push the needle into the wood. Mean DR and FR values from profiles of the entire length of the poles were compared between treated and untreated samples.

3. Results and Discussion

3.1. Moisture Content

Variation in mean MC over the length of the poles is presented in Figure 4a. Due to severe decay (Figure 5), it was not possible to measure the MC for underground parts of untreated poles with a resistance-type moisture meter. A higher MC was observed at the ground-level, top, and underground parts of treated and untreated poles. The MC in treated poles was around 40% in both the underground and top parts of the poles. Despite the top parts of the poles being covered in paint to eliminate moisture absorption because of precipitation, the protective layer wore out over time. The mean MC of the top parts was between 23%–31% in the untreated and 30%–46% in the treated poles (Figure 4a).

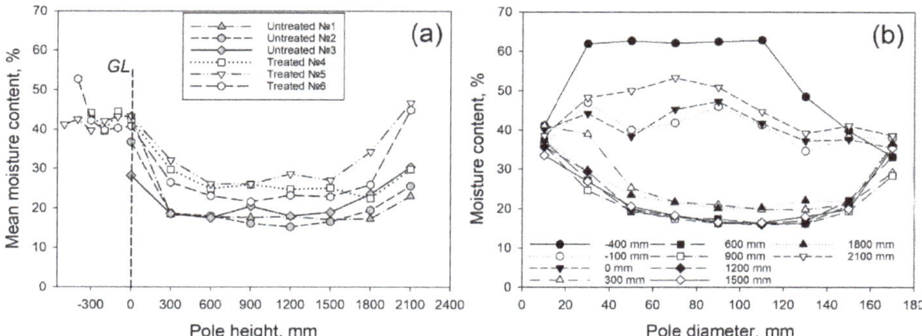

Figure 4. MC obtained from disc samples. (**a**) Mean MC over length of tested poles. Note: *GL*, ground level (0 mm). (**b**) MC variation along diameter of treated pole №3 at different heights in relation to *GL*.

Figure 5. Underground sections of treated (top) and untreated (bottom) poles.

Wood treated with preservative salts shows lower electrical resistance than untreated wood with the same MC [40]. Brischke and Lampen [41] showed that the accuracy of resistance-based MC measurements is not negatively affected by impregnation with inorganic salts. However, they found that accuracy decreased at an MC that was above fiber saturation.

The MC gradient in the radial direction differed between the underground, middle, and top parts of the poles (Figure 4b). The outer layers of above-ground and underground parts of the poles showed a similar MC of about 35%. This indicated that these parts were susceptible to a significant level of decay. MC decreased closer to the pith in the middle parts of the poles, but was higher near the pith in underground and top parts of the poles (Figure 4b). However, variation in wood MC along the length and diameter of the poles might have been affected by seasonal or current weather conditions. The data shown here are therefore just a snapshot of current conditions.

MC influences the mechanical [42] and cutting (drilling) properties of wood [43,44]. It was reported by Sharapov et al. [39] that the influence of MC on DR and FR depended on the rotational frequency and feed rates of the drill bit. These factors can both increase and decrease DR and FR. For Type 1 and the applied speed parameters, the feed rate per major cutting edge of the drill bit was 0.5 mm [45]. A significant influence of MC on DR could be expected given a lower DR is observed when MC is above fiber saturation. The effect of wood MC on DR when using Type 2 was not investigated. However, given the results for Type 1 [39], the impact of MC on DR should be lower (with a rotational frequency of 2500 min^{-1} and a feed rate of 1 m/min).

3.2. Typical Drilling Resistance Profiles

We present typical DR and FR profiles of treated and untreated poles tested 600 mm above ground level with Type 1 in Figure 6. The oscillation of both datasets (Figure 6) was related to different densities of the early wood and latewood portions of the poles.

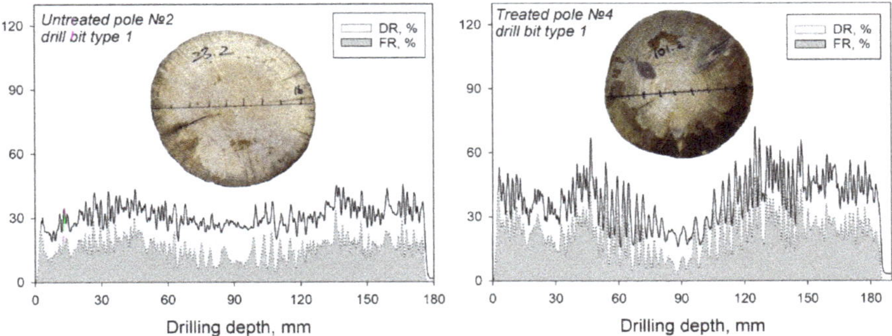

Figure 6. Typical drilling-resistance (DR) and feeding-resistance (FR) profiles obtained from untreated and treated poles using drill bit Type 1.

Poles were not fully impregnated in the radial direction. Impregnated layers showed a darker color compared to the untreated wood that was closer to the pith. The border between impregnated and untreated wood was more prominent for parts of the poles with a high MC (Figure 7b). DR and FR in regions close to the pith and DR variation within annual layers were different in untreated and treated poles (Figure 6). This may be attributed to imperfect drilling that was not in a precise radial direction in untreated poles (Figure 6). However, the outer layers of the treated wood generally showed a higher DR and FR than corresponding layers in untreated poles.

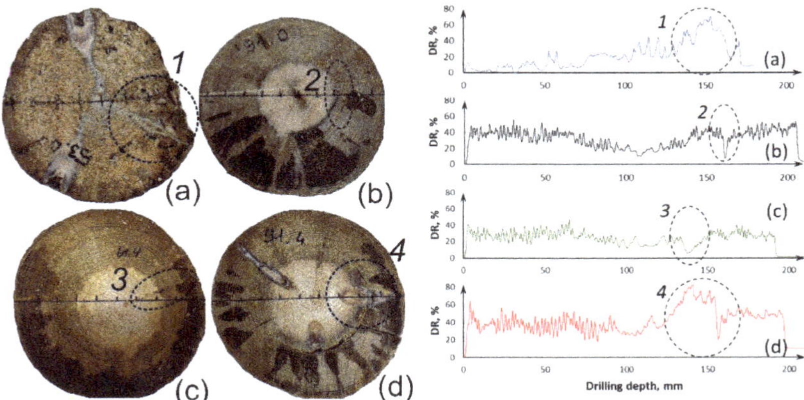

Figure 7. Cross-cut ends of poles and their DR profiles. (**a**) Untreated pole №3 at ground level. Note: 1, section with wood knot (see also Figure 9). (**b**) Treated pole №5 at ground level. Note: 2, region of cracked wood; (**c**) Treated pole №4 at 1200 mm from ground level. Note: 3, region of cracked wood (see Figure 9). (**d**) Treated pole №5 at 1200 mm from ground level. Note: 4, region with wood knot and cracked wood (see also Figure 8).

Many authors have reported that the treatment of wood with waterborne preservatives can affect its mechanical properties [33,46]: modulus of elasticity (MOE) is usually unaffected, maximum crushing strength is usually unaffected or slightly increased, modulus of rupture (MOR) is often reduced by up to 20%, and energy-related properties are usually reduced by up to 50%. However, the effect of preservative treatments on the elasto-mechanical properties of wood depends on the impregnation process and type of preservative. Ulunam et al. [47] showed that larch (*Larix decidua*) and black pine (*Pinus nigra*) treated with Korasit-KS (a preservative) demonstrated minor differences in static compression and bending strength after one year of weathering. Cutting resistance of Scots pine (*P. sylvestris* L.) treated with Korasit KS2 was higher compared to untreated wood in a laboratory setting when using a frame saw [48]. Based on a model for the prediction of cutting power, Chuchała and Orłowski [48] concluded that power consumption in the band-sawing of treated wood can be up to 50% higher compared to that of untreated wood. The higher DR and FR of treated wood can be explained by an increase in friction forces during drilling (cutting).

Typical internal defects detected in treated and untreated poles are presented in Figure 7. Untreated poles were severely degraded in the ground-level and underground parts (Figure 5). The DR and FR of degraded untreated wood were low. However, wood knots in the decayed wood were not degraded. This was evident in DR and FR data, presented in Figure 7a. The maximum DR of wood knots in the decayed wood of untreated poles was close to the DR of wood knots in treated poles (Figure 7a,d). Drilling of severely decayed wood that contains wood knots can result in a higher mean DR and FR. This therefore may not reflect the actual degree of destruction of the entire structure under inspection. Other internal defects, such as radial (Figure 7c) or annual ring (Figure 7b,d) cracks, were detected in drill-penetration paths.

3.3. Mean Drilling Resistance and Feeding Resistance

Variations in mean DR and FR across the length of treated and untreated poles are shown in Figure 8. Preservative treatment had a significant influence on wood durability in the underground part of the poles. Mean DR and FR were remarkably reduced in the underground parts of untreated poles with both drill bit types. The reduction in DR for Type 1 was more prominent than in Type 2 (Figure 8).

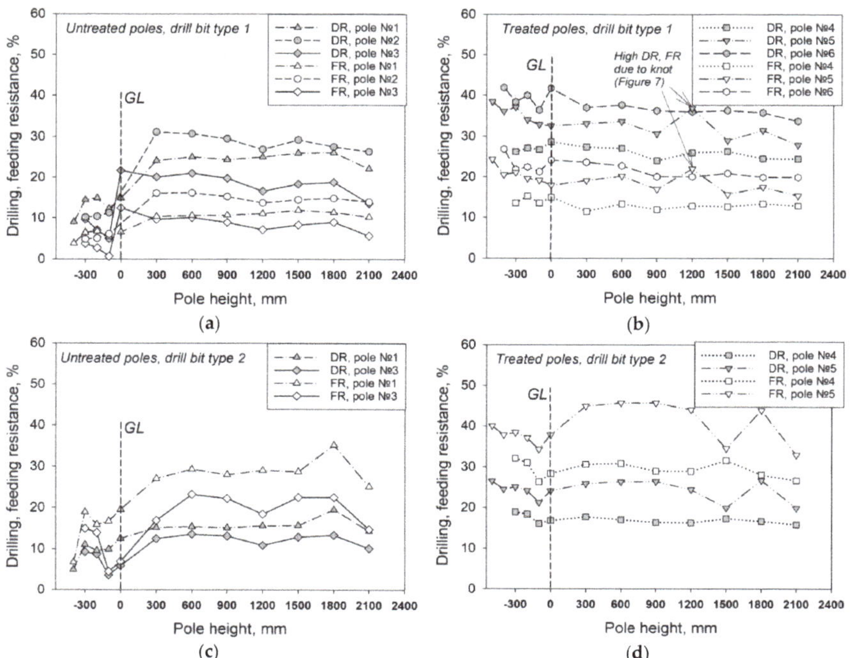

Figure 8. Mean drilling resistance (DR) and feeding resistance (FR) across length of tested poles. (**a**) untreated poles, drill bit type 1, (**b**) treated poles, drill bit type 1, (**c**) untreated poles, drill bit type 2, (**d**) treated poles, drill bit type 2. Horizontal dashed line, ground level (GL). Negative pole-height values indicate DR and FR data points observed for underground parts.

Differences in the design of the cutting part of drill bits had significant influence on mean DR and FR. Using the drill bit with one major cutting edge led to higher FR values compared to DR values. Different rotational frequencies and feed rates were used for both drill-bit types due to the high energy consumption of drills with one major cutting edge. Owing to its higher amplitude, FR is most often used to preselect the feed rate for drill bit Type 2. However, DR is a more accurate parameter for the prediction of the density and mechanical properties of wood [24]. The expected advantages of using a drill bit with one major cutting edge were not evident. The lower DR amplitude for the drill bit with two major cutting edges might be offset by an increased feed rate. DR and FR were generally not reduced in the underground parts of treated poles for both drill bits (Figure 5).

In general, mean DR and FR were higher in the above-ground parts of the impregnated poles (Figure 8). However, this may have been because the lower-density 3 m poles that were cut from the upper part of 9 m logs were used as untreated reference poles. This hypothesis was partly confirmed by the smooth decline in mean DR along the length of the poles from the small to large end (Figure 8). To confirm the hypothesis that DR and FR are significantly affected by preservative treatments, defect-free specimens need to be tested after conditioning at normal climate.

To represent DR measurements along poles in one plane in the same drilling direction, DR and FR data are presented as contour plots (Figure 9). DR and FR data profiles were interpolated between measurements across the length of the poles, including underground parts.

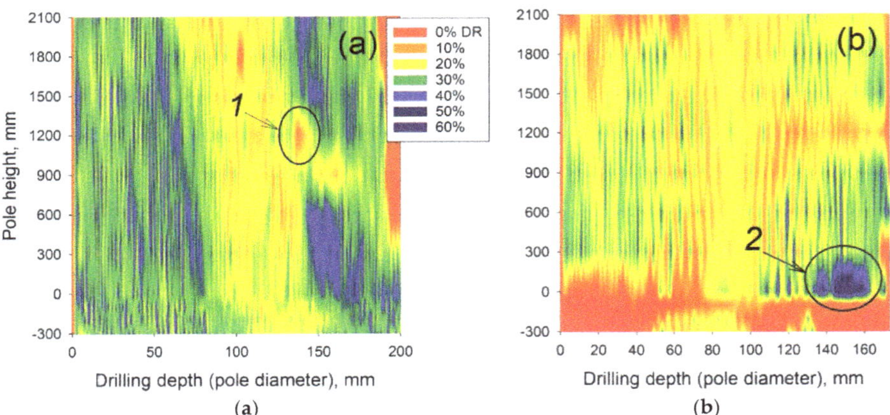

(a) (b)

Figure 9. Contour plots including all DR data obtained from (**a**) treated pole №5 and (**b**) untreated pole №3. Note: 0 on Y axis, ground level; 1, low DR in cracked wood; 2, high DR in a wood knot (see Figure 7).

Vertical colored lines corresponded to early wood and latewood portions as presented for individual DR profiles in Figure 6. This form of data presentation shows the distribution of variations in the mechanical properties of wood due to density differences, knots, cracks, decay, and other defects. As can be seen in Figure 9a, the central part of the treated pole, which was not impregnated with the preservative, had lower DR compared to the treated outer (sapwood) parts. This area might have been erroneously characterized as decayed. The DR of untreated poles (presented as a contour plot in Figure 9b) was more consistent in middle parts of the poles and significantly reduced in underground parts of the poles. This was likely due to decay.

4. Conclusions

MC of wood varied between approximately 15% and 60% in the radial and axial directions in treated and untreated poles. This can have a significant impact on DR and FR measurements. DR is a potential method for the development of standard tests for the in-situ or in-vitro assessment and monitoring of the quality of preservative treatment or wood durability. Internal defects in poles can both decrease and increase DR and FR. This should be considered when assessing the condition of wooden structures. The technological benefits of using a drill bit with one major cutting edge instead of a standard drill bit with a center-spiked tip and two major cutting edges have not been established. Contour graphs using DR measurements obtained from one element of a wooden structure are recommended for the presentation and analysis of the internal conditions of wood. Further studies should focus on the effect of different wood preservatives on DR measurements using defect-free specimens at different MCs. Similarly, DR may be used to assess the effect of modifying agents, adhesives, and other chemicals on the condition of wood and timber.

Author Contributions: E.S. and C.B. conceived and designed the experiments and wrote the original draft of the manuscript; E.S. performed the experiments and analyzed the data; H.M. reviewed and edited the manuscript. All authors have read and agreed to the published version of the manuscript.

Funding: This study was supported by the German Academic Exchange Service, DAAD (ID: 57447934), and the Ministry of Science and Higher Education of the Russian Federation (№ 5.8394.2017/8.9; 37.13394.2019/13.2).

Acknowledgments: The authors gratefully acknowledge Susanne Bollmus and Antje Gellerich (University of Goettingen, Wood Biology and Wood Products) for providing the test material. Carl Scholl GmbH, Cologne, is acknowledged for the impregnation of the poles. Kurt Obermeier GmbH and Co. KG, Bad Berleburg are acknowledged for providing the wood preservative.

Conflicts of Interest: The authors declare no conflict of interest.

References

1. Dana, H.J. Pole Soundness Tester. U.S. Patent 2,359,030, 13 November 1945.
2. Kamm, W.F.G.; Voss, S. Process and Device for Determining the Internal Condition of Trees or Wooden Components. U.S. Patent 4,671,105, 9 January 1987.
3. Rinn, F.; Schweingruber, F.H.; Schar, E. Resistograph and X-ray density charts of wood comparative evaluation of drill resistance profiles and X-ray density charts of different wood species. *Holzforschung* **1996**, *50*, 303–311. [CrossRef]
4. Tannert, T.; Anthony, R.; Kasal, B.; Kloiber, M.; Piazza, M.; Riggio, M.; Rinn, F.; Widmann, R.; Yamaguchi, N. In-situ assessment of structural timber using semi-destructive techniques. *Mater. Struct.* **2014**, *47*, 767–785. [CrossRef]
5. Schimleck, L.; Dahlen, J.; Apiolaza, L.A.; Downes, G.; Emms, G.; Evans, R.; Moore, J.; Pâques, L.; Van den Bulcke, J.; Wang, X. Non-destructive evaluation techniques and what they tell us about wood property variation. *Forests* **2019**, *10*, 728. [CrossRef]
6. Eckstein, D.; Sass, U. Measurements of drill resistance on broadleaved trees and their anatomical interpretation. *Holz Roh Werks.* **1994**, *52*, 279–286. [CrossRef]
7. Schwarze, F.W.M.R. *Diagnosis and Prognosis of the Development of Wood Decay in Urban Trees*; ENSPEC Pty Ltd: Rowville, Australia, 2008; pp. 1–336.
8. Henriques, D.F.; Nunes, L.; Machado, J.S.; Brito, J. Timber in buildings: Estimation of some properties using Pilodin® and Resistograph®. In Proceedings of the International Conference on Durability of Building Materials and Components, Porto, Portugal, 12–15 April 2011; pp. 1–8.
9. Sharapov, E.; Brischke, C.; Militz, H.; Smirnova, E. Effects of white rot and brown rot decay on the drilling resistance measurements in wood. *Holzforschung* **2018**, *72*, 905–913. [CrossRef]
10. Görlacher, R.; Hättich, R. Untersuchung von altem Konstruktionsholz-Die Bohrwiderstandsmessung. *Bauen Mit Holz* **1990**, *6*, 455–459.
11. Ceraldi, C.; Mormone, V.; Russo-Ermolli, E. Resistographic inspection of ancient timber structures for the evaluation of mechanical characteristics. *Mater. Struct.* **2001**, *34*, 59–64. [CrossRef]
12. Park, C.Y.; Kim, S.J.; Lee, J.J. Evaluation of specific gravity in post member by drilling resistance test. *Mokchae Konghak* **2006**, *34*, 1–9.
13. Bouffier, L.; Charlot, C.; Raffin, A.; Rozenberg, P.; Kremer, A. Can wood density be efficiently selected at early stage in maritime pine (*Pinus pinaster* Ait.)? *Ann. For. Sci.* **2008**, *65*, 106–113. [CrossRef]

14. Zhang, H.; Guo, Z.; Su, J. Application of a drill resistance technique for rapid determining wood density. Progress of Machining Technology. *Key Eng. Mater.* **2009**, *407*, 494–499. [CrossRef]
15. Acuña, L.; Basterra, L.A.; Casado, M.; López, G.; Ramon-Cueto, G.; Relea, E.; Martínez, C.; González, A. Application of resistograph to obtain the density and to differentiate wood species. *Mater. Constr.* **2011**, *61*, 451–464. [CrossRef]
16. Faggiano, B.; Grippa, M.R.; Marzo, A.; Mazzolani, F.M. Experimental study for non-destructive mechanical evaluation of ancient chestnut timber. *J. Civ. Struct. Health Monit.* **2011**, *1*, 103–112. [CrossRef]
17. Zhang, H.; Zhu, L.; Sun, Y.; Wang, X.; Yan, H. Determining modulus of elasticity of ancient structural timber. *Adv. Mater. Res.* **2011**, *217*, 407–412. [CrossRef]
18. Kloiber, M.; Tippner, J.; Hrivnak, J. Mechanical properties of wood examined by semi-destructive devices. *Mater. Struct.* **2014**, *47*, 199–212. [CrossRef]
19. Morales-Conde, M.J.; Rodríguez-Liñán, C.; Saporiti-Machado, J. Predicting the density of structural timber members in service. The combine use of wood cores and drill resistance data. *Mater. Constr.* **2014**, *64*, 1–11. [CrossRef]
20. Oliveira, J.T.S.; Wang, X.; Vidaurre, G. Assessing specific gravity of young Eucalyptus plantation trees using a resistance drilling technique. *Holzforschung* **2017**, *71*, 137–145. [CrossRef]
21. Karlinasari, L.; Danu, M.I.; Nandika, D.; Tujaman, M. Drilling resistance method to evaluate density and hardness properties of resinous wood of agarwood (*Aquilaria malaccensis*). *Wood Res.* **2017**, *6*, 683–690.
22. Downes, G.M.; Lausberg, M.; Potts, B.M.; Pilbeam, D.L.; Bird, M.; Bradshaw, B. Application of the IML Resistograph to the infield assessment of basic density in plantation eucalypts. *Aust. For.* **2018**, *81*, 177–185. [CrossRef]
23. Fundova, I.; Funda, T.; Wu, H.X. Non-destructive wood density assessment of Scots pine (*Pinus sylvestris* L.) using Resistograph and Pilodyn. *PLoS ONE* **2018**, *13*, e0204518. [CrossRef]
24. Sharapov, E.; Brischke, C.; Militz, H.; Smirnova, E. Prediction of modulus of elasticity in static bending and density of wood at different moisture contents and feed rates by drilling resistance measurements. *Eur. J. Wood Wood Prod.* **2019**, *77*, 833–842. [CrossRef]
25. Brashaw, B.K.; Vatalaro, R.J.; Wacker, J.P.; Ross, R.J. *Condition Assessment of Timber Bridges 1. Evaluation of a Micro-Drilling Resistance Tool*; General Technical Report FPL-GTR-159; USDA Forest Service, Forest Products Laboratory: Madison, WI, USA, 2005; p. 8.
26. Kappel, R.; Mattheck, C. Inspection of timber construction by measuring drilling resistance using Resistograph F300-S. *WIT Trans. Built Environ.* **2003**, *66*, 825–834.
27. Branco, J.M.; Sousa, H.S.; Tsakanika, E. Non-destructive assessment, full-scale load-carrying tests and local interventions on two historic timber collar roof trusses. *Eng. Struct.* **2017**, *140*, 209–224. [CrossRef]
28. Frontini, F. In situ evaluation of a timber structure using a drilling resistance device. Case study: Kjøpmannsgata 27, Trondheim (Norway). *Int. Wood Prod. J.* **2017**, *8*, 14–20. [CrossRef]
29. Imposa, S.; Mele, G.; Corrao, M.; Coco, G.; Battaglia, G. Characterization of decay in the wooden roof of the S. Agata Church of Ragusa Ibla (Southeastern Sicily) by means of sonic tomography and resistograph penetration tests. *Int. J. Archit. Herit.* **2014**, *8*, 213–223. [CrossRef]
30. Nowak, T.P.; Jasienko, J.; Hamrol-Bielecka, K. In situ assessment of structural timber using the resistance drilling method–Evaluation of usefulness. *Constr. Build. Mater.* **2016**, *102*, 403–415. [CrossRef]
31. Gezer, E.D.; Ali Temiz, A.; Yüksek, T. Inspection of wooden poles in electrical power distribution networks in Artvin, Turkey. *Adv. Mater. Sci. Eng.* **2015**, *2015*, 1–11. [CrossRef]
32. Reinprecht, L.; Šupina, P. Comparative evaluation of inspection techniques for impregnated wood utility poles: Ultrasonic, drill-resistive, and CT-scanning assessments. *Eur. J. Wood Prod.* **2015**, *73*, 741. [CrossRef]
33. Winandy, J.E. Effects of waterborne preservative treatment on mechanical properties: A review. In Proceedings of the American Wood-Preservers' Association, New York, NY, USA, 21–24 May 1995; USDA Forest Service: Washington, DC, USA, 1995; Volume 914, pp. 17–33.
34. Ross, R.J.; Brashaw, B.K.; Wang, X.; White, R.H.; Pellerin, R.F. *Wood and Timber Condition Assessment Manual*, 2nd ed.; General Technical Report FPL-GTR-234; USDA Forest Service, Forest Products Laboratory: Madison, WI, USA, 2014; p. 92.
35. Brischke, C.; Rapp, A.O. Dose–response relationships between wood moisture content, wood temperature and fungal decay determined for 23 European field test sites. *Wood Sci. Technol.* **2008**, *42*, 663–677. [CrossRef]

36. Mattheck, C.; Bethge, K.; Albrecht, W. How to read the results of resistograph M. *Arboric. J.* **1997**, *21*, 331–346. [CrossRef]
37. Lin, C.J.; Wang, S.Y.; Lin, F.C.; Chiu, C.M. Effect of moisture content on the drill resistance value in Taiwania plantation wood. *Wood Fiber Sci.* **2003**, *35*, 234–238.
38. Ukrainetz, N.K.; O'Neill, G.A. An analysis of sensitivities contributing measurement error to Resistograph values. *Can. J. For. Res.* **2010**, *40*, 806–811. [CrossRef]
39. Sharapov, E.; Brischke, C.; Militz, H.; Smirnova, E. Combined effect of wood moisture content, drill bit rotational speed and feed rate on drilling resistance measurements in Norway spruce (*Picea abies* (L.) Karst.). *Wood Mater. Sci. Eng.* **2018**, 1–7. [CrossRef]
40. James, W.L. *Electric Moisture Meters for Wood*; General Technical Report FPL-GTR-6; USDA Forest Service, Forest Products Laboratory, Department of Agriculture: Madison, WI, USA, 1988; p. 17.
41. Brischke, C.; Lampen, S.C. Resistance based moisture content measurements on native, modified and preservative treated wood. *Eur. J. Wood Wood Prod.* **2014**, *72*, 289–292. [CrossRef]
42. Gerhards, C.C. Effect of the moisture content and temperature on the mechanical properties of wood and analysis of immediate effects. *Wood Fiber Sci.* **1982**, *14*, 4–36.
43. Franz, N.C. *An Analysis of the Wood-Cutting Process*; University of Michigan Press: Ann Arbor, MI, USA, 1958; pp. 1–166.
44. Ivanovskii, E.G. *Rezanie Drevesiny [Wood Cutting]*; Lesnaya Promyshlennost [Forest Industry]: Moscow, Russia, 1975; pp. 1–200.
45. Sharapov, E.; Brischke, C.; Militz, H.; Toropov, A. Impact of drill bit feed rate and rotational frequency on the evaluation of wood properties by drilling resistance measurements. *Int. Wood Prod. J.* **2019**, 1–11. [CrossRef]
46. Humar, M.; Krzisnik, D.; Brischke, C. The effect of preservative treatment on mechanical strength and structural integrity of wood. In Proceedings of the International Research Group on Wood Protection; IRG/WP 15-30667, Viña del Mar, Chile, 10–14 May 2015.
47. Ulunam, M.; Özalp, M.; Sofuoğlu, S.D.; Çerçioğlu, M. Changes in technological properties of black pine and larex woods impregnated with olive oil and Korasit-KS under open air conditions. In Proceedings of the 2nd International Conferences on Science and Technology; Life Science and Technology (ICONST-2019), Prizren, Kosovo, 26–30 August 2019; pp. 198–206.
48. Chuchała, D.; Orłowski, K.A. Forecasting values of cutting power for the sawing process of impregnated pine wood on band sawing machine. *Mechanik* **2018**, *91*, 766–768. [CrossRef]

Article

White-Rot Fungi Control on *Populus* spp. Wood by Pressure Treatments with Silver Nanoparticles, Chitosan Oligomers and Propolis

María Milagrosa Casado-Sanz [1], Iosody Silva-Castro [2], Laura Ponce-Herrero [3], Pablo Martín-Ramos [4,*], Jesús Martín-Gil [2] and Luis Acuña-Rello [1]

1 Laboratorio de Tecnología de la Madera, Departamento de Ingeniería Agrícola y Forestal, Universidad de Valladolid, Avenida de Madrid 44, 34004 Palencia, Spain; mariamilagrosa.casado@uva.es (M.-M.C.-S.); maderas@iaf.uva.es (L.A.-R.)

2 Laboratorio de Tecnología Ambiental, Departamento de Ingeniería Agrícola y Forestal, Universidad de Valladolid, Avenida de Madrid 44, 34004 Palencia, Spain; iodosody@gmail.com (I.S.-C.); mgil@iaf.uva.es (J.M.-G.)

3 Sustainable Forest Management Research Institute, University of Valladolid—INIA, Avenida de Madrid 57, 34004 Palencia, Spain; laura.ponce@alumnos.uva.es

4 EPS, Instituto Universitario de Investigación en Ciencias Ambientales de Aragón (IUCA), Universidad de Zaragoza, Carretera de Cuarte s/n, 22071 Huesca, Spain

* Correspondence: pmr@unizar.es; Tel.: +34-974292668

Received: 10 September 2019; Accepted: 1 October 2019; Published: 7 October 2019

Abstract: There is growing interest in the development of non-toxic, natural wood preservation agents to replace conventional chemicals. In this paper, the antifungal activities of silver nanoparticles, chitosan oligomers, and propolis ethanolic extract were evaluated against white-rot fungus *Trametes versicolor* (L.) Lloyd, with a view to protecting *Populus* spp. wood. In order to create a more realistic in-service type environment, the biocidal products were assessed according to EN:113 European standard, instead of using routine in vitro antimicrobial susceptibility testing methods. Wood blocks were impregnated with the aforementioned antifungal agents by the vacuum-pressure method in an autoclave, and their biodeterioration was monitored over 16 weeks. The results showed that treatments based on silver nanoparticles, at concentrations ranging from 5 to 20 ppm, presented high antifungal activity, protecting the wood from fungal attack over time, with weight losses in the range of 8.49% to 8.94% after 16 weeks, versus 24.79% weight loss in the control (untreated) samples. This was confirmed by SEM and optical microscopy images, which showed a noticeably higher cell wall degradation in control samples than in samples treated with silver nanoparticles. On the other hand, the efficacy of the treatments based on chitosan oligomers and propolis gradually decreased over time, which would be a limiting factor for their application as wood preservatives. The nanometal-based approach is thus posed as the preferred choice for the industrial treatment of poplar wood aimed at wood-based engineering products (plywood, laminated veneer lumber, cross-laminated timber, etc.).

Keywords: decay fungi; nanomaterials; natural protectors; poplar; wood preservatives

1. Introduction

Populus spp., part of the Salicaceae family, are amongst the most frequently cultivated trees for industrial purposes. These fast-growing species have a significant economic impact, given that their wood is used for the production of pulp, panels, and many other commercial applications [1–3], in addition to being very important from an environmental perspective [4]. According to EN 350:2016, poplar wood is non-durable, and some studies have shown that it is highly susceptible to *Trametes versicolor* [5,6]. The fungal hyphae propagate through the woody elements, vessels, and tracheids of the sapwood, feeding on the constituents of the cell wall (lignin and structural carbohydrates) through the secretion of enzymes capable of metabolizing these structures [7]. This causes weight and mechanical resistance losses, color changes, and an increase in moisture content.

In recent decades, significant efforts have been devoted to the evaluation of natural products for wood protection applications, since they represent an alternative to traditionally used chemical compounds, which have toxic effects on humans and on the environment [8]. Amongst them, renewable polymers have attracted intense industrial interest, and chitosan in particular has especially promising application prospects [9]. Other products, such as propolis [10,11] and metallic nanoparticles [12] have also been the focus of increasing attention.

Various nano-sized inorganic materials based on metals or metal oxides have been assayed for wood protection, including silver, gold, zinc, copper, boron, titanium, tin, silicon, and cerium [12–16]. Among these, silver has shown the highest toxicity against bacterial and fungal growth [17–22]. However, it should be clarified that those studies were conducted against other wood-staining fungi and solely in agar plates. Only Dorau et al. [17] used wood blocks (as in this study), although they tested the activity of ionic silver instead of silver nanoparticles (AgNPs) against three brown-rot fungi (*Postia placenta, Tephrophana palustris* and *Gloeophyllum trabeum*). It is worth noting that the toxicity mechanisms of these two silver forms differ, as some biological effects—including the formation of reactive oxygen species (ROS) and extensive membrane damage—have been observed more severely for AgNPs than for ionic Ag+ [23].

Chitosan is a linear aminopolysaccharide biopolymer [24] obtained by the alkaline deacetylation of chitin extracted from the exoskeleton of crustaceans and from the cell walls of some fungi and algae. It stands out for its antimicrobial properties against algae, yeasts, some bacteria, and fungi [25]. With regards to the latter, chitosan not only inhibits their growth (fungistatic), but may also act as a fungicide at high concentrations [26]. Moreover, it has other very interesting properties, including biocompatibility, high biodegradability, bioactivity, and non-toxicity in humans [24,27]. Chitosan can be chemically and/or enzymatically modified [28] to prepare chitosan oligomers (pentamers and heptamers), which present enhanced antifungal behaviors [21,29–31].

Propolis is a resinous substance collected and transformed by bees (*Apis mellifera* L.) for use in honeycomb construction and repair [32]. The use of propolis as an antifungal agent is well known, and its effectiveness has been demonstrated, for instance, against *Candida* spp. [33,34]. Partially purified propolis extracts have been investigated for wood protection due to their antimicrobial effects against yeasts, molds, bacteria, and parasites [10,32]. Propolis' activity against wood-decay fungi has been demonstrated in previous works [21,30,31].

The aim of this study was to compare the antifungal capacity of these three products (i.e., AgNPs, chitosan oligomers (CO), and propolis (P)), considered harmless to human health and the environment, against *T. versicolor*, assaying different concentrations on *Populus* spp. wood. Treatments were carried out by vacuum-pressure impregnation in an autoclave, according to EN:113 standard. The novelty of this study lies both in the use of a more realistic assessment method of the biocidal efficacy of the candidate wood preservatives (using wood blocks instead of agar plates) and in the fungus against which the products were assayed.

2. Materials and Methods

2.1. Reagents and Biological Material

Medium-molecular-weight chitosan (60–130 kDa; CAS No. 9012-76-4) with 90% deacetylation was acquired from Hangzhou Simit Chemical Technology Co. (Hangzhou, China). Propolis, with a content of ca. 10% *w/v* of polyphenols and flavonoids, came from the Duero river basin region (Burgos, Spain). Silver nitrate (CAS number 7761-88-8), sodium citrate (CAS 6132-04-3), acetic acid (CAS 64-19-7), and potassium methoxide (CAS 865-33-8) were supplied by Merck Millipore (Darmstadt, Germany).

The Wood Technology Laboratory, Universidad de Valladolid (Spain) supplied 432 wood blocks ($50 \times 25 \times 15$ mm^3) of *Populus × euramericana* I-214 clone. The selected isolate of *T. versicolor* L. Lloyd 1920 (DSM 3086 strain, CECT 863A) was supplied by the Spanish Type Culture Collection (Valencia, Spain) and was cultivated on malt agar (supplied by Scharlau, Barcelona, Spain). Test specimen supports were sterilized by the steam method using an Autester ST DRY PV-II 30 LA autoclave (JP Selecta, Barcelona, Spain).

2.2. Synthesis of the Antifungal Solutions

Silver nanoparticles were synthetized by a sonication method; 50 mL of silver nitrate (50 mM) was mixed with 50 mL of sodium citrate (30 mM) as a reducing agent, and the solution was heated to 90 °C until it turned from colorless to pale yellow, which then became more intense. The yellowish solution was sonicated for 3–5 min with a probe-type UIP1000hdT ultrasonicator (Hielscher, Teltow, Germany; 1000 W, 20 kHz), and finally it was stabilized for at least 24 h in a refrigerator at 5 °C [35]. The resulting AgNPs, characterized by transmission electron microscopy (TEM) with a JEOL (Akishima, Tokyo, Japan) JEM-FS2200 HRP microscope, were spherical in shape with variable sizes, ranging from 10 to 30 nm.

Chitosan oligomers were prepared by oxidative degradation. Commercial medium molecular-weight (MW) chitosan was first dissolved in acetic acid (1%) under constant stirring at 60 °C for 2 h, then the pH was adjusted from 4.5 to 6.5 with potassium methoxide, and the molecular weight was finally reduced by adding hydrogen peroxide (0.3 M) to the solution, under stirring for 1 h at the same temperature. Chitosan oligomers of a viscosity-average MW of ca. 2000 Da were obtained [30] according to the experimental procedure proposed by Costa et al. [36].

Propolis ethanolic extract was prepared by grinding the resin and adding it to an hydroalcoholic solution 7:3 (v/v), followed by stirring for 72 h at room temperature and filtration with a stainless-steel 220 mesh to remove insoluble particles [37].

Every solution was tested at four different concentrations (Table 1). These concentrations were chosen on the basis of previously published work [30]. It should be clarified that, in contrast with that study (in which the wood block samples were only dipped in the preservative solution, and hence the adhesion properties of the bioagents played a major role), the vacuum-pressure impregnation method used herein allowed to use chitosan oligomers instead of medium MW chitosan, taking advantage of their superior minimum inhibitory values and lower viscosity.

Table 1. Concentrations of the wood preservative substances used in the impregnation treatments.

Silver Nanoparticles (ppm)	Chitosan Oligomers (mg/mL)	Propolis (mg/mL)
20	80	40
15	40	20
10	20	10
5	10	5

2.3. Vacuum-Pressure Treatments and Antifungal Tests

The vacuum-pressure treatment and the subsequent antifungal efficacy assessment were carried out based on a modified EN:113 standard, with six replicates for each solution (i.e., antifungal product), concentration, and exposure period, accounting for a total of 288 blocks. The vacuum-pressure cycle can be briefly described as follows: wood blocks were first introduced into a stainless-steel autoclave, and a vacuum was created and maintained for 15 min. The selected product was then introduced by pressure difference, and pressure was increased to 6 kg·cm^{-2} for 2 h; pressure was finally removed, and the wood blocks were extracted from the remaining liquid. After impregnation, wood blocks were left at 20 °C and 65% relative humidity (RH) for 4 weeks in a Medline Scientific (Oxon, UK) conditioning chamber.

Antifungal activity tests were carried out by placing two treated wood blocks and one control block into a Kolle flask that contained a one-week-old fungal culture of *T. versicolor*. The samples were incubated in a culture chamber (Incubat 2000944) at 22 °C and 70% RH for 16 weeks. Every 4 weeks, six samples were taken out to measure the fungal attack by the weight loss. Wood blocks were carefully cleaned and dried at 103 ± 2 °C before the final weight was recorded. The weight loss in each sample was calculated according to Equation (1):

$$\text{Weight loss } (\%) = (M_0 - M_f)/M_f \cdot 100 \quad (1)$$

where M_0 is the oven-dry weight of the sample before impregnation and prior to exposure, and M_f is the oven-dry weight of samples after exposure to fungus. Since the concentrations of the wood preservatives used in this study were very low (in the order of ppm), it was assumed that the dry weight of the impregnated wood was the same as the initial oven-dry weight.

2.4. Wood Degradation Monitoring

Wood degradation was monitored using optical microscopy and scanning electron microscopy (SEM). For the former, 30-μm microtome cuts were taken from the sample using a Leica (Wetzlar, Germany) microtome apparatus. These were analyzed with a Leica DMLM transmission optical microscope. The microtomes were also subjected to SEM examination at the facilities of the Microscopy Unit of the Parque Científico UVa (Spain), using a FE-SEM Leica LEO-1530 microscope. For SEM analyses, samples were adhered to the holder with double-sided carbon tape and covered with gold with a Quorum Emitech K575X sputter coater (Quorum Technologies, Ashford, Kent, UK). To confirm the composition of the AgNPs impregnated into the samples, an FEI QUANTA 200 FEG (Thermo-Fisher Scientific, Waltham, MA, USA) SEM microscope equipped with a Genesis XM 4i energy dispersive X-ray microanalysis unit was used to obtain EDX elemental maps of non-sputter-coated samples.

2.5. Statistical Analyses

All the statistical analyses were performed using R software (v. 3.4.4; R Development Core Team, 2018). Data from 288 treated samples, corresponding to 48 different individual groups (3 preservation agents × 4 concentration values × 4 exposure periods) × 6 replicates, in addition to 144 control samples, were analyzed. Prior to the analyses, the assumptions of independence, normality, and homoscedasticity were checked for all groups. Since all the data met the normality requirement (checked with a Shapiro–Wilks test) and the homoscedasticity requirement (checked with a Bartlett's test), ANOVA was used. Bootstrapping and robust homogenous groups were used.

3. Results

Table 2 shows the average percentage of weight loss for the untreated samples (control) and for the samples treated with various concentrations of silver nanoparticles after exposure to *T. versicolor* for 4, 8, 12, and 16 weeks. The results of the analysis of variance verified that there were significant differences ($p < 0.01$) between the weight loss of the treatment and control samples. A long-term protective effect of the silver nanoparticles was observed during the development of the experiment, since the weight losses for all treatment concentrations oscillated between 8% and 10%, with no statistically significant differences, regardless of the exposure time. On the other hand, controls showed a progressive increase in weight loss over time: 13.7%, 17.5%, 23.8%, and 24.8% for the 4-, 8-, 12-, and 16-week exposure times, respectively.

Table 2. Weight loss after different incubation periods for the silver nanoparticles treatments.

Exposure Period (weeks)	AgNPs Concentration (ppm)	Weight Loss (%) *	Shapiro–Wilk Test *p*-Value	Bartlett *p*-Value	ANOVA *p*-Value	Homogeneous Groups †			
	Control	13.75 ± 1.21	0.0532						X
	5	8.80 ± 0.09	0.8964					c	
4	10	8.67 ± 0.09	0.8290	0.6538	2.71×10^{-6}		b	c	
	15	8.52 ± 0.19	0.7587				b		
	20	8.33 ± 0.19	0.9515			a			
	Control	17.55 ± 2.26	0.7000						X
	5	8.93 ± 0.11	0.6224					c	
8	10	8.73 ± 0.18	0.9775	0.9059	0.00135		b	c	
	15	8.59 ± 0.19	0.7082			a	b		
	20	8.43 ± 0.18	0.8408			a			
	Control	23.86 ± 1.95	0.5846				X		
	5	8.85 ± 0.21	0.6889			a			
12	10	8.74 ± 0.13	0.6406	0.7605	0.0542	a			
	15	8.59 ± 0.19	0.9706			a			
	20	8.43 ± 0.19	0.9854			a			
	Control	24.79 ± 5.21	0.4149					X	
	5	8.94 ± 0.16	0.7192				b		
16	10	8.81 ± 0.14	0.5510	0.511	0.0291		b		
	15	8.80 ± 0.13	0.7146				b		
	20	8.49 ± 0.22	0.8678			a	b		

* Expressed as average ± robust confidence interval. † Values denoted with identical letters do not differ significantly.

The weight loss results for the chitosan-oligomers-based treatment, assayed at different concentrations, are summarized in Table 3. As in the case of AgNPs, antifungal behavior was observed even at low concentrations, since all the treatments showed significant differences from the control treatment. In the first sampling, after 4 weeks of exposure, the highest concentration (80 mg·mL^{-1}) showed the best antifungal activity (5% weight loss), showing statistically significant differences vs. the other concentrations (with weight losses in the 7.9%–11.6% range). Nonetheless, the effectiveness of the chitosan treatment clearly decreased over time and, after 12 weeks, there were no significant differences in weight loss between the different protective agent concentrations (all were in the 17.5%–22.3% range).

Table 3. Effect of chitosan oligomers concentration and fungus exposure time on the protection of poplar wood against *Trametes versicolor*.

Exposure Period (weeks)	Chitosan Oligomers Concentration (mg/mL)	Weight Loss (%) *	Shapiro–Wilk Test *p*-Value	Bartlett *p*-Value	ANOVA *p*-Value	Homogeneous Groups †			
	Control	14.54 ± 0.82	0.0621						X
	10	11.58 ± 1.70	0.6750					c	
4	20	9.34 ± 0.78	0.7082	0.1393	2.71×10^{-7}		b	c	
	40	7.91 ± 0.87	0.3403				b		
	80	4.96 ± 0.63	0.3632			a			
	Control	19.43 ± 3.34	0.5128					X	
	10	12.33 ± 0.75	0.6924				b		
8	20	9.34 ± 0.78	0.6589	0.9059	0.00135		b		
	40	9.93 ± 0.81	0.6982			a	b		
	80	8.92 ± 1.03	0.5524			a			
	Control	25.86 ± 2.50	0.6602				X		
	10	22.25 ± 1.72	0.9546			a			
12	20	21.36 ± 2.16	0.7794	0.8604	0.0542	a			
	40	20.12 ± 1.84	0.6710			a			
	80	17.44 ± 2.47	0.8068			a			
	Control	32.14 ± 3.05	0.7002					X	
	10	29.95 ± 4.25	0.6900				b		
16	20	21.36 ± 2.16	0.5467	0.8871	0.0291	a	b		
	40	23.28 ± 4.27	0.8115			a	b		
	80	19.35 ± 3.34	0.8773			a			

* Expressed as average ± robust confidence interval. † Values denoted with identical letters do not differ significantly.

The antifungal activity of the propolis ethanolic extract on wood blocks exposed to *T. versicolor* for 4, 8, 12, and 16 weeks (Table 4) was more dependent on the concentration as a function of time than the other two protective agents evaluated in this work. A clear relationship between propolis concentration and wood weight loss was observed, as they were inversely proportional in every sampling period (e.g., in the first sampling, the weight loss decreased from of 13.7% to 9.8% as propolis concentration was increased from 5 to 40 mg/mL). Although all propolis ethanolic extract concentrations showed significant differences vs. the control, a gradual decrease in the antifungal activity of these treatments over time was evidenced. The weight loss in the first 8 weeks was higher than that obtained for chitosan (14.3% for propolis vs. 10.1% for chitosan), but after 16 weeks the weight loss values were comparable for the two treatments (21.0% vs. 23.5%).

Table 4. Effect of treatment concentration of propolis ethanolic extract and time of fungus exposure (sampling) on the protection of poplar wood against *T. versicolor*.

Exposure Period (weeks)	Propolis Concentration (mg/mL)	Weight Loss (%) *	Shapiro–Wilk Test *p*-Value	Bartlett *p*-Value	ANOVA *p*-Value	Homogeneous Groups †
	Control	14.212 ± 2.065	0.3225			X
	5	13.742 ± 0.650	0.8964			c
4	10	12.116 ± 0.565	0.4721	0.863	7.91×10^{-8}	b
	20	11.113 ± 0.627	0.9877			b
	40	9.775 ± 0.458	0.9794			a
	Control	19.897 ± 1.801	0.505			X
	5	17.870 ± 1.383	0.829			c
8	10	15.416 ± 0.955	0.9422	0.5844	5.68×10^{-7}	b
	20	12.587 ± 1.561	0.8152			a
	40	11.115 ± 0.877	0.6586			a
	Control	21.036 ± 3.141	0.770			X
	5	19.892 ± 1.620	0.7587			b
12	10	18.639 ± 1.321	0.8894	0.5511	0.713	b
	20	17.413 ± 1.053	0.8444			a
	40	16.288 ± 0.810	0.7739			a
	Control	28.795 ± 2.841	0.2954			X
	5	24.934 ± 2.133	0.9515			c
16	10	22.456 ± 1.614	0.8911	0.9257	0.00014	b c
	20	19.439 ± 1.664	0.7038			b
	40	17.105 ± 1.690	0.2817			a

* Expressed as average ± robust confidence interval. † Values denoted with identical letters do not differ significantly.

The optical microscopy images confirmed that the wood blocks exposed to the attack of *T. versicolor* without any protective treatment (control) suffered a progressive degradation over time, with weight loss values ranging from 13.7%–14.5% after 4 weeks to 24.8%–38.1% after 16 weeks. Figure 1 shows the degradation of vessel elements and fibers on untreated wood upon exposure to *T. versicolor* for 16 weeks, whereas in the case of the sample treated with AgNPs the observed biodeterioration was very small. A micrograph of a healthy poplar wood sample with no degradation is shown for comparison purposes.

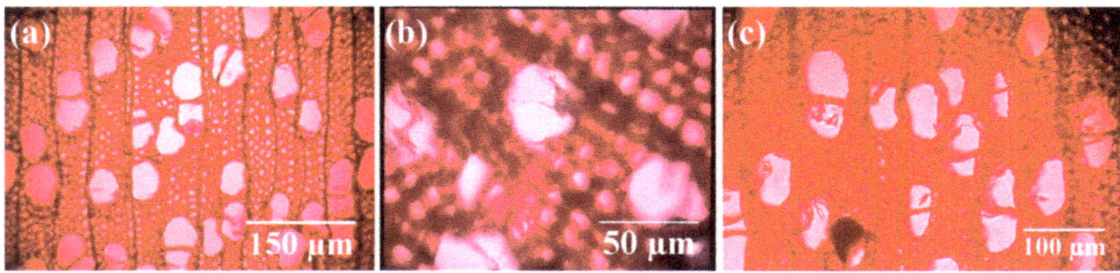

Figure 1. Optical micrographs of (**a**) undecayed control wood sample; (**b**) decayed control wood sample; and (**c**) sample treated with silver nanoparticles (AgNPs) (20 ppm), subjected to the action of *T. versicolor* for 16 weeks.

The SEM images (Figure 2) also confirmed that the fungal attack caused severe damage to the wood cell wall after 16 weeks of exposure in control samples (Figure 2a). The damage was reduced in the samples treated with propolis (Figure 2b), suggesting that propolis had a protective activity.

However, the best wood preservation was attained for silver nanoparticles: as shown in Figure 2c, the cell wall was practically intact. In Figure 2d, at 5000× magnification, silver nanoparticles can be observed on the cell walls. The composition of AgNPs was confirmed by EDX elemental mapping.

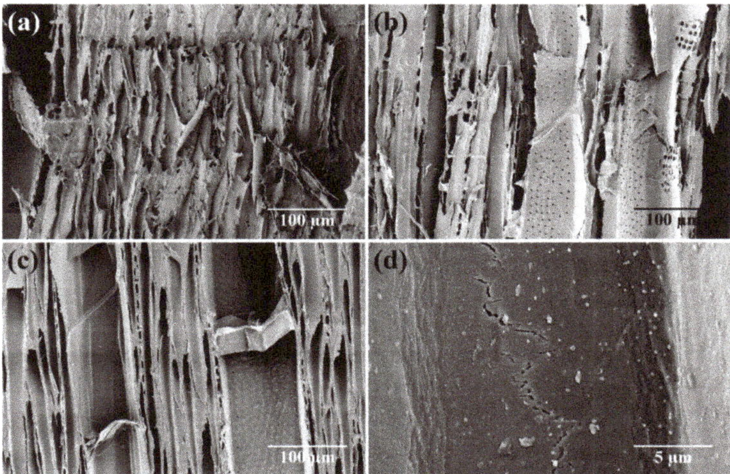

Figure 2. SEM images of wood samples after 16 weeks of exposure to *T. versicolor*: (**a**) control samples, (**b**) samples treated with propolis (40 mg/mL), (**c**) samples treated with silver nanoparticles (20 ppm), and (**d**) zoomed-in view of silver nanoparticles on the cell wall.

4. Discussion

The optical microscopy images showed that wooden blocks exposed to *T. versicolor* attack without any protective treatment (control) suffered a significant degradation of the cell wall, thus confirming the high susceptibility of poplar wood to white-rot fungus [6]. This was further evidenced in the SEM micrographs, which also showed that the fungal attack caused severe damage to the wood cell wall after 16 weeks of exposure in control samples: the middle lamella was degraded and the fibers were separated due to lignin degradation [38].

In relation to the antifungal effectiveness of the pressure treatment with silver nanoparticles, the obtained results were similar to those of Akhtari et al. [39]. These authors evaluated the effectiveness of aqueous solutions of silver nanoparticles against *T. versicolor*, albeit at much higher concentrations (400 ppm), and reported a weight loss of only 2.1% for the treated wood 4 weeks after the inoculation. The low weight losses of around 8.5% for the pieces treated with nanosilver are in agreement with Clausen et al. [13] and Matsunaga et al. [14], who suggested that the advantageous behavior of nanometal-based wood protectors would arise from the slow and controlled release of the bioactive metal ions into the wood cells.

Apropos of the observed decrease in the antifungal activity of chitosan oligomers over time, it was also reported by Alfredsen et al. [26], who noted that the protective nature of chitosan in the long term is lower than those of other commercial fungicides, thus confirming that organic compounds are more prone to biodegradation [8]. According to Larnøy et al. [40], the decrease in the effectiveness of chitosan over time could be due to fungal degradation, as it would be affected by the action of enzymes excreted by fungi in cellulose degradation. Other studies that investigated the applicability of chitosan oligomers as a wood protection agent also observed a gradual loss of its antifungal activity over time [30,40,41].

A gradual decrease in the antifungal activity of propolis treatments over time was also noticed, which may be ascribed either to biodegradability or to a lower retention of the propolis ethanolic extract solution in wood, both of which are very common problems in organic biocides [8].

Consequently, although the use of these two organic bioagents may be improved by adding stabilizers to the solution, the nanometal-based approach would be the preferred option for industrial applications.

5. Conclusions

On the basis of biocidal efficacy tests conducted on wood blocks, more realistic than routine antimicrobial susceptibility tests conducted in agar plates, it was evidenced that the three protective agents assayed in the vacuum-pressure treatments (AgNPs, chitosan, and propolis) showed a protective effect on *Populus* spp. wood. However, the one based on AgNPs featured the highest effectiveness against *T. versicolor* over time: even at the lowest concentration (5 ppm), it minimized the biodeterioration of poplar wood for over 4 months, albeit higher concentrations (of up to 20 ppm) were more effective. SEM images confirmed the effectiveness of the impregnation with nanosilver and evidenced the severe cell wall degradation by the fungus on untreated samples. Chitosan oligomers and propolis-based treatments also showed results with significant differences vs. the non-treated wood, but a noticeable decrease in their effectiveness over the 16 weeks was observed, pointing to either biodegradation or adsorption issues (to which all organic bioactive agents are susceptible). Hence, AgNPs can be put forward as effective protective agents for poplar-wood-based engineering products.

Author Contributions: Conceptualization, M.M.C.-S., J.M.-G., and L.A.-R.; Formal analysis, P.M.-R. and L.A.-R.; Funding acquisition, M.M.C.-S.; Investigation, M.M.C.-S., I.S.-C., L.P.-H., and J.M.-G.; Methodology, M.M.C.-S., P.M.-R., J.M.-G., and L.A.-R.; Resources, M.M.C.-S., J.M.-G., and L.A.-R.; Supervision, M.M.C.-S., J.M.-G., and L.A.-R.; Validation, I.S.-C., L.P.-H., and L.A.-R.; Visualization, P.M.-R.; Writing—original draft, M.M.C.-S., I.S.-C., L.P.-H., and P.M.-R.; Writing—review & editing, P.M.-R. and L.A.-R.

Funding: This study was funded by JUNTA DE CASTILLA Y LEÓN under project VA258P18, with FEDER co-funding.

Acknowledgments: P.M.-R. acknowledges the support of Universidad de Zaragoza under project UZ2019-TEC-07. I.S.-C. would like to gratefully acknowledge the financial support of the National Council for Science and Technology (CONACYT) of Mexico, through PhD Scholarship ref. no. 329975. The authors also wish to thank Melissa Boyd for kindly revising the use of English language.

Conflicts of Interest: The authors declare no conflict of interest.

References

1. Karimi, A.; Taghiyari, H.R.; Fattahi, A.; Karimi, S.; Ebrahimi, G.; Tarmian, A. Effects of wollastonite nanofibers on biological durability of poplar wood (*Populus nigra*) against *Trametes versicolor*. *BioResources* **2013**, *8*, 4134. [CrossRef]
2. Casado, M.; Acuña, L.; Basterra, L.-A.; Ramón-Cueto, G.; Vecilla, D. Grading of structural timber of *Populus × euramericana* clone I-214. *Holzforschung* **2012**, *66*, 633. [CrossRef]
3. Todaro, L.; Russo, D.; Cetera, P.; Milella, L. Effects of thermo-vacuum treatment on secondary metabolite content and antioxidant activity of poplar (*Populus nigra* L.) wood extracts. *Ind. Crop. Prod.* **2017**, *109*, 384–390. [CrossRef]
4. Kollert, W.; Borodowski, E.D. Situación de las Salicáceas en el Mundo. In Proceedings of the Jornadas de Saliáceas 2014—IV Congreso Internacional de Salicáceas en Argentina, Buenos Aires, Argentina, 18–21 March 2014; p. 10.
5. Diaz, B.; Murace, M.; Peri, P.; Keil, G.; Luna, L.; Otaño, M.Y. Natural and preservative-treated durability of *Populus nigra* cv *Italica* timber grown in Santa Cruz Province, Argentina. *Int. Biodeterior. Biodegrad.* **2003**, *52*, 43–47. [CrossRef]
6. Xing, J.-Q.; Ikuo, M.; Wakako, O. Natural resistance of two plantation woods *Populus × canadensis* cv. and *Cunninghamia lanceolata* to decay fungi and termites. *For. Stud. China* **2005**, *7*, 36–39. [CrossRef]
7. Pandey, K.K.; Pitman, A.J. FTIR studies of the changes in wood chemistry following decay by brown-rot and white-rot fungi. *Int. Biodeterior. Biodegrad.* **2003**, *52*, 151–160. [CrossRef]

8. Singh, T.; Singh, A.P. A review on natural products as wood protectant. *Wood Sci. Technol.* **2011**, *46*, 851–870. [CrossRef]
9. Thakur, V.K.; Thakur, M.K. Recent advances in graft copolymerization and applications of chitosan: A review. *ACS Sustainable Chem. Eng.* **2014**, *2*, 2637–2652. [CrossRef]
10. Quiroga, E.N.; Sampietro, D.A.; Soberon, J.R.; Sgariglia, M.A.; Vattuone, M.A. Propolis from the northwest of Argentina as a source of antifungal principles. *J. Appl. Microbiol.* **2006**, *101*, 103–110. [CrossRef]
11. Torlak, E.; Sert, D. Antibacterial effectiveness of chitosan–propolis coated polypropylene films against foodborne pathogens. *Int. J. Biol. Macromol.* **2013**, *60*, 52–55. [CrossRef]
12. Kartal, S.N.; Green, F.; Clausen, C.A. Do the unique properties of nanometals affect leachability or efficacy against fungi and termites? *Int. Biodeterior. Biodegrad.* **2009**, *63*, 490–495. [CrossRef]
13. Clausen, C.A.; Yang, V.W.; Arango, R.A.; Green, F., III. Feasibility of nanozinc oxide as a wood preservative. *Proc. Am. Wood Prot. Assoc.* **2009**, *105*, 255–260.
14. Matsunaga, H.; Kiguchi, M.; Evans, P.D. Microdistribution of copper-carbonate and iron oxide nanoparticles in treated wood. *J. Nanopart. Res.* **2008**, *11*, 1087–1098. [CrossRef]
15. Marzbani, P.; Mohammadnia-afrouzi, Y. Investigation on leaching and decay resistance of wood treated with nano-titanium dioxide. *Adv. Environ. Biol.* **2014**, *8*, 974–979.
16. Nair, S.; Pandey, K.K.; Giridhar, B.N.; Vijayalakshmi, G. Decay resistance of rubberwood (*Hevea brasiliensis*) impregnated with ZnO and CuO nanoparticles dispersed in propylene glycol. *Int. Biodeterior. Biodegrad.* **2017**, *122*, 100–106. [CrossRef]
17. Dorau, B.; Arango, R.; Green, F. *An Investigation into the Potential of Ionic Silver as a Wood Preservative, Proceedings from the Woodframe Housing Durability and Disaster Issues Conference, Las Vegas, NV, USA, 4–6 October 2004*; Forest Products Society: Las Vegas, NV, USA, 2004; pp. 133–145.
18. Velmurugan, N.; Kumar, G.G.; Han, S.S.; Nahm, K.S.; Lee, Y.S. Synthesis and characterization of potential fungicidal silver nano-sized particles and chitosan membrane containing silver particles. *Iran. Polym. J.* **2009**, *18*, 383–392.
19. Kaur, P.; Thakur, R.; Choudhary, A. An in vitro study of the antifungal activity of silver/chitosan nanoformulations against important seed borne pathogens. *Int. J. Sci. Technol. Res.* **2012**, *1*, 83–86.
20. Narayanan, K.B.; Park, H.H. Antifungal activity of silver nanoparticles synthesized using turnip leaf extract (*Brassica rapa* L.) against wood rotting pathogens. *Eur. J. Plant Pathol.* **2014**, *140*, 185–192. [CrossRef]
21. Silva-Castro, I.; Martín-García, J.; Diez, J.J.; Flores-Pacheco, J.A.; Martín-Gil, J.; Martín-Ramos, P. Potential control of forest diseases by solutions of chitosan oligomers, propolis and nanosilver. *Eur. J. Plant Pathol.* **2017**, *150*, 401–411. [CrossRef]
22. Kim, S.W.; Jung, J.H.; Lamsal, K.; Kim, Y.S.; Min, J.S.; Lee, Y.S. Antifungal effects of silver nanoparticles (AgNPs) against various plant pathogenic fungi. *Mycobiology* **2018**, *40*, 53–58. [CrossRef]
23. Ivask, A.; ElBadawy, A.; Kaweeteerawat, C.; Boren, D.; Fischer, H.; Ji, Z.; Chang, C.H.; Liu, R.; Tolaymat, T.; Telesca, D.; et al. Toxicity mechanisms in *Escherichia coli* vary for silver nanoparticles and differ from ionic silver. *ACS Nano* **2013**, *8*, 374–386. [CrossRef] [PubMed]
24. Bin Ahmad, M.; Lim, J.J.; Shameli, K.; Ibrahim, N.A.; Tay, M.Y. Synthesis of silver nanoparticles in chitosan, gelatin and chitosan/gelatin bionanocomposites by a chemical reducing agent and their characterization. *Molecules* **2011**, *16*, 7237–7248. [CrossRef] [PubMed]
25. Goycoolea, F. Monografía XXVIII: Nanotecnología Farmacéutica. Available online: https://www.analesranf.com/index.php/mono/article/view/990/1024 (accessed on 7 October 2019).
26. Alfredsen, G.; Eikenes, M.; Militz, H.; Solheim, H. Screening of chitosan against wood-deteriorating fungi. *Scand. J. For. Res.* **2011**, *19*, 4–13. [CrossRef]
27. Xia, W.; Liu, P.; Zhang, J.; Chen, J. Biological activities of chitosan and chitooligosaccharides. *Food Hydrocoll.* **2011**, *25*, 170–179. [CrossRef]
28. Rabea, E.I.; Badawy, M.E.T.; Stevens, C.V.; Smagghe, G.; Steurbaut, W. Chitosan as antimicrobial agent: Applications and mode of action. *Biomacromolecules* **2003**, *4*, 1457–1465. [CrossRef] [PubMed]
29. Badawy, M.E.I.; Rabea, E.I. Potential of the biopolymer chitosan with different molecular weights to control postharvest gray mold of tomato fruit. *Postharvest Biol. Technol.* **2009**, *51*, 110–117. [CrossRef]
30. Silva-Castro, I.; Casados-Sanz, M.; Alonso-Cortés, A.; Martín-Ramos, P.; Martín-Gil, J.; Acuña-Rello, L. Chitosan-based coatings to prevent the decay of *Populus* spp. wood caused by *Trametes versicolor*. *Coatings* **2018**, *8*, 415. [CrossRef]

31. Silva-Castro, I.; Diez, J.; Martín-Ramos, P.; Pinto, G.; Alves, A.; Martín-Gil, J.; Martín-García, J. Application of bioactive coatings based on chitosan and propolis for *Pinus* spp. protection against *Fusarium circinatum*. *Forests* **2018**, *9*, 685. [CrossRef]
32. Burdock, G.A. Review of the biological properties and toxicity of bee propolis (propolis). *Food Chem. Toxicol.* **1998**, *36*, 347–363. [CrossRef]
33. Marcucci, M.C. Propolis: Chemical composition, biological properties and therapeutic activity. *Apidologie* **1995**, *26*, 83–99. [CrossRef]
34. Oryan, A.; Alemzadeh, E.; Moshiri, A. Potential role of propolis in wound healing: Biological properties and therapeutic activities. *Biomed. Pharmacother.* **2018**, *98*, 469–483. [CrossRef] [PubMed]
35. Matei, P.M.; Martin-Ramos, P.; Sanchez-Bascones, M.; Hernandez-Navarro, S.; Correa-Guimaraes, A.; Navas-Gracia, L.M.; Rufino, C.A.; Ramos-Sanchez, M.C.; Martin-Gil, J. Synthesis of chitosan oligomers/propolis/silver nanoparticles composite systems and study of their activity against *Diplodia seriata*. *Int. J. Polym. Sci.* **2015**, *2015*, 864729. [CrossRef]
36. Costa, C.N.; Teixeira, V.G.; Delpech, M.C.; Souza, J.V.S.; Costa, M.A.S. Viscometric study of chitosan solutions in acetic acid/sodium acetate and acetic acid/sodium chloride. *Carbohydr. Polym.* **2015**, *133*, 245–250. [CrossRef] [PubMed]
37. Araujo-Rufino, C.; Fernandes-Vieira, J.; Martín-Ramos, P.; Silva-Castro, I.; Fernandes-Correa, M.; Matei, P.M.; Sánchez-Báscones, M.; Ramos-Sánchez, M.C.; Martín-Gil, J. Synthesis of chitosan oligomers composite systems and study of their activity against *Bipolaris Oryzae*. *J. Mater. Sci. Eng. Adv. Technol.* **2016**, *13*, 29–52.
38. Schwarze, F.W.M.R. Wood decay under the microscope. *Fungal Biol. Rev.* **2007**, *21*, 133–170. [CrossRef]
39. Akhtari, M.; Arefkhani, M. Study of microscopy properties of wood impregnated with nanoparticles during exposed to white-rot fungus. *Agric. Sci. Dev.* **2013**, *2*, 116–119.
40. Larnøy, E.; Eikenes, M.; Militz, H. Evaluation of factors that have an influence on the fixation of chitosan in wood. *Wood Mater. Sci. Eng.* **2006**, *1*, 138–145. [CrossRef]
41. El-Gamal, R.; Nikolaivits, E.; Zervakis, G.I.; Abdel-Maksoud, G.; Topakas, E.; Christakopoulos, P. The use of chitosan in protecting wooden artifacts from damage by mold fungi. *Electron. J. Biotechnol.* **2016**, *24*, 70–78. [CrossRef]

Article

Impact of Water Holding Capacity and Moisture Content of Soil Substrates on the Moisture Content of Wood in Terrestrial Microcosms

Christian Brischke * and Friedrich L. Wegener

Wood Biology and Wood Products, University of Goettingen, Buesgenweg 4, D-37077 Goettingen, Germany; f.wegener@stud.uni-goettingen.de

* Correspondence: christian.brischke@uni-goettingen.de; Tel.: +49-551-3929514

Received: 20 May 2019; Accepted: 30 May 2019; Published: 4 June 2019

Abstract: Terrestrial microcosms (TMCs) are frequently used for testing the durability of wood and wood-based materials, as well as the protective effectiveness of wood preservatives. In contrary to experiments in soil ecology sciences, the experimental setup is usually rather simple. However, for service life prediction of wood exposed in ground, it is of imminent interest to better understand the different parameters defining the boundary conditions in TMCs. This study focused, therefore, on soil–wood–moisture interactions. Terrestrial microcosms were prepared from the same compost substrate with varying water holding capacities (WHCs) and soil moisture contents (MC_{soil}). Wood specimens were exposed to 48 TMCs with varying WHCs and MC_{soil}. The wood moisture content (MC_{wood}) was studied as well as its distribution within the specimens. For this purpose, the compost substrate was mixed with sand and peat and its WHC was determined using two methods in comparison, i.e., the "droplet counting method" and the "cylinder sand bath method" in which the latter turned out advantageous over the other. The MC_{wood} increased generally with rising MC_{soil}, but WHC was often negatively correlated with MC_{wood}. The distance to water saturation S_{soil} from which MC_{wood} increased most intensively was found to be wood-species specific and might, therefore, require further consideration in soil-bed durability-testing and service life modelling of wood in soil contact.

Keywords: decay; ENV 807; soft rot test; soil moisture content; use class 4 (UC4)

---

1. Introduction

For determining the durability of wood or the protective effectiveness of wood preservatives against soft rot fungi and other soil-inhabiting micro-organisms, terrestrial microcosms (TMCs) can be used. For this purpose, natural top soil or a fertile loam-based horticultural soil should be used and various requirements need to be fulfilled with respect to the soil substrate.

It is well known that many parameters affect the decay activity of soils [1–5]. Therefore, it is recommended to consider more than one in-ground field test site for durability testing of wood and more than one soil substrate for laboratory studies using TMCs [2,6–9].

Consequently, for standardized test protocols several parameters are more or less strictly defined. For instance, according to the European standard CEN/TS 15083-2 [10], the following soil-related boundary conditions need to be assured:

- pH 6–8
- no added agrochemical
- water holding capacity (WHC): 25%–60%
- natural soils: peat or top 50 mm removed and not taken from a depth below 200 mm

- soil collected in moist conditions
- soil passed through a sieve of nominal aperture size 12.5 mm
- storage of soil prior to use only in closed moisture proof containers
- thorough mixing of soil before use
- horticultural soil which was sterilized during its preparation needs to be replenished with 20% natural soil
- soil not used before

Finally, a moisture content of the soil (MC_{soil}) equivalent to 95% of its WHC is required and the TMC should be stored at 27 °C ± 2 °C and 70% ± 5% relative humidity (RH) during the whole period of exposure in a dark room.

A previous study by Wälchli [11] showed that MC_{wood} decreased with both decreasing MC_{soil} and WHC as determined for two different soils and five different MC_{soil}. However, mass loss (ML) by decay of untreated and differently copper–chromium–boron (CCB)-treated Scots pine sapwood was neither correlated with MC_{wood} nor with MC_{soil}. Similarly, Mieß [12] found an increase in MC_{wood} with rising MC_{soil} in three different soil types and for different untreated and modified timbers. Furthermore, she found a gradient of MC_{wood} in untreated wood from the highest MC_{wood} in the bottom part and lowest MC_{soil} in the top part of the buried test stakes. In contrast, a remarkable 20% of the Scots pine sapwood specimens showed the highest MC_{wood} in the top or central part of the specimens. Mieß [12] suggested that the MC_{wood} gradients were the consequence of vertical gradients of MC_{soil}, which were differently severe due to the different soil wetting and re-drying regimes. It is further likely that the gradients were the consequence of ML gradients along the stake-shaped specimens, because the MC_{wood} of the different specimen segments had been determined not before the end of the test after 17 weeks of incubation when significant ML had already occurred.

Gray [13] performed durability tests in TMCs using different soils at different MC_{soil} and found that the highest ML occurred at an MC_{soil} between 108% and 148% of the WHC of the respective soil. The highest MC_{wood} after harvesting was found at an MC_{soil} between 120% and 218% of its WHC referring to an MC_{soil} at approximately 40% in all soil types used. Thus, ML increased with increasing MC_{soil}, but found an optimum, which was, however, far beyond the recommended 95% WHC. Again, MC_{wood} data are needed to obtain a set perspective, since they refer to the different severely decayed specimens after harvesting.

In summary, it becomes evident that both WHC and MC_{soil} influence MC_{wood} and ML through fungal decay, and do seemingly interact. Clear relationships between the three moisture-related parameters have not yet been established.

Others [6,12,14,15] previously demonstrated that all three rot types, i.e., brown, white, and soft rot, occur in TMCs complemented by tunneling, erosion, and cavity bacteria. However, neither MC_{wood} nor MC_{soil} seemed to limit their occurrence. Solely, soft rot apparently copes better with very high moisture contents, which are not favorable for brown and white rot fungi. Nevertheless, soft rot fungi can degrade wood in a rather large moisture range. They are early colonizers, so-called "ruderal organisms" [16], which, in contrast to basidiomycetes ('combative organisms'), are rarely able to take over a substrate [17].

The WHC of soil substrates can vary remarkably, and therefore, it needs to be determined before each test. In both standards, CEN/TS 15083–2 [10] as well as ENV 807 [18], a suitable method for determining the WHC of soil is described: the so-called "droplet counting method". The method is based on determining the ability of a sample of a test substrate to retain water against the pull of a vacuum pump, as a measure of its WHC. However, the method is rather laborious and time consuming. Furthermore, the standards lack a definition of the vacuum that needs to be applied, wherefore one might question the reproducibility of the test results.

Within this study, we conducted comparative WHC measurements on a series of different mixtures of compost and silica sand using the "droplet counting method" and an alternative method according to ISO 11268-2 [19], where wet soil samples are allowed to drain on a sand bath. Based on this comparison of methods, TMCs should be prepared representing soil substrates of varying WHCs and MC_{soil}. The overall objective of this study was to establish relationships between WHC, MC_{soil}, and the resulting MC_{wood} of different wood species after exposure in the TMC.

2. Materials and Methods

2.1. Soil Substrates

Three soil substrates were used to prepare TMCs of defined water holding capacities (WHCs). The basis substrate was a compost produced by the University of Goettingen from horticultural waste (i.e., leaf litter, grass, cut softwoods, and hardwoods, sand). To lower its WHC, silica sand (grain size > 0.2 mm) was added; to increase its WHC, peat (moderately-to-severely decomposed high-moor peat (H3–H8), total nitrogen 0.35%, magnesium 0.15%, organic substance 30%) was added. Both peat and compost were passed through a sieve of a nominal aperture size 8.5 mm. The soil moisture content (MC_{soil}) and the WHC were determined according to the "droplet counting method" and the "cylinder sand bath method".

2.2. Determination of the soil Moisture Content (MC_{soil})

Soil samples of 7–64 g (depending on the soil density) were taken for determining the soil moisture content (MC_{soil}). Three replicate samples were taken, weighed to the nearest 0.01 g, oven-dried at 103 °C, and weighed again. MC_{soil} was calculated as follows:

$$MC_{soil} = \frac{m_{soil,\ wet} - m_{soil,0}}{m_{soil,0}} \times 100\ [\%] \tag{1}$$

where MC_{soil} is the soil moisture content, in %; $m_{soil,wet}$ is the wet soil mass, in g; $m_{soil,0}$ is the oven-dry soil mass, in g.

2.3. Determination of the Water Holding Capacity (WHC) of Soil

2.3.1. "Droplet Counting Method"

A small quantity of water was added to soil samples of 200 g, the substrate was mixed well, and the operation was repeated until the soil particles stuck to another (crumb structure). Further, 25 mL water were added, mixed well, and allowed to stand for 2 h. A coarse filter paper was placed in the bottom of a Buchner funnel (100 mm diameter) and moistened to seal the filter paper to the funnel. The prepared test sample was transferred into the funnel and spread evenly. The bottom of the Buchner funnel was covered by soil substrate to a height of at least 10 mm. Suction was applied using a vacuum pump until no more than five drops of water per minute were being withdrawn from the sample, increasing the suction slowly to avoid perforation of the filter paper. The sample was transferred to an aluminum container of known mass and weighed. The container was oven-dried at 103 °C ± 2 °C and weighed again. The *WHC* of $n = 5$ peat samples and $n = 3$ sand and compost samples was determined and calculated as follows:

$$WHC = \frac{m_{soil,\ saturated} - m_{soil,0}}{m_{soil,0}} \times 100\ [\%] \tag{2}$$

where *WHC* is the water holding capacity, in %; $m_{soil,saturated}$ is the soil mass at saturation, in g; $m_{soil,0}$ is the oven-dry soil mass, in g.

2.3.2. "Cylinder Sand Bath Method"

Soil was inserted into polyethylene cylinders with 4 cm diameters. The bottoms of the cylinders were covered with a fine polymer grid and filter paper (MN 640 W 70 mm). All cylinders were placed in a vat for 3 h, which was filled with water to a height 1 cm above the soil filling height of 7 cm. After soaking the soil in water, the cylinders were placed on a water-saturated sand bath for 2 h to allow the unbound water to drain. The soil samples were then weighed wet, oven-dried at 103 °C ± 2 °C, and the WHC of the soil was calculated according to Equation (2) analogously to the "droplet counting method".

2.4. Preparation of Mixed Soil Substrates

For a comparison of the "Droplet counting method" and the "cylinder sand bath method" and for establishing a regression between mixing ratios of the different soil substrates and their resulting WHC, a total of 22 soil substrate mixtures were prepared as summarized in Table 1.

Table 1. Mixing ratios of soil substrates for water holding capacity (WHC) tests. Percentage is based on the oven-dried mass.

Percentage Compost (%)	**100**	**95**	**90**	**80**	**70**	**60**	**50**	**40**	**30**	**20**	**10**
Percentage silica sand/peat (%)	0	5	10	20	30	40	50	60	70	80	90

For preparing mixed soil substrates, the following equation was used:

$$m_{soil\ x,target,\ wet} = m_{target,\ total,\ 0} \times \left(\frac{a_{target,0}}{100}\right) \times \left(\frac{100}{100 - MC_{soil\ x}}\right)\ (g) \tag{3}$$

where $m_{soil\ x.target,\ wet}$ is the target mass of the wet substrate x, in g; $m_{target,\ total,\ 0}$ is the target oven-dry mass of the total mix, in g; $a_{target,0}$ is the target percentage of substrate x based on the oven-dry mass, in %. $MC_{soil\ x,}$ is the soil moisture content of substrate x, in %.

2.5. Terrestrial Microcosms (TMCs)

Miniature terrestrial microcosms were prepared in polypropylene containers of 110 (height) × 110 × 80 mm^3 and a volume of 500 mL. In total, 48 different substrates were filled in the containers each to a height of 100 mm. The combinations of the parameters WHC and MC_{soil} are summarized in Table 2, where the latter is expressed as (%WHC). The containers were weighed to the nearest 0.01 g, closed with a lid, and their total mass maintained over a period of three weeks.

The following regression functions were used (see also Section 3.1):

$$WHC_{mix:compost-sand} = -0.586 \times a_{target,\ sand,\ 0} + 80.81\ (\%) \tag{4}$$

$$WHC_{mix:compost-turf} = 2.499 \times a_{target,\ turf,\ 0} + 87.15\ (\%) \tag{5}$$

where $WHC_{mix:compost\text{-}sand}$ is the water holding capacity of compost mixed with silica sand, in %; $WHC_{mix:compost\text{-}peat}$ is the water holding capacity of compost mixed with peat, in %; $a_{target,\ sand,\ 0}$ is the target percentage of sand based on the oven-dry mass, in %; $a_{target,\ peat,\ 0}$ is the target percentage of peat based on the oven-dry mass, in %.

Table 2. Soil moisture content MC_{soil} (%) for combinations of target WHC [1] and target MC_{soil} expressed as (%WHC).

MC_{soil} (%WHC) [1]	**WHC (%) [1]**							
	21.8 [2]	**30.0**	**40.0**	**50.0**	**60.0**	**70.0**	**80.0**	**90.0**
30.0	6.5	9.0	12.0	15.0	18.0	21.0	24.0	27.0
50.0	10.9	15.0	20.0	25.0	30.0	35.0	40.0	45.0
70.0	15.3	21.0	28.0	35.0	42.0	49.0	56.0	63.0
80.0	17.4	24.0	32.0	40.0	48.0	56.0	64.0	72.0
95.0	20.7	28.5	38.0	47.5	57.0	66.5	76.0	85.5
120.0	26.2	36.0	48.0	60.0	72.0	84.0	96.0	108.0

[1] WHC determined according to the "cylinder sand bath method" according to ISO 11268-2 [19]. [2] WHC of pure silica sand was 21.8% and consequently the lowest WHC achieved.

The WHC of the substrates used for TMCs were determined exclusively according to ISO 11268-2 [19]. The basic substrate was compost. Silica sand and peat were added according to the regression obtained by comparative WHC measurements as described in Section 2.4 (Figure 1).

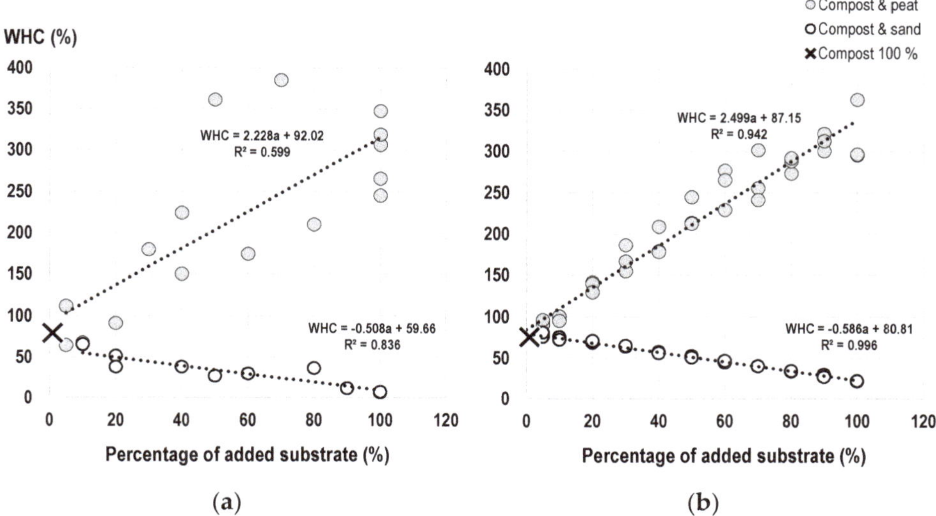

Figure 1. Interrelationship between the percentage of added sand and peat and the WHC of the substrate mixtures: (**a**) WHC determined according to the "droplet counting method" [10]. (**b**) WHC determined according to the "cylinder sand bath method" [19].

The soil mixtures used for the TMCs are summarized in Table 3.

Table 3. WHC of different mixtures of compost with sand and peat.

	WHC (%) [1]							
	21.8 [2]	30.0	40.0	50.0	60.0	70.0	80.0	90.0
Percentage compost (%)	0.0	13.2	30.3	47.4	64.5	81.6	98.6	98.6
Percentage sand (%)	100.0	86.8	69.7	52.6	35.5	18.5	1.4	0.0
Percentage peat (%)	0.0	0.0	0.0	0.0	0.0	0.0	0.0	1.4

[1] WHC determined according to the "cylinder sand bath method" according to ISO 11268-2 [19]. [2] WHC of pure silica sand was 21.8% and consequently the lowest WHC achieved.

2.6. Preparation and Exposure of Wood Specimens

Specimens of 5 × 10 × 100 (ax.) mm^3 were prepared from Scots pine sapwood (*Pinus sylvestris* L.), Douglas fir heartwood (*Pseudotsuga menziesii* Franco), English oak heartwood (*Quercus robur* L.), and European beech (*Fagus sylvatica* L.). All specimens were free from defects such as cracks, decay, and discoloration. For each of the 48 combinations of WHC and MC_{soil}, $n = 5$ replicate specimens of each species were prepared, which corresponded to a total of 960 specimens.

In total, 48 soil substrates, i.e., combinations of WHC and MC_{soil}, were prepared and each was used to fill two containers (miniature TMCs). Wood specimens were conditioned at 20 °C/65% RH until constant mass before soil exposure. Afterwards, ten wood specimens were buried to 4/5 of their length in each container and exposed for three weeks. The MC_{soil} was maintained by adding water about every third day if needed.

2.7. Determination of the Wood Moisture Content (MC_{wood})

Specimens from selected TMCs were used to determine the MC_{wood} distribution within the specimens. Therefore, after harvest, the specimens were cleaned from adhering soil particles, cut into five segments of 20 mm length using ratchet scissors, weighed, dried, and weighed again in each step separately. Segment-wise MC_{wood} was determined on specimens after exposure in TMC with substrates of 30, 60, and 90% WHC, and a MC_{soil} of 30, 75, and 95% of its WHC.

$$MC_{wood} = \frac{m_{wood,\,wet} - m_{wood,\,0}}{m_{wood,0}} \times 100\ (\%) \tag{6}$$

where MC_{wood} is the wood moisture content, in %; $m_{wood,wet}$ is the wet mass of the wood specimen, in g; $m_{wood,0}$ is the oven-dry wood mass, in g.

2.8. Statistical Analysis

Regression functions between different variables were established using the method of least squares to achieve the best fit. Statistical differences between collectives were considered significant at a probability of error less than 5% according to a modified Student *t*-test (Welch test).

3. Results and Discussion

3.1. Water Holding Capacity (WHC) of Soil Mixtures

The WHC of the three initial substrates was highest for peat, followed by compost and sand according to both methods applied (Table 4).

Table 4. WHC (%) of the initial soil substrates determined according to the "droplet counting method" and the "cylinder sand bath method". Standard deviation in parentheses.

	WHC (%) [1]		
	Silica Sand	**Compost**	**Peat**
Droplet counting method	6.7 (0.3)	78.9 (1.3)	296.6 (40.9)
Cylinder sand bath method	21.8 (0.4)	75.5 (1.1)	318.3 (38.8)

[1] Number of replicate samples was $n = 3$ and $n = 5$ for peat tested according to the droplet counting method.

According to both methods, the WHC of the different soil mixtures was linearly correlated with the percentage of added sand or peat, respectively. With increasing percentage of sand and decreasing percentage of peat, the WHC decreased (Figure 1).

It became evident that: (1) single WHC values scattered more and (2) the regression between substrate ratios and WHC was less pronounced when using the "droplet counting method". Furthermore, the following advantages of the "cylinder sand bath method" over the "droplet counting method" became apparent:

- Less time consumption: 7 min/sample compared to more than 60 min/sample using the "droplet counting method".
- Consistency of the test setup: Time for counting droplets varied between 6 and 60 min/sample. Intervals between single droplets varied partly drastically, especially when testing substrates of high WHC. In contrast, up to several hundred cylinders can be used in parallel and the duration of test was always constant.
- Clear description of setup: Simple, less expensive, and well described setup. In contrast, the description of the "droplet counting method" suffers from some vagueness: size of Buchner funnel, applied vacuum, and type of filter paper are not specified, but likely affect the test results.
- Independency from sample size and volume: In contrast, the given sample thickness according to the "droplet counting method" led to strongly varying mass of the soil sample in the Buchner

funnel and affected the WHC, as shown for peat in Figure 2. With increasing sample size (= sample mass) the WHC increased.

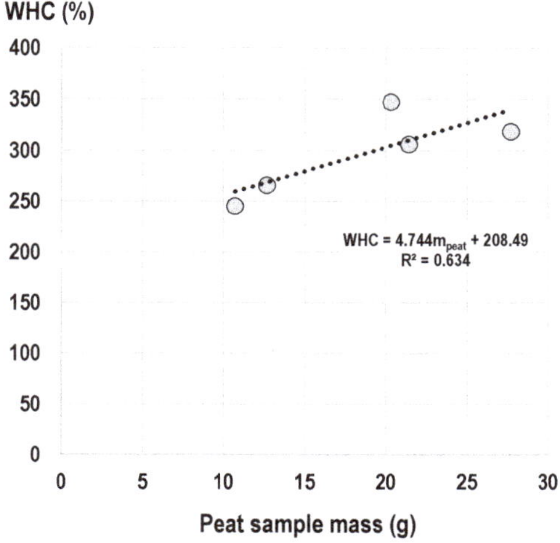

Figure 2. Interrelationship between the WHC of peat according to the 'droplet counting method' (CEN/TS 15083-2, 2005) and the mass of the peat sample.

Generally, it was observed that substrates with very high WHC, such as peat, require longer wetting periods than specified by the standards, i.e., 1–2 h according to CEN/TS 15083-2 [10] and 3 h according to ISO 11268-2 [19]. After 3 h of submersion, the peat was still not fully water saturated when oven-dried before, which consequently led to an underestimation of its WHC.

The WHC determined according to both methods were highly correlated, especially for WHC below 200%, as shown in Figure 3. Therefore, and regarding its numerous advantages, in the following, all WHC measurements were conducted using the "cylinder sand bath method".

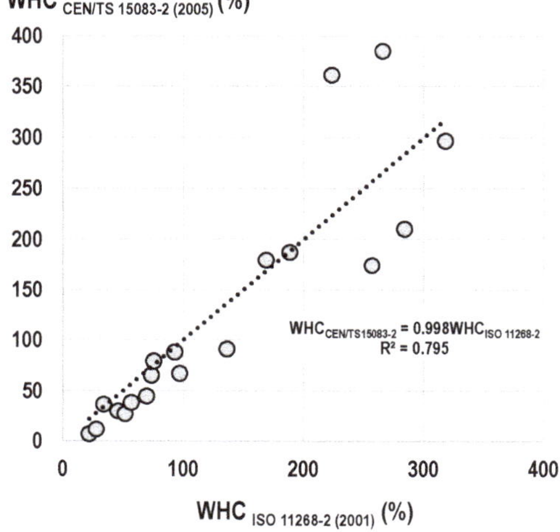

Figure 3. Interrelationship between the WHC of different substrate mixtures according to the "droplet counting method" [10] and the "cylinder sand bath method" [19].

3.2. Impact of WHC and MC_{soil} on the Moisture Content of Wood (MC_{wood}) Exposed in TMCs

After three weeks of exposure in different TMCs, the average MC_{wood} was highest in Scots pine sapwood (88%), followed by English oak (75%), Beech (67%), and Douglas fir (48%). In general, MC_{wood} increased with increasing MC_{soil}, but was strongly dependent on the WHC of the soil. The lower the WHC, the higher was the MC_{wood} at a given MC_{soil}, which coincided with previous findings [12]. Lower WHC in this study corresponded with a higher percentages of silica sand, which can only physically absorb water in contrast to organic soil substrates such as compost soil and peat, which also form chemical bonds with water [20]. The capacity to bind water is therefore higher in organic substrates which restricts the amount of available water which potentially wets the wood specimens.

This effect became especially prominent when considering MC_{soil} as a percentage of the WHC of the soil as illustrated in Figure 4. The lower MC_{soil} (%WHC), the lower the MC_{wood} was at a given WHC of the substrate in the TMCs. However, it also became apparent that the effect of MC_{soil} became more pronounced at higher WHCs, i.e., the range of MC_{wood} between 30% and 120% MC_{soil} expressed as percentage of its WHC was higher by up 2 factors in substrates of high WHC (120%) compared to those with very low WHC (30%).

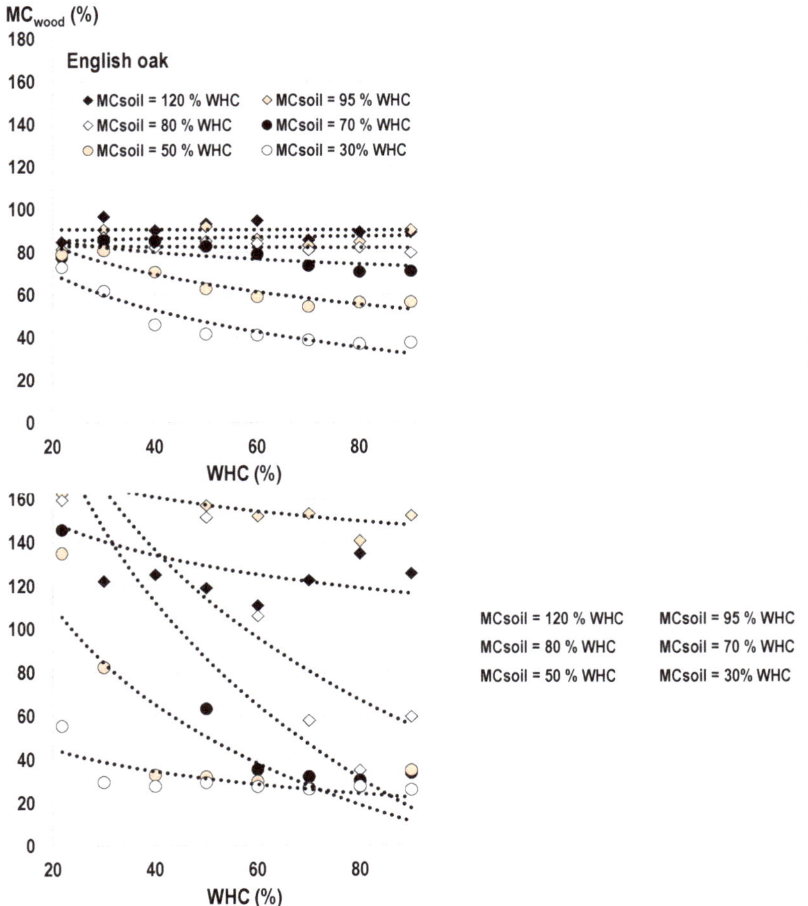

Figure 4. The interrelationship between wood moisture content (MC_{wood}) and WHC for different MC_{soil} expressed as a percentage of the WHC of the TMC (regression functions are shown in Table 5).

Table 5. Regression functions for fitting curves shown in Figure 5 (y = wood moisture content MC_{wood}; x = water holding capacity WHC).

MC_{soil} (%WHC)	English Oak	Beech	Scots Pine Sapwood	Douglas Fir
120	$-0.20\ln(x) + 91.20$	$6.81\ln(x) + 75.78$	$-16.06\ln(x) + 220.01$	$-21.77\ln(x) + 214.39$
95	$1.56\ln(x) + 80.56$	$-8.52\ln(x) + 125.38$	$-21.77\ln(x) + 214.39$	$-3.25\ln(x) + 74.33$
80	$-0.65\ln(x) + 84.98$	$-26.67\ln(x) + 173.92$	$-99.34\ln(x) + 502.65$	$-14.01\ln(x) + 106.82$
70	$-8.17\ln(x) + 110.07$	$-30.11\ln(x) + 175.84$	$-116.8\ln(x) + 543.26$	$-19.14\ln(x) + 121.88$
50	$-20.27\ln(x) + 144.18$	$-22.67\ln(x) + 131.58$	$-66.53\ln(x) + 310.55$	$-15.69\ln(x) + 97.79$
30	$-25.01\ln(x) + 144.87$	$-20.20\ln(x) + 113.81$	$-14.83\ln(x) + 89.25$	$-13.02\ln(x) + 81.44$

The difference between MC_{wood} results achieved after three weeks of in-soil exposure was surprisingly small between Beech and English oak heartwood, because the latter is known to take up water more slowly due to the formation of tyloses in the vessels. In contrast, Beech wood—apart from false heartwood which was excluded in this study— usually takes up liquid water very easy, although its vessel diameters are much smaller compared to the early wood vessels of English oak. Similarly, the maximum MC_{wood} of Douglas fir heartwood was in the same range of that of Scots pine sapwood when exposed in soil at an MC_{soil} of 120% WHC. Solely, at a lower MC_{soil}, the more permeable Scots pine sapwood showed higher MC_{wood} compared to the refractory heartwood of Douglas fir. In summary, it became evident that already after a short exposure period of three weeks in wet soil, wood anatomy-induced differences in moisture uptake diminished confirming previous findings [12].

To further illustrate the interdependency between WHC and MC_{soil} and their effect on MC_{wood}, the distance to water saturation of the soil substrate (S_{soil}) was determined according to Equation (7) and correlated with MC_{wood} (Figure 5).

$$S_{soil} = MC_{soil} - WHC\ (\% - \text{points}) \quad (7)$$

where S_{soil} is the distance to water saturation of the soil substrate, in %-points; MC_{soil} is the soil moisture content, in %; *WHC* is the water holding capacity, in %.

Generally, with increasing S_{soil}, the MC_{wood} increased as well, but followed wood species-specific curves with differently steep increases and a plateau at S_{soil} = 0%, i.e., soil water saturation. For English oak, Beech, and Douglas fir, the MC_{wood} stayed approximately constant at S_{soil} > 0%. Solely, for Scots pine sapwood, MC_{wood} dropped significantly after exceeding the saturation point, although unlimited uptake of liquid water was expected to be provided above this threshold.

The distance to water saturation from which MC_{wood} remarkably rose differed also between wood species and was approximately at −55%-points for English oak, −25%-points for Beech and Douglas fir, and −20%-points for Scots pine sapwood. Scots pine sapwood showed also by far the highest increase in MC_{wood} with increasing S_{soil}, i.e., between 32% and up to 180% MC_{wood} between −20 and 0% S_{soil}.

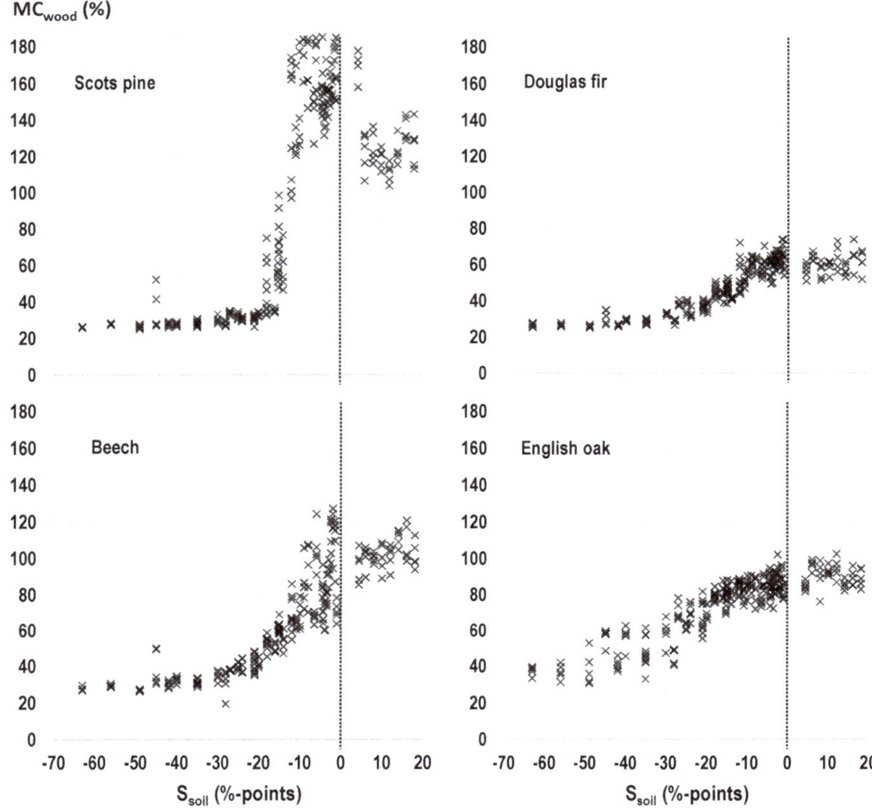

Figure 5. The interrelationship between the distance to water saturation of the soil substrates (S_{soil}) and the wood moisture content (MC_{wood}).

3.3. Moisture Content Gradients in Buried Wood Specimens

The MC_{wood} in specimens buried to 4/5 of their length in TMCs showed partly drastic gradients from high moisture content in the bottom to less in the upper part, which was not buried (Figures 6–9). Solely, Scots pine sapwood specimens exposed at high MC_{soil} (95%WHC) showed barely significant gradients, but very high MC_{wood} in all parts of the specimen. Similarly, deviating MC_{wood} gradients were reported by Mieß [12] for Scots pine sapwood specimens. As expected, generally, the highest difference in MC_{wood} was found between the upper segments and upper next segments.

The MC_{wood} in the upper segments of English oak and Douglas fir specimens was in the range of their equilibrium moisture content (EMC) at fiber saturation. In contrast, the upper segments of Scots pine sapwood and Beech specimens showed MC_{wood} up to 180 and 80% respectively, indicating strong capillary water transport along the specimen axis.

The MC_{wood} gradient in the specimens was positively correlated with MC_{soil} (%WHC). However, two types of MC gradients became apparent: (1) increasing MC_{wood} from the upper to the next segment, but rather constant MC_{wood} below (e.g., English oak at WHC = 30%), and (2) a steady increase of MC_{wood} from the upper to the bottom segment (e.g., Beech and Scots pine sapwood).

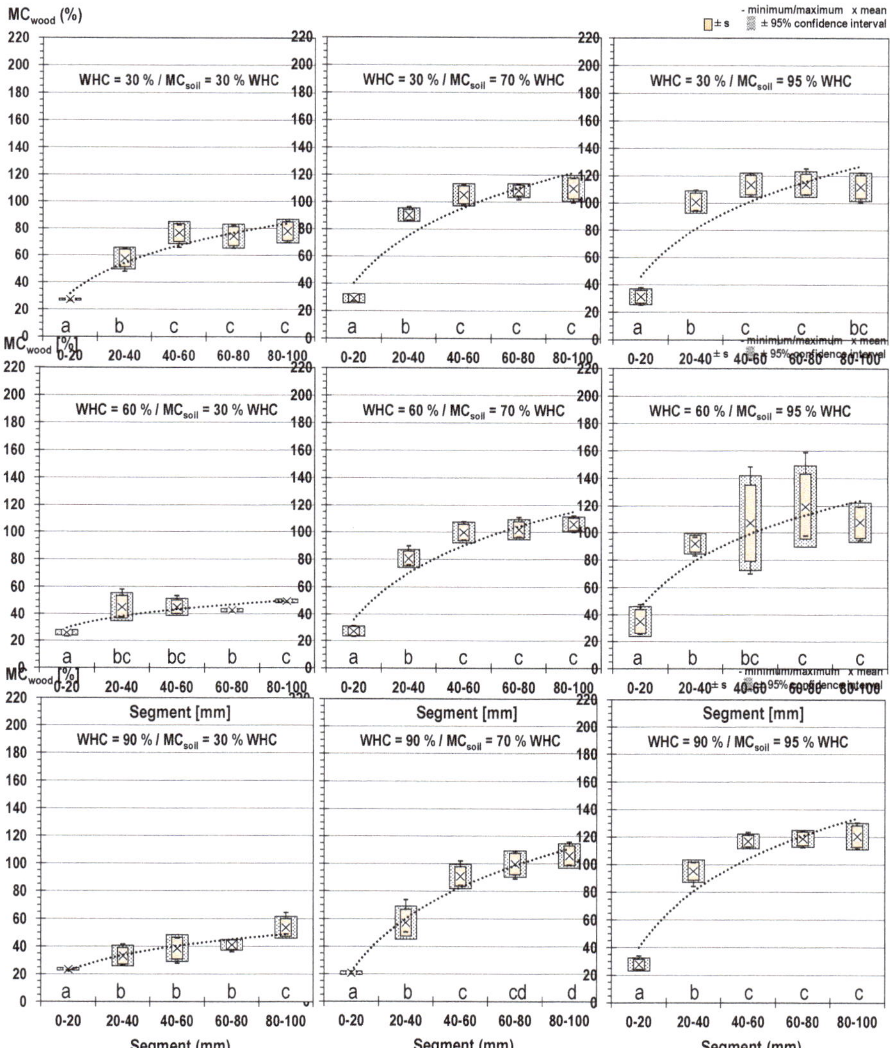

Figure 6. Distribution of MC_{wood} in English oak specimens buried to 4/5 of their length (20–100 mm) in different TMCs. Different letters indicating significant differences between groups at $p < 5\%$ according to a Student *t*-test for non-paired samples.

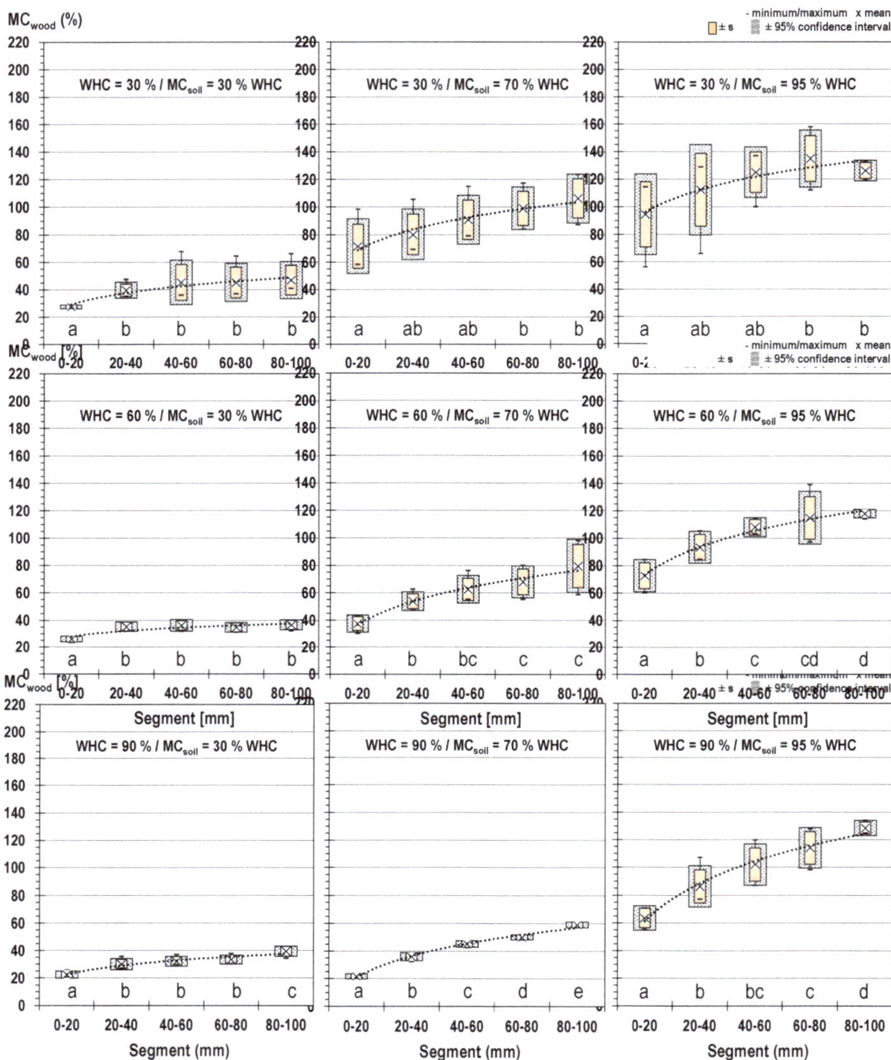

Figure 7. Distribution of MC_{wood} in Beech specimens buried to 4/5 of their length (20–100 mm) in different TMCs. Different letters indicating significant differences between groups at $p < 5\%$ according to a Student *t*-test for non-paired samples.

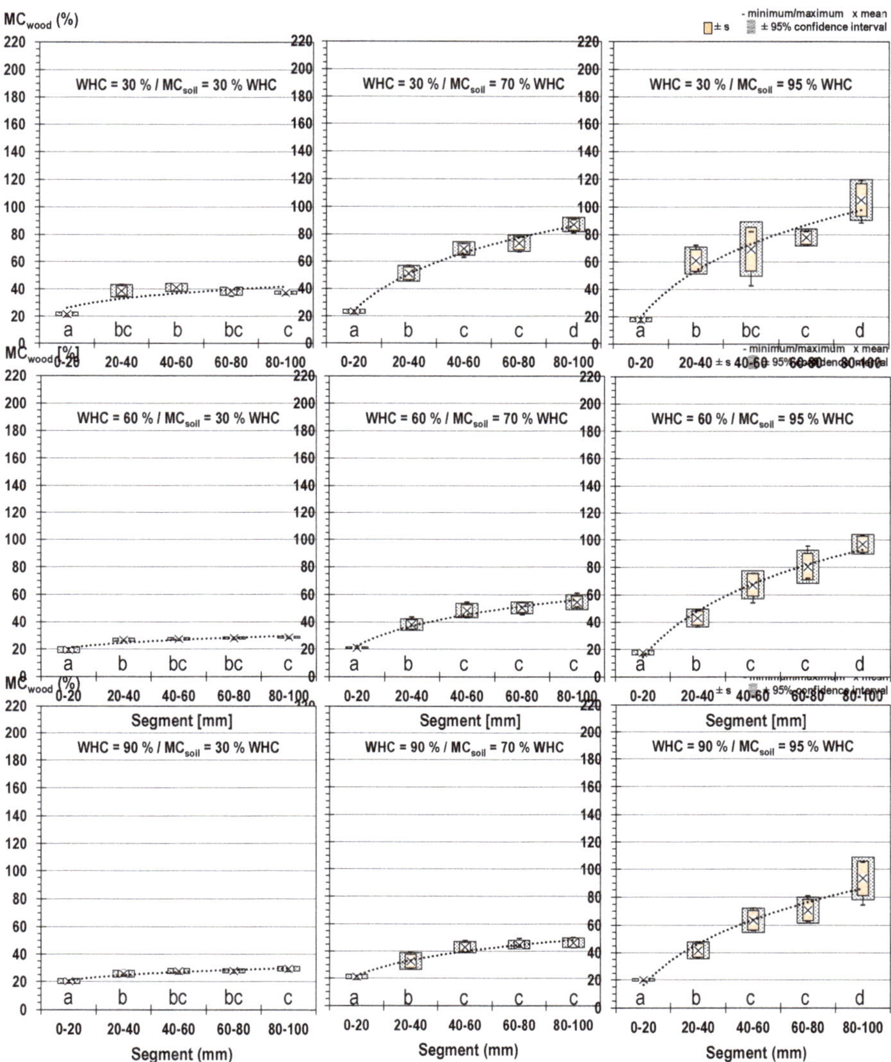

Figure 8. Distribution of MC_{wood} in Douglas fir specimens buried to 4/5 of their length (20–100 mm) in different TMCs. Different letters indicating significant differences between groups at $p < 5\%$ according to a Student t-test for non-paired samples.

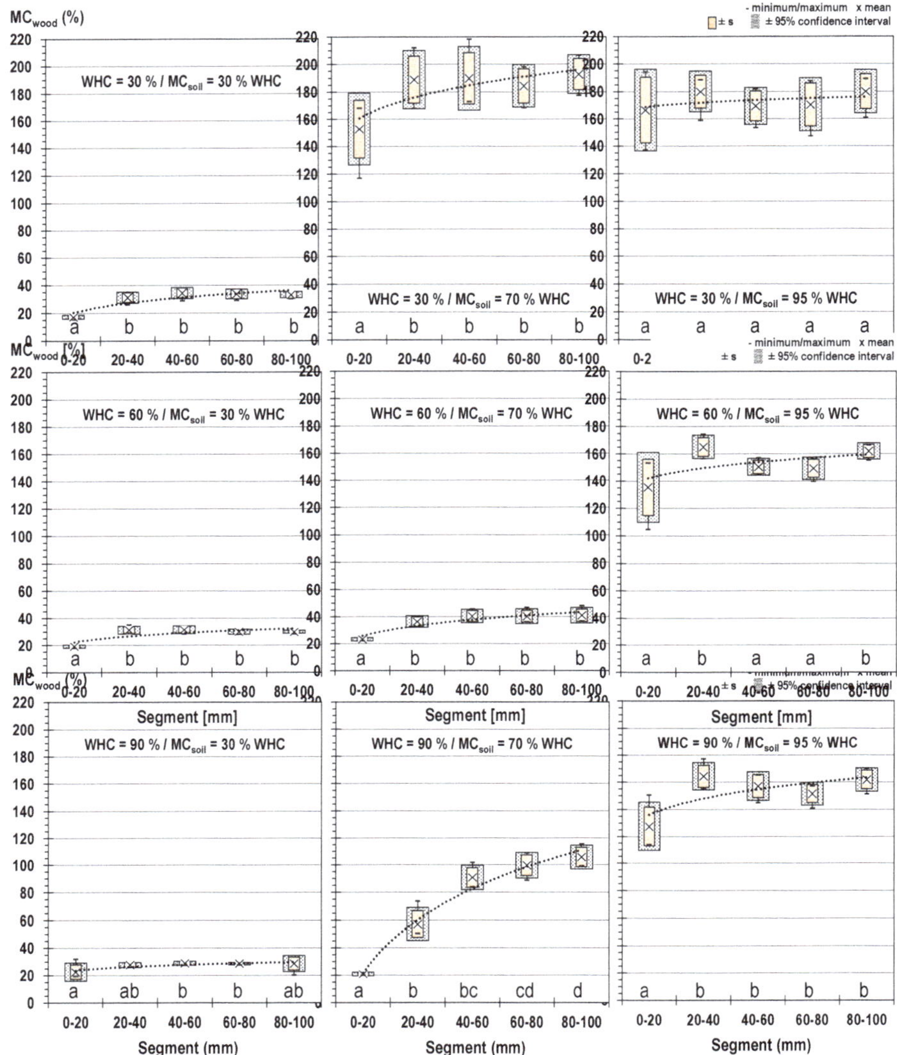

Figure 9. Distribution of MC_{wood} in Scots pine sapwood specimens buried to 4/5 of their length (20–100 mm) in different TMCs. Different letters indicating significant differences between groups at $p < 5\%$ according to a Student *t*-test for non-paired samples.

4. Conclusions

The findings from this laboratory study on the soil–wood–moisture interactions in terrestrial microcosms led us to the following conclusions:

- The more advantageous "Cylinder sand bath method" should consequently be seen as an adequate alternative for the "Droplet counting method", which turned out disadvantageous regarding practical applicability, reproducibility, and reliability.

- Water holding capacityvalues obtained from both test methods applied seemed to be easily transferable to each other. It is, therefore, recommended to replace the "droplet counting method" with the "cylinder sand bath method".
- The average MC_{wood} of specimens buried in TMCs increased with rising MC_{soil}, but WHC was often negatively correlated with MC_{wood}.
- The distance to water saturation S_{soil} appeared as a more predictive measure for MC_{wood}.
- With increasing S_{soil} the MC_{wood} increased but followed wood species-specific curves with differently steep increase and a plateau at S_{soil} = 0%.
- The distance to water saturation S_{soil} from which MC_{wood} increased most intensively was found to be wood-species specific and might, therefore, require further consideration in soil-bed durability testing. Thus, S_{soil} can likely be used to establish moisture conditions which are favorable for a specific decay type, i.e., brown, white or soft rot.
- The segment-wise determination of MC_{wood} revealed that a combination of low MC_{soil} and high WHC of the soil can easily lead to moisture conditions which are not favorable, neither for fungal decay in general, nor for soft rot decay in particular. They should, therefore, be avoided in durability testing, but might be of interest for the service life prediction of wood exposed to soil.

Based on the findings from this study, further experiments have been initiated to examine the effect of soil–wood–moisture interactions on fungal decay, in particular soft rot decay. In addition, the outstanding role of soil and wood temperature on decay in constantly wet wood will be further investigated.

Author Contributions: F.W. together with C.B. were mainly responsible for the conceptualization, methodology used, data evaluation, data validation, and formal analysis. Investigations and data curation were conducted by F.W. The original draft of this article was prepared by C.B. who was also responsible for the review and editing process of this article. C.B. and F.W. oversaw the visualization.

Funding: Parts of the research were funded in the frame of the research project 'CEMWOGEO—Cement coating of wood for geotechnical applications'—Funding Program for Renewable Resources—by the German Federal Ministry of Food and Agriculture. Fachagentur Nachwachsende Rohstoffe e.V. (FNR)—Project reference number: 22007617.

Acknowledgments: The authors gratefully acknowledge support with the TMC experiments from Brendan Marais.

Conflicts of Interest: The authors declare no conflict of interest.

References

1. Wälchli, O. Influence of the Content of Organic Matter of Soil on the Degradation of Wood by Soft Rot Fungi. In Proceedings of the IRG Annual Meeting, Nancy, France, 29–30 September 1970. IRG/WP 70–27, 7 pp.
2. Edlund, M.-L.; Evans, F.G.; Henriksen, K.; Nilsson, T. Testing Durability of Treated Wood According to EN 252-Interpretation of Data from Nordic Test Fields. In Proceedings of the IRG Annual Meeting, Tromsø, Norway, 18–22 June 2006. IRG/WP 06–20341, 17 pp.
3. Wakeling, R. Is Field Test Data from 20 × 20mm Stakes Reliable? Effects of Decay Hazard, Decay Type and Preservative Depletion Hazard. In Proceedings of the IRG Annual Meeting, Tromsø, Norway, 18–22 June 2006. IRG/WP 06–20327, 20 pp.
4. Augusta, U. Untersuchung der natürlichen Dauerhaftigkeit wirtschaftlich bedeutender Holzarten bei Verschiedener Beanspruchung im Außenbereich. Ph.D. Thesis, University Hamburg, Hamburg, Germany, 2007.
5. Brischke, C.; Olberding, S.; Meyer, L.; Bornemann, T.; Welzbacher, C.R. Intra-site variability of fungal decay on wood exposed in ground contact. *Int. Wood Prod. J.* **2013**, *4*, 37–45. [CrossRef]
6. Edlund, M.-L.; Nilsson, T. Testing the durability of wood. *Mat. Struct.* **1998**, *31*, 641–647. [CrossRef]
7. Westin, M.; Alfredsen, G. Durability of Modified Wood in UC3 and UC4-Results from Lab Natural Durability in soil-bed Assay. In Proceedings of the IRG Annual Meeting, Queenstown, New Zealand, 8–12 May 2011. IRG/WP 11–40562, 9 pp.

8. Meyer, L.; Brischke, C.; Melcher, E.; Brandt, K.; Lenz, M.T.; Soetbeer, A. Durability of English oak (*Quercus robur* L.)—Comparison of decay progress and resistance under various laboratory and field conditions. *Int. Biodeter. Biodegr.* **2014**, *86*, 79–85. [CrossRef]
9. EN 252. *Wood Preservatives. Field Test Methods for Determining the Relative Protective Effectiveness in Ground Contact*; European Committee for Standardisation: Brussels, Belgium, 2015.
10. CEN/TS 15083-1. *Durability of Wood and Wood-Based Products-Determination of the Natural Durability of Solid Wood against Wood-Destroying Fungi, Test Methods-Part 2: Soft Rotting Micro-Fungi*; European Committee for Standardisation: Brussels, Belgium, 2005.
11. Wälchli, O. Soft Rot–Soil Burial Tests–Influence of the Water Content of the Soil on Wood Decay. In Proceedings of the IRG Annual Meeting, West-Berlin, Germany, 26–28 October 1972. IRG/WP 72–212, 14 pp.
12. Mieß, S. Einfluß des Wasserhaushaltes auf Abbau und Fäuletypen von Holz in Terrestrischen Mikrokosmen. Diploma thesis, University Hamburg, Hamburg, Germany, 1997.
13. Gray, S.M. Effect of Soil Type and Moisture Content on Soft Rot Testing. In Proceedings of the IRG Annual Meeting, Avignon, France, 25–30 May 1986. IRG/WP 86–2270, 27 pp.
14. Nilsson, T.; Daniel, G. Decay Types Observed in Small Stakes of Pine and Alstonia scholaris Inserted in Different Types of Unsterile Soil. In Proceedings of the IRG Annual Meeting, Rotorua, New Zealand, 13–19 May 1990. IRG/WP 90-1443, 8 pp.
15. Edlund, M.-L.; Nilsson, T. Performance of copper and non-copper based wood preservatives in terrestrial microcosms. *Holzforschung* **1999**, *53*, 369–375. [CrossRef]
16. Rayner, A.D.; Boddy, L. *Fungal Decomposition of Wood. Its biology and ecology*; John Wiley & Sons Ltd.: Bath, UK, 1988.
17. Eaton, R.A.; Hale, M.D. *Wood: Decay, Pests and Protection*; Chapman and Hall Ltd.: London, UK, 1993.
18. ENV 807. *Wood Preservatives-Determination of the Effectiveness against Soft Rotting Micro-Fungi and Other Soil Inhabiting Micro-Organisms*; European Committee for Standardisation: Brussels, Belgium, 2001.
19. ISO 11268-2. *Soil Quality-Effects of Pollutants on Earthworms-Part 2: Determination of Effects on Reproduction of Eisenia fetida/Eisenia andrei*; International Organization for Standardization (ISO): Geneve, Switzerland, 2012.
20. Essington, M.E. *Soil and Water Chemistry: An Integrative Approach*; CRC press: Boca Raton, FL, USA, 2015.

Article

Biological Durability of Sapling-Wood Products Used for Gardening and Outdoor Decoration

Christian Brischke *, Lukas Emmerich, Dirk G.B. Nienaber and Susanne Bollmus

Wood Biology and Wood Products, University of Goettingen, D-37077 Goettingen, Germany; lukas.emmerich@uni-goettingen.de (L.E.); dirk.nienaber94@gmail.com (D.G.B.N.); susanne.bollmus@uni-goettingen.de (S.B.)

* Correspondence: christian.brischke@uni-goettingen.de

Received: 28 November 2019; Accepted: 13 December 2019; Published: 17 December 2019

Abstract: Sapling-wood products from different wood species such as willow (*Salix* spp. L.) and Common hazel (*Corylus avellana* L.) are frequently used for gardening and outdoor decoration purposes. Remaining bark is suggested to provide additional biological durability. Even for temporary outdoor use it seemed questionable that durability of juvenile sapwood can provide acceptably long service lives of horticultural products. Therefore, sapling-wood from seven European-grown wood species was submitted to laboratory and field durability tests. In field tests, specimens with and without bark were tested in comparison and submitted to differently severe exposure situations, i.e., in-ground contact, and above-ground situations with and without water trapping. All materials under test were classified 'not durable' independently from any potential protective effect of remaining bark, which contradicted their suitability for outdoor applications if multi-annual use is desired.

Keywords: basidiomycetes; fungal decay; horticulture; juvenile wood; resistance; sapwood

1. Introduction

The biological durability of most European-grown wood species is often insufficient for outdoor applications and wood needs protection either by design, wood modification, or wood preservation as pointed out in prEN 460 (Durability of wood and wood-based products—Natural durability of solid wood—Guide to the durability requirements for wood to be used in hazard classes [1]). Nevertheless, more recently, some wood species were advertised and customized for gardening and outdoor decoration purposes, although their durability is generally considered low, but had been rarely studied systematically. Among those, Common hazel (*Corylus avellana* L.) and different willow species (*Salix* spp. L.) are suggested for climb supports for clamberers [2,3], paling and woven fences [4], fascines, screens, flower bed edgings, raised beds, and other decoration items (Figure 1). Frequently, such products are manufactured from sapling-wood (here defined as roundwood from stems of less than 10 years of age) and bark is not removed since it is suggested to provide additional biological durability.

Goat willow (*Salix caprea*) is classified as 'non-durable' (DC 5, [5]), although some previous studies indicated slightly higher durability [6], i.e., DC 4. The durability of Common hazel is not classified within EN 350 (Durability of wood and wood-based products—Testing and classification of the durability to biological agents of wood and wood-based materials [5]). However, sapling-wood is juvenile wood. The latter has previously been reported to be less durable than adjacent mature wood for different wood species [7–11]. In addition, sapling-wood is exclusively sapwood, which is per definition 'non-durable' according to EN 350 [5], independent from the wood species. Consequently, it is hypothesized that sapling-wood is the least durable kind of xylem and manufacturing sapling-wood products for outdoor use appears questionable, especially when ground contact is proposed such as for fences and fascines.

Figure 1. Examples of using Common hazel wood for gardening, outdoor decoration, and historical use for stabilization of soil. Left: climb support for roses; Center: miniature woven fence; Right: reproduction trench during World War I in Belgium.

Bark tissue of various wood species is known for containing substantial amounts of extractives which can have inhibitory effects on fungal growth and wood degradation. Bark extractives such as different organic acids, tannins, and alkaloids have, therefore, been used to improve the biological durability of wood and other lignocellulosic products as previously reported by different authors [12–15]. Bark itself has been used for application where biological durability is requested, that is, such as for roofing [16,17], boats [18,19], and mulch [20]. Recommendations to leave bark on sapling-wood products could therefore be meaningful, although it contains rather thin layers of bast and thick layers of secondary bark. The latter contains usually significantly more extractives than bast, but the ratio as well as the total amount of extractives is tree species specific [21]. Apart from potential biocidal or inhibitory effects of extractives, bark might serve as a chemo-mechanical barrier for moisture and can protect the wood tissue beneath from wetting for instance due to hydrophobic substances (e.g., suberin, resin acids) or the formation of thyloses. In contrary, re-drying of once wetted wood is inhibited by bark layers as well.

This study aimed at examining the natural biological durability of sapling-wood comprehensively. Therefore, sapling-wood from seven European-grown wood species was submitted to laboratory and field durability tests. In field tests, specimens with and without bark were tested in comparison and submitted to differently severe exposure situations, i.e., in-ground contact and above-ground situations with and without water trapping, to fully reflect the anticipated in-use conditions of gardening products available in specialized trade.

2. Materials and Methods

2.1. Wood Specimens

For laboratory decay resistance tests, sapling-wood was sampled from young trees (less than 10 years old) of English oak (*Quercus robur*, Oak), Common hazel (*Corylus avellana*, Hazel), Black cherry (*Prunus serotina*), White willow (*Salix alba*, Willow), European beech (*Fagus sylvatica*, Beech), Silver birch (*Betula pendula*, Birch), Rowan (*Sorbus aucuparia*), and Scots pine (*Pinus sylvestris*) at different stands in Lower Saxony, Germany. For laboratory decay tests, specimens with a length of 50 ± 1 mm and an average diameter (without bark) between 17.3 and 20.9 mm. The target diameter of the sapling-wood specimens at a given length of 50 mm was d_{target} = 21.9 mm according to Equation (1). However, the diameter of the collected samples varied around the target value as shown in Table 1.

$$d_{target} = \sqrt{\frac{A_{CEN\ TS\ 15083-1}}{\pi}} \cdot 2\ (\mathrm{mm}) \tag{1}$$

where, d_{target} is the target diameter of the sapling-wood specimens, in mm; $A_{CEN\ TS\ 15083\text{-}1}$ is the cross-sectional specimen area acc. to CEN/TS 15083-1 (2005) = 375 mm^2.

The bark was peeled off immediately after cutting the trees. In addition, sapwood specimens of 15 × 25 × 50 mm^3 were cut from Beech, Hazel, Willow, Birch, and Scots pine according to CEN/TS 15083-1 (Durability of wood and wood-based products—Determination of the natural durability of solid wood against wood-destroying fungi, test methods—Part 1: Basidiomycetes [22]). For each test fungus, 16 replicate specimens were used. Further detailed information about the specimens is summarized in Table 1. For the different field tests, sapling-wood was sampled from the same species as listed above, but separate sets of specimens were prepared with and without bark. The length of the field test specimens was 500 mm, the mid-length diameter varied between 15 and 50 mm (Table 1). For each test set-up, 10 replicates with and without bark were exposed, resulting in 60 specimens per each wood species.

Table 1. Wood species and specimen mid-length diameter used in laboratory decay tests (n = 16), in-ground field tests (n = 10), above-ground tests (n = 10), and above-ground sponge tests (n = 10).

Wood Species	Botanical Name	Diameter (mm)						
		Lab Decay Test	In-Ground Field Test		Above-Ground Field Test		Above-Ground Sponge Field Test	
		Without Bark	With Bark	Without Bark	With Bark	Without Bark	With Bark	Without Bark
English oak	*Quercus robur*	17–22	20–38	20–29	18–36	12–32	18–33	15–37
European beech	*Fagus sylvatica*	16–18	20–37	23–38	18–35	11–36	20–40	18–39
Common hazel	*Corylus avellana*	17–21	21–33	17–35	17–37	18–36	15–39	18–37
Black cherry	*Prunus serotina*	17–22	17–33	17–34	14–37	14–37	14–28	15–40
Rowan	*Sorbus aucuparia*	16–25	22–31	18–33	20–33	14–34	22–33	17–30
White willow	*Salix alba*	18–22	15–50	21–44	14–45	13–42	12–34	13–36
Silver birch	*Betula pendula*	16–20	20–50	22–39	22–43	21–33	17–35	19–37
Scots pine	*Pinus sylvestris*	18–22	n.a.	n.a.	n.a.	n.a.	n.a.	n.a.

2.2. Durability Test with Basidiomycete Monocultures

Laboratory decay resistance tests were conducted according to a modified CEN/TS 15083-1 [22] protocol as follows: all specimens were oven-dried at 103 ± 2 °C for 48 h, weighed to the nearest 0.001 g, and afterwards conditioned at 20 °C/65% relative humidity (RH) until constant mass. After sterilization in an autoclave at 121 °C and 2.4 bar for 20 min, two specimens of the same species were placed on fungal mycelium in a Kolle flask. To avoid direct contact between wood and overgrown malt agar (4%) stainless steel washers were used. The incubation time was 16 weeks. The following test fungi were used: *Coniophora puteana* = (Schum.:Fr.) P. Karsten BAM Ebw. 15 and *Trametes versicolor* = (L.:Fr.) Pilat CTB 863A. After incubation, the specimens were cleaned from adhering mycelium, weighed to the nearest 0.001 g, and mass loss (ML) calculated according to Equation (2).

$$ML_f = \frac{m_{0,i} - m_{0,f}}{m_{0,i}} \cdot 100\ (\%) \quad (2)$$

where, $m_{0,i}$ is the oven-dry mass before incubation, in g; $m_{0,f}$ is the oven-dry mass after incubation, in g.

2.3. Field Durability Tests

Specimens with and without bark, each of 500 mm length, were exposed outdoors in three different settings. Firstly, specimens were buried to half of their length in the loamy soil on the in-ground field test site at the University of Goettingen (51°33′34.6″N 9°57′19.1″E) in September 2017. At the Goettingen test site brown, white, and soft rot decay occur. To avoid the growth of grass and other plants a horticultural water permeable textile sheet was placed on the soil. Secondly, specimens were placed horizontally on aluminum L-profiles (4 × 30 × 60 mm) with a distance of 10 mm to each other. In a third test setting specimens were wrapped with cellulose sponges (thickness: 5 mm, width: 100 mm), which were fixed with two cable strips at the center of the specimens and served for water trapping (Figure 2).

(a) (b)

Figure 2. Field tests with sapling wood specimens. (**a**): in-ground exposure. (**b**): specimens with and without bark, partly wrapped with a cotton sponge.

Decay was assessed every 6 months (results during first 1.5 years of exposure are reported). Therefore, the specimens were evaluated according to EN 252 (Field test method for determining the relative protective effectiveness of a wood preservative in ground contact [23]) using a pick-test where a pointed knife was pricked into the specimens and backed out again. The fracture characteristics of the splinters as well as depth and appearance of decay were assessed visually, and referred to the evaluation scheme according to EN 252 [23] (Table 2). Due to varying cross-sectional areas of the specimens and their circular shape, the rating system had to be adjusted. Therefore, based on the depth of decay the minimum intact cross-sectional area was determined according to Equation (3). The latter was assigned to the five different rating steps according to EN 252 [23], based on the percentage minimum remaining intact cross-sectional area A_{intact} as shown in Table 2. A_{intact} and the corresponding adapted decay rating according to EN 252 [23] were determined for each specimen separately, considering the individual mid-length diameters of the specimens.

$$A_{intact} = \frac{\left(\frac{d_i}{2} - s_{decay}\right)^2 \cdot \pi}{\left(\frac{d_i}{2}\right)^2 \cdot \pi} \ (\%) \quad (3)$$

where, A_{intact} is the minimum remaining intact cross-sectional area, in %; d_i is the initial diameter of specimen, in mm; s_{decay} is the maximum depth of decay in mm.

Table 2. Decay rating scheme according to EN 252 [23], corresponding maximum depth of decay s_{decay}, and remaining minimum intact cross-sectional area A_{intact}.

Rating	Description	s_{decay}	A_{intact}	A_{intact}
		(mm)	(mm^2)	(%)
0	No attack	0	1250	100
1	Slight attack	1	1104	88
2	Moderate attack	3	836	67
3	Severe attack	5	600	48
4	Failure	50	0	0

3. Results and Discussion

3.1. Durability against Basidiomycetes

The average mass loss (ML_f) caused by *C. puteana* was between 32% and 50%, and between 24% and 48% after incubation with *T. versicolor* (Figure 3). The median ML_f caused by *C. puteana* was well above 30% corresponding to durability class 5 (DC 5, 'not durable') according to CEN/TS 15083-1 [21] and EN 350 [5]. Sapwood of any wood species is 'not durable' as defined in EN 350 [5], although several studies showed that sapwood of different wood species showed less than 5% median ML_f in laboratory decay tests according to CEN/TS 15083-1 [22] or similar test protocols such as Atlas cedar (*Cedrus atlantica (Endl.) Manetti*, [24]), Red maple (*Acer rubrum* L., [25]), Douglas fir (*Pseudotsuga menziesii* Franco., [26]) and different other American conifers [27]. However, in most cases where durability was assigned better than DC 5 might be related (1) to its within-species variation and (2) to varying virulence of the respective test fungus with respect to discrete ML_f boundaries for the different DC according to CEN/TS 15083-1 [22].

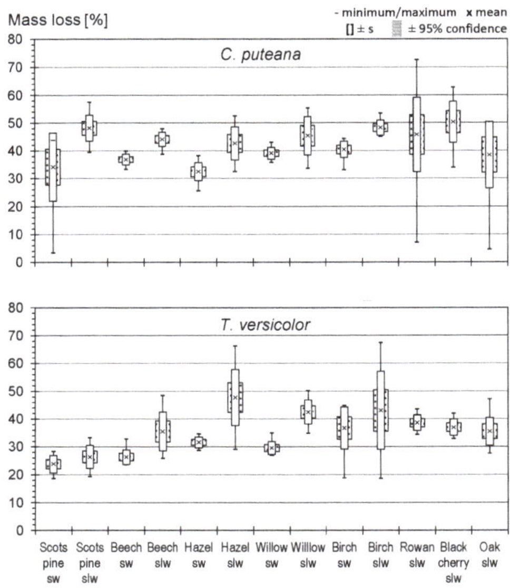

Figure 3. Mass loss (ML_f) of sapwood (sw) and sapling-wood (slw) specimens after 16 weeks of incubation with *Coniophora puteana* and *Trametes versicolor*.

In most cases, ML_f of sapling-wood was significantly higher compared to sapwood of the same wood species, which underpins the hypothesis that juvenile wood, which has still not undergone any heartwood formation, can be considered less durable than regular sapwood.

3.2. Durability against Fungal Decay in Field Tests

All in-ground specimens failed after only 1 year of exposure (Figure 4), which coincides with decay rates determined for beech wood in a test field in Hannover, Germany, where the average service life of standard graveyard test specimens was between 0.6 and 0.9 years [28]. In-ground decay was initiated generally faster in specimens without bark, but different decay rates between sets with and without bark were equalized during the second half-year of exposure.

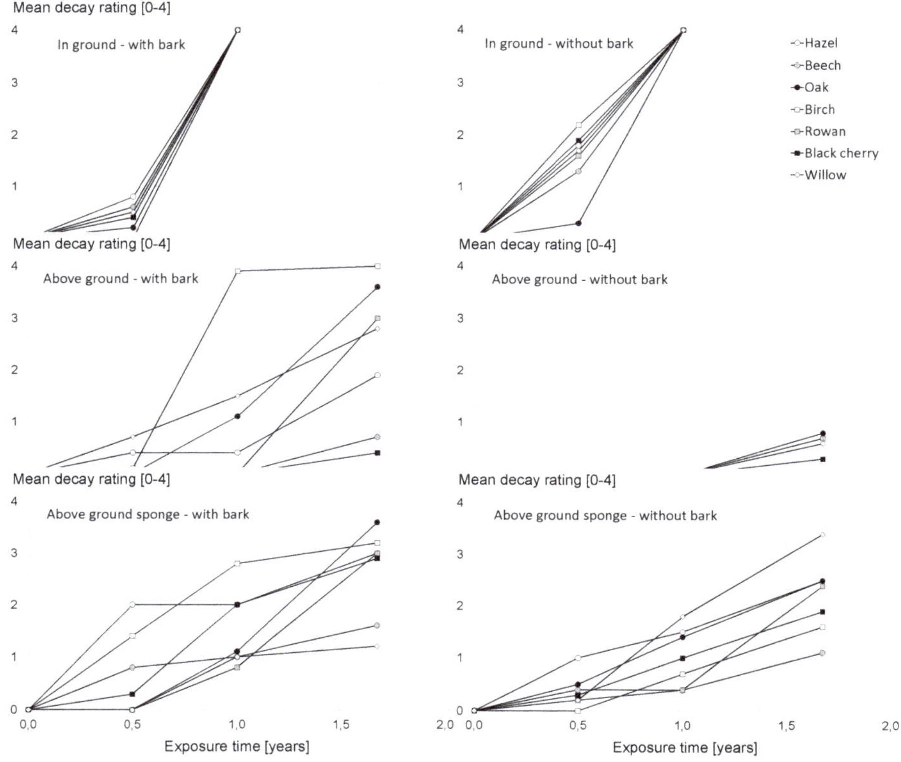

Figure 4. Mean decay rating of field test sapling-wood specimens with and without bark in-ground and above-ground exposed as single specimens with and without sponge wrapping for moisture trapping.

Specimens exposed 1 m above ground showed first signs of decay at least after 1.7 years of exposure, and earlier in most cases. Specimens with bark decayed slightly faster than those where bark had been removed before exposure. Wrapping a sponge around the center of the specimens led to higher decay rates only in specimens without bark. Similar acceleration measures were previously applied to L-joint specimens by Van Acker and Stevens [29] and led to increased decay rates compared to specimens without water capturing. In contrast, within this study permanent moistening of the bark might have improved the performance of the bark envelope around the cylindrical specimens. Especially on Beech, severe flaking of bark was observed, as shown in Figure 5, which then led to an increased formation of cracks. Consequently, the expected negative effect of wetting on the durability of the specimens was superposed by positively affecting the integrity of the protective bark layer. In specimens without bark, decay was significantly accelerated by wrapping sponges.

(a) (b)

Figure 5. Visual appearance of above-ground specimens after 6 months of exposure. (**a**): severe flake off of bark on Beech specimens with sponge wrapping. (**b**): formation of cracks in Hazel specimens without bark.

After 2 years of exposure, all above-ground specimens showed at least moderate decay, most specimens were severely decayed or had failed. Provided that mechanical strength is not an issue for the respective application, the service life of all sapling-wood specimens was less than 1 year when exposed in ground, and less than 2 years when exposed above ground. The serviceability of fences or load-bearing items such as climbing supports would have been lost earlier. Differences in decay rate between wood species were statistically insignificant, which can also be related to their general low durability and thus to short service lives of all tested materials.

4. Conclusions

Sapling-wood from seven different tree species turned out to be not durable, both in laboratory and different field tests. Against advertisement promises and as expected from studies on the durability of sapwood and juvenile heartwood of various wood species, the durability of sapling-wood, which is considered juvenile sapwood, was even lower than common sapwood. Protective effects of remaining bark were not observed. It is therefore expected that service lives of gardening and outdoor decoration accessories made from sapling-wood are generally low. In particular, under Central European climatic conditions the service life is barely 2 years in above-ground and 1 year in in-ground contact. The effect of additional water capturing seemed to be negligibly small due to the very low durability of the material itself.

Author Contributions: C.B. was mainly responsible for the conceptualization, methodology used, data evaluation, data validation, and formal analysis. Investigations and data curation were conducted by D.G.B.N. and L.E. The original draft of this article was prepared by C.B., S.B. and L.E. who was also responsible for the review and editing process of this article. C.B. and L.E. did care for the visualization.

Funding: This research received no external funding.

Acknowledgments: Martin Rosengren, Mirko Küppers, and Rainer Henke are acknowledged for providing sapling-wood test material and corresponding sapwood portions.

Conflicts of Interest: The authors declare no conflict of interest.

References

1. prEN 460. *Durability of Wood and Wood-Based Products—Natural Durability of Solid Wood—Guide to the Durability Requirements for Wood to Be Used in Hazard Classes*; European Committee for Standardization: Brussels, Belgium, 2018.
2. Uglow, J. *A Little History of British Gardening*; Random House: Munich, Germany, 2012.
3. Larkcom, J. *Creative Vegetable Gardening*; Hachette UK: London, UK, 2017.

4. Buch, C.; Kasielke, T.; Mörtl, B. *Corylus avellana*—Gewöhnliche Hasel, Haselstrauch (Betulaceae). *Jahrb. Boch. Bot. Ver.* **2016**, *7*, 197–211.
5. EN 350. *Durability of Wood and Wood-Based Products—Testing and Classification of the Durability to Biological Agents of Wood and Wood-Based Materials*; European Committee for Standardization: Brussels, Belgium, 2016.
6. Meyer-Veltrup, L.; Brischke, C.; Alfredsen, G.; Humar, M.; Flæte, P.O.; Isaksson, T.; Larsson Brelid, P.; Westin, M.; Jermer, J. The combined effect of wetting ability and durability on outdoor performance of wood: Development and verification of a new prediction approach. *Wood Sci. Technol.* **2017**, *51*, 615–637. [CrossRef]
7. Gierlinger, N.; Jacques, D.; Grabner, M.; Wimmer, R.; Schwanninger, M.; Pâques, L.E. Colour of larch heartwood and relationships to extractives and brown-rot decay resistance. *Trees* **2004**, *18*, 102–108. [CrossRef]
8. Bhat, K.M.; Thulasidas, P.K.; Florence, E.M.; Jayaraman, K. Wood durability of home-garden teak against brown-rot and white-rot fungi. *Trees* **2005**, *19*, 654–660. [CrossRef]
9. Dünisch, O.; Richter, H.G.; Koch, G. Wood properties of juvenile and mature heartwood in *Robinia pseudoacacia* L. *Wood Sci. Technol.* **2010**, *44*, 301–313. [CrossRef]
10. Latorraca, J.V.; Dünisch, O.; Koch, G. Chemical composition and natural durability of juvenile and mature heartwood of *Robinia pseudoacacia* L. *An. Acad. Bras. Cienc.* **2011**, *83*, 1059–1068. [CrossRef] [PubMed]
11. Blohm, J.H. Holzqualität und Eigenschaften des juvenilen und adulten Holzes der Douglasie (*Pseudotsuga menziesii* (Mirb.) Franco) aus Süddeutschen Anbaugebieten. Ph.D. Thesis, University of Hamburg, Hamburg, Germany, 2015.
12. Doi, S.; Kurimoto, Y. Durability of sugi (*Cryptomeria japonica* D. Don) bark against wood decay fungi and a subterranean termite. *Eur. J. Wood Wood Prod.* **1998**, *56*, 178. [CrossRef]
13. Nemli, G.; Gezer, E.D.; Yıldız, S.; Temiz, A.; Aydın, A. Evaluation of the mechanical, physical properties and decay resistance of particleboard made from particles impregnated with *Pinus brutia* bark extractives. *Bioresour. Technol.* **2006**, *97*, 2059–2064. [CrossRef] [PubMed]
14. Harun, J.; Labosky, P. Antitermitic and antifungal properties of selected bark extractives. *Wood Fiber Sci.* **2007**, *17*, 327–335.
15. Tascioglu, C.; Yalcin, M.; de Troya, T.; Sivrikaya, H. Termiticidal properties of some wood and bark extracts used as wood preservatives. *BioResources* **2012**, *7*, 2960–2969.
16. Lines, R. Man's use of birch—Past and present. *Proc. R. Soc. Edinb. Sect. B Biol. Sci.* **1984**, *85*, 203–213. [CrossRef]
17. Gottesfeld, L.M.J. The importance of bark products in the aboriginal economies of northwestern British Columbia, Canada. *Econ. Bot.* **1992**, *46*, 148–157. [CrossRef]
18. Gidmark, D. *The Birchbark Canoe Makers of Lac Barriere*; Algonquian Papers-Archive; 19th Algonquian Conference: Maniwaki, Quebec, 1988; Volume 19.
19. Raffan, J.; Franks, C.E.S. Bark, skin & cedar: Exploring the canoe in Canadian experience. *Am. Rev. Can. Stud.* **2000**, *30*, 393.
20. Skroch, W.A.; Powell, M.A.; Bilderback, T.E.; Henry, P.H. Mulches: Durability, aesthetic value, weed control, and temperature. *J. Environ. Hortic.* **1992**, *10*, 43–45.
21. Herrick, F.W. Chemistry and utilization of western hemlock bark extractives. *J. Agric. Food Chem.* **1980**, *28*, 228–237. [CrossRef]
22. CEN/TS 15083-1. *Durability of Wood and Wood-Based Products—Determination of the Natural Durability of Solid Wood against Wood-Destroying Fungi, Test Methods—Part 1: Basidiomycetes*; European Committee for Standardization: Brussels, Belgium, 2005.
23. EN 252. *Field Test Method for Determining the Relative Protective Effectiveness of a Wood Preservative in Ground Contact*; European Committee for Standardization: Brussels, Belgium, 2015.
24. Brunetti, M.; De Capua, E.L.; Macchioni, N.; Monachello, S. Natural durability, physical and mechanical properties of Atlas cedar (*Cedrus atlantica* Manetti) wood from Southern Italy. *Ann. For. Sci.* **2001**, *58*, 607–613. [CrossRef]
25. Anagnost, S.E.; Smith, W.B. Comparative decay of heartwood and sapwood of Red maple. *Wood Fiber Sci.* **1997**, *29*, 189–194.
26. Brischke, C.; Gellerich, A.; Homann, P. Impact of sapwood portions on the durability of adjacent heartwood of *Pinus sylvestris*, *Pseudotsuga menziesii* and *Quercus robur*: Part 1: Laboratory studies. In Proceedings of the IRG Annual Meeting, IRG/WP 18-10922, Johannesburg, South Africa, 29 April–3 May 2018; p. 15.

27. Humphrey, C.J. Laboratory tests on the durability of American woods: I. Flask tests on conifers. *Mycologia* **1916**, *8*, 80–92. [CrossRef]
28. Stirling, R.; Alfredsen, G.; Brischke, C.; De Windt, I.; Francis, L.P.; Frühwald Hansson, E.; Humar, M.; Jermer, J.; Klamer, M.; Kutnik, M.; et al. Global survey on durability variation—On the effect of the reference species. In Proceedings of the IRG Annual Meeting, IRG/WP 16-20573, Lisbon, Portugal, 15–19 May 2016; p. 26.
29. Van Acker, J.; Stevens, M. Biological durability of wood in relation to end-use—Part 2: The use of an accelerated outdoor L-joint performance test. *Holz Roh Werkst.* **2003**, *61*, 125–132. [CrossRef]

Article

Initial *Rhodonia placenta* Gene Expression in Acetylated Wood: Group-Wise Upregulation of Non-Enzymatic Oxidative Wood Degradation Genes Depending on the Treatment Level

Martina Kölle [1,*], Rebecka Ringman [2] and Annica Pilgård [1,2]

[1] TUM School of Life Sciences Weihenstephan, Technical University of Munich, 85354 Freising, Germany; annica.pilgard@ri.se

[2] RISE, Research Institutes of Sweden, Bioeceonmy, Box 857, 501 15 Borås, Sweden; rebecka.ringman@ri.se

* Correspondence: koelle@hfm.tum.de

Received: 25 November 2019; Accepted: 5 December 2019; Published: 7 December 2019

Abstract: Acetylation has been shown to delay fungal decay, but the underlying mechanisms are poorly understood. Brown-rot fungi, such as *Rhodonia placenta* (Fr.) *Niemelä, K.H. Larss. & Schigel*, degrade wood in two steps, i.e., oxidative depolymerization followed by secretion of hydrolytic enzymes. Since separating the two degradation steps has been proven challenging, a new sample design was applied to the task. The aim of this study was to compare the expression of 10 genes during the initial decay phase in wood and wood acetylated to three different weight percentage gains (WPG). The results showed that not all genes thought to play a role in initiating brown-rot decay are upregulated. Furthermore, the results indicate that *R. placenta* upregulates an increasing number of genes involved in the oxidative degradation phase with increasing WPG.

Keywords: brown-rot fungi; oxidative degradation; fenton degradation; acetylation; scots pine; *Pinus sylvestris*; *Postia placenta*

---

1. Introduction

Wood is an attractive building material that is biodegradable, renewable, stores carbon [1,2], and can serve as an alternative to concrete and steel due to its mechanical properties [3,4]. Biodegradability is a disadvantage when wood is used in outdoor construction [5,6], if no adequate preservative method is applied. Hence, wood protection is necessary to prolong the service life of wooden products made of nondurable wood. Traditionally, wood is treated with copper-based preservatives, however, ecological and health issues surrounding copper-based preservatives have led to restrictions and regulations [7–9]. An alternative approach to traditional preservatives is wood modification, which alters the properties of the wood to enhance its resistance to degradation [8]. Wood can be either physically (heat-treated wood) or chemically (for example wood acetylation and furfurylation) modified. Acetylation is a widely studied wood modification method [8,10–13]. During acetylation, hydroxyl groups in the wood are replaced by acetyl groups [1,8], leading to reduced hygroscopicity and volumetric swelling which bulks the cell walls and reduces water absorption [8,14–16]. Several studies have found, that a weight percent gain (WPG) of 10% leads to limited decay inhibition, whereas a WPG of 20% significantly increases decay resistance [8,17–20].

The exact mechanisms behind the decay resistance in acetylated wood is not known. However, recent results indicate that diffusion of fungal depolymerizing agents through the wood cell wall may be inhibited due to decreased equilibrium moisture content [8,21–25]. Hunt et al. [26] reported a decreased diffusion of K^+ ions with an increasing level of acetylation, which supports the theory that acetylation inhibits diffusion of hydrolytic reductants due to moisture exclusion [25]. Zelinka et al. [24,27,28] hypothesized in their studies that an interrupted moisture network inhibits diffusion. This conclusion was supported by Beck et al. [16], who characterized moisture in acetylated wood with low-field nuclear magnetic resonance relaxometry (LF-NMR).

Thirty-two percent of the described fungi in the world belong to the Basidiomycota [29]. Wood decaying fungi, belonging to the Basidiomycota, are filamentous fungi and can be divided into two groups, white-rot and brown-rot fungi [30–32]. However, because decay mechanisms are much more diverse than assumed, such a distinction may be too simplistic [33]. Although more species of white-rot fungi are known, approximately 80% of the wood decay fungi found in wooden constructions belong to the brown-rot subgroup [34–37]. Because of their preference for softwoods, brown-rot fungi are the main recyclers of lignocellulose in Northern Hemisphere coniferous forests, and decay associated with brown-rot is reportedly the most destructive type [5,38,39]. Brown-rot decay rapidly leads to significant strength loss through depolymerization of the cellulose and hemicellulose fractions through non-enzymatic oxidative degradation processes [40–44], followed by the secretion of hydrolyzing enzymes [44–46]. Zhang et al. found, in 2016, that the two decay phases are in fact spatially separated and that the presence of sugars solubilized during oxidative decay triggers the transition from oxidative to enzymatic degradation [47].

Brown-rot wood degradation has been studied mainly in the organism *Gloeophyllum trabeum* (Pers.) *Murrill* and verified to some degree in *Rhodonia placenta* (Fr.) *Niemelä, K.H. Larss. & Schigel* (also known as *Postia placenta*), *Coniophora puteana* (Schumach.) *P. Karst.*, and *Serpula lacrymans* (Wulfen) *J. Schröt*. The details regarding the non-enzymatic oxidative degradation phase still remains unknown. The current theory is that during decay brown-rot fungi secrete oxalic acid, which diffuses into the lumen, where it functions as a chelator to sequester Fe^{3+} [46,48,49]. Fe^{2+} is formed through reduction by hydroquinones, and H_2O_2 is believed to be formed through a reaction between hydroquinones and oxygen [50,51]. Hydrogen peroxide and Fe^{2+} react to hydroxyl radicals, which polymerize cellulose and hemicellulose, and modifies lignin [45,46,52]. This process solubilizes sugars, which can diffuse through the cell walls into the lumen, to become accessible to cellulases and hemicellulases [42,53].

Zhang et al. [47] found that thirty-three genes likely associated with redox processes were upregulated during early stages of decay, and 21 of them may be involved in generating H_2O_2 and Fe^{2+}. Martinez et al. [54] identified a putative quinone reductase in *R. placenta* that is believed to recover iron reductants (hydroquinones) [50,54–56] and is also putatively mediating the reduction of iron chelators (oxalic acid) [55]. *R. placenta* has only one quinone reductase (QRD Ppl124517), compared with the two in *Gloeophyllum trabeum*, which have been shown to have different functions; one is involved in wood degradation and the other in stress defense [57]. *R. placenta* quinone reductase belongs to the Carbohydrate-Active enZYmes (CAZy) family AA6 and is likely involved in oxidative processes and the formation of intracellular enzymes for fungal protection. Mueckler et al. [58] suggested that Ppl44553 is a quinate transporter (PQT), belonging to the major facilitator superfamily domain. This domain includes sugar transporters, which bind and transport a variety of carbohydrates, organic alcohols, and acids. A major part of this family catalyzes sugar transport [59]. The exact function of the *R. placenta* PQT remains unclear. Laccases appears to play an important role by oxidizing methoxyhydroquinones into semiquinones that reduce Fe^{2+} [60] and are believed to be effective producers of H_2O_2 [61]. Four putative laccases have been found in *R. placenta* [54] and one of them is included in this present study (Lac1 Ppl111314). Extracellular H_2O_2 has also been suggested to be produced by copper radical oxidases (Cro), gluco-oligosaccharide oxidases, and glucose-methanol-choline (GMC) oxidoreductases, including alcohol oxidases (AlOx) and glucose oxidases (GOx) [54,62,63]. Alcohol oxidase is upregulated in the presence of cellulose and wood [38,54]. *R. placenta* has three genes encoding alcohol oxidases, belonging to the CAZy family AA3, AlOx1 Ppl44331, AlOx2 Ppl129158, and AlOx3 Ppl118723. Copper radical oxidase (Cro1 Ppl56703) belongs to the CAZy family AA5. Cro1, together with glucose oxidase 2 (GOx2 Ppl108489), are oxidoreductases involved in the H_2O_2 production [47]. Brown-rot fungi accumulate significant amounts of oxalic acid, which has been shown to be involved in their wood decaying system [64]. Oxaloacetate dehydrogenase (CyOx Ppl112832) is presumed to be involved in oxalic acid synthesis, which is important for sequestering and reducing Fe^{3+} to Fe^{2+} [13]. Glyoxylate dehydroxygenase (GlyD Ppl121561) is likely involved in oxalic acid synthesis, by catalyzing the production of oxalate through oxidation of glyoxalate [65]. It is also involved in the glyoxylate cycle [66]. Munir et al. [66,67] purified a GlyD that catalyzes dehydrogenation of glyoxylate to oxalate in the presence of cytochrome c. Munir et al. [66] also found that GlyD exhibited a strong correlation with the biosynthesis of oxalic acid and fungal growth.

Previous studies (mostly on *R. placenta*) have shown that brown-rot fungi are not killed when growing on acetylated wood and can express genes needed for both the non-enzymatic oxidative degradation process and the enzymatic wood degradation process [11,13,68–70]. Even though the fungi express the genes needed for degradation, the wood remains intact for a prolonged period of time as compared with untreated wood materials. Since the fungi express genes involved in wood degradation in modified wood, before mass loss occurs, at similar levels as in untreated wood where degradation has begun, it is clearly shown that the fungi are attempting to degrade the wood but fail [11,13,69–71]. Despite previous research, it is still not known which regulatory mechanisms in the brown-rot degradation machinery are prevalent during degradation of acetylated wood. In order to improve the decay resistance of acetylated wood, it is of crucial importance to understand the decay mechanisms of incipient decay.

To investigate and improve wood modification methods, degradation tests are often used, for example, the miniblock test [72]. *Rhodonia placenta* (Fr.) M.J. *Larsen & Lombard* is used as a model fungus for standardized degradation tests in Europe (FPRL280) and the USA (MAD-698R) (AWPA E10-16, 1991; EN 113, 1996). Previous comparisons of gene expression when growing on untreated and modified wood supply hints on fungal behavior [11,24,25,73]. Most studies used only a small number of biological replicates and often encountered difficulties differentiating the non-enzymatic oxidative degradation phase from the enzymatic degradation phase. Zhang et al. presented in 2016 a method making it possible to separate the two phases. They used wood wafers (with the largest area being the cross-section) placed in an upward position on previously inoculated feeder strips. Fungi were allowed to grow up along the wafers forming a clear hyphal front. This method has previously never been used on modified wood.

The aim of this study was to investigate the differences in gene expression of 10 different genes in *R. placenta* during the non-enzymatic oxidative degradation phase, when grown on wafers of untreated wood and wood acetylated to three different levels.

2. Material and Methods

2.1. Wood Samples

Wood boards from Scots pine sapwood (*Pinus sylvestris* L.) were cut into wafers ($80 \times 18 \times 2.5$ mm^3), with the largest area being the cross section [74]. The samples were dried at 103 °C for 24 h and their dry weights were determined. Dried wafers were acetylated for 15 min, 30 min, and 60 min to achieve three different acetylation levels; 10 (AC10), 15 (AC15), and 20 (AC20) weight percent gain (WPG) (Table 1). Then, 50 samples (only 25 for AC10) were put in a glass flask (1 L) and a vacuum was attached for 30 min. Then, 50 mL of acetic anhydride was injected, followed by 50 mL of pyridine. After vacuuming for another minute, the samples were incubated at room temperature for 3 h, followed by lowering the flask into an 80 °C water bath for the corresponding reaction time. The reaction was stopped by washing the samples twice with ice-cold acetone and twice with an acetone-water mixture with a ratio of 1:1. Samples were rinsed in distilled water several times for 3 days, vacuum-impregnated with water, to remove all accessible chemicals, and dried again before the weight gain was measured. All samples were packed and autoclaved before the decay test.

Table 1. Mean weight percent gain (WPG) and standard deviation of the wood samples.

Mean WPG	Standard Deviation
10	0.18
15	0.13
20	0.14

2.2. Decay Test

R. placenta FPRL 280 (Fr.) was used for the decay test previously described by Zhang et al. [47]. As growth medium, 50 g of soil, 25 g of sand, 20 g of vermiculite, and 45 mL of water were mixed for each specimen container. The jars were autoclaved. Feeder strips of *P. sylvestris* sapwood were autoclaved, and three per glass were placed on the soil medium and inoculated with agar plugs from pregrown 4% malt agar plates. For each treatment, untreated (UT), 10% acetylated (AC10), 15% acetylated (AC15), and 20% acetylated (AC20), 16 glasses (n = 16) with three samples each were prepared. After the feeder strips were completely overgrown, one wood wafer was placed on each feeder strip. The jars were stored in a climate chamber at 22 °C and 70% relative humidity. Samples were harvested according to the height of the hyphal front (3/4 of the wood wafer overgrown). Only samples with even hyphal growth, around the whole sample, were included in the test. Since the cross section was the largest area, the mycelia growing up the sample could easily grow into the sample through the wood cell lumen. With this sample design, it was possible to assure that the fungi had reached the same level inside the sample as outside the sample [47]. Sections with a size of five millimeters, including the hyphal front, were cut out of the wafers, immediately frozen in liquid nitrogen and stored at −80 °C awaiting further analysis. The section size was chosen due to an expected growth rates of about 2.5 mm/day, with the first 5 mm representing approximately a 48 h window, during which the non-enzymatic degradation phase has been shown to take place [47]. Additional tests were done to obtain data on mass loss and growth rates for all treatments. For these tests a different set of samples was used.

2.3. RNA Purification and cDNA Synthesis

All 5 mm sections from one jar (three samples) were pooled into one biological replicate. A Mixer Mill MM 400 (Retsch GmbH, Haan, Germany) using one 1.5 cm steel ball and 30 Hz for 2 min was used to produce wood powder. Containers, beads, and samples were frozen with liquid nitrogen. Sixty milligrams of each sample were taken for RNA purification. Total RNA was extracted using a MasterPureTM RNA Purification Kit (Lucigen, Middleton, USA) [24]. RNA was converted to cDNA using TaqMan Reverse Transcription Reagents using Oligo d(T)16 (Applied Biosystems, Foster City, USA) with 10 times the standard dNTP concentration.

2.4. Quantitative Real-Time Polymerase Chain Reaction

A rotor-gene SYBRGreen polymerase chain reaction (PCR) kit (Qiagen, Hilden, Germany) was used according to the manufacturer's protocol for quantitative real-time PCR. Each sample was run with three technical replicates. β-tubulin was used as an endogenous control [13]. A list with all used primers including the number of the Joint Genome Institute (JGI) can be found in Table 2. The ten genes used in this study were chosen because of their assumed importance during the non-enzymatic degradation phase. This selection was based on previous findings [11,13,47,68–70,75–79]. Most importantly, the ten genes included in this paper have never been studied with this sample design before. Rotor-gene Q series software (Version 2.3.1, Qiagen, Hilden, Germany) was used to evaluate the runs. Expression levels for each technical replicate of the target genes (Tg) were calculated and normalized to the endogenous control according to the formula [80]:

$$\text{Expression level} = 10^4 \times 2^{C_t\beta t - C_t Tg} \quad (1)$$

Table 2. Primer sequences and JGI number of the target genes.

Gene	JGI no.	Primer Sequence
β-tubulin (bT)	113871	CAGGATCTTGTCGCCGAGTAC/ CCTCATACTCGCCCTCCTCTT
Quinone oxidoreductase (QRD)	124517	CGACGACAAGCCCAACAAG/ GATGACGATGATGGCGATTTTAGG
Alcohol oxidase 1 (AlOx1)	44331	GGAGGTACAGACGGACGAAC/ AGAGTCGACGACACCGTTCT
Alcohol oxidase 2 (AlOx2)	129158	TACTCGACGGCCCTCACTAT/ CCGCTTGAGACTGAACACTG
Alcohol oxidase 3 (AlOx3)	118723	ACACCAAGGAGGACGACGAG/ GACGAGCAAGGCAGACGAGTA
Putative quinate transporter (PQT)	44553	ACTGACCTTTTGCGCAGACT/ CAATGTTGATTGTGGCGAAC
Laccase (Lac)	111314	CGGTGCTCTTGGCCACTTAG/ CCATTGGTTATGGGCAGCTC
Copper radical oxidase (Cro1)	56703	CCTACCAGCTGCTTCCTGAC/ AACGTTCGGCTGTATGAACC
Glucose oxidase (GOx2)	108489	GTCCGCTCTAACGTTGCTTC/ CCGGCGTTATTGGAGAGATA
Glyoxylate dehydrogenase (GlyD)	121561	CGGAGCTGGACCTTTGTTAC/ GCGCGAAGGCAAATCTAATA
Oxaloacetate dehydrogenase (CyOx)	112832	AAGGCGTTCTTCGAGGTCAT/ AAAGCAGCAACCCGAGAAG

A mean concentration was calculated for each sample (n = 11–16) and normalized. Significance ($p < 0.05$) was calculated using the Student's t-test. Expression levels of the four different treatment levels (UT, AC10, AC15 and AC20) were compared.

2.5. Statistical Evaluation

A simple linear regression in Excel (Version 2016) was constructed for each of the four treatments for growth rate calculations.

3. Results

3.1. Mass Loss and Growth Rates

Average mass loss data and linear regression values of the growth test for all treatments are supplied in Table 3. Significantly lower mass loss was found in treated samples as compared with that in untreated samples. In AC20, a negative mass loss was detected.

Table 3. Mass loss and growth rates of all treatments during the period of colonizing the wood wafers.

Treatment	Mass Loss (%)	Growth Rate [mm/day]	Growth Rate R^2 Values
untreated	9.36	0.23	0.9955
10% acetylated	1.63	0.24	0.9955
15% acetylated	0.25	0.22	0.91
20% acetylated	−0.16	0.25	0.9897

The growth rates of *R. placenta* growing on untreated samples, as well as on acetylated samples, can be seen in Table 3. These correlate well with the findings of Zhang et al. (2016) [47]. However,

R. placenta growing on AC15 behaved differently with a slower and more uneven growth rate as compared with the other acetylation levels.

3.2. Expression of Target Genes

3.2.1. Upregulated Genes

The results of all upregulated genes can be found in detail in Figure 1. Note that the figure shows logarithmic values and that the *y*-axis scale varies between the different genes.

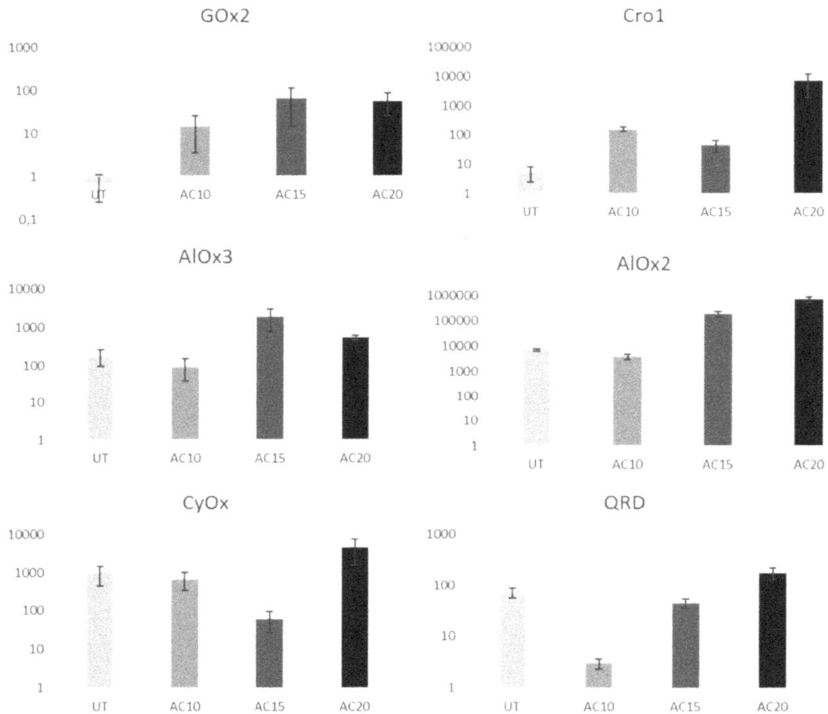

Figure 1. Logarithmic values for gene expression for upregulated genes (UT, untreated; AC10, 10 WPG acetylated; AC15, 15 WPG acetylated; and AC20, 20 WPG acetylated samples; GOx2, glucose oxidase 2; Cro1, copper radical oxidase 1; AlOx3 & AlOx2, alcohol oxidases 2 & 3; CyOx, oxaloacetate dehydrogenase; QRD, quinone oxidoreductase).

Expression levels for AlOx2 on acetylated samples were significantly upregulated as were those for AlOx3 (except for AC10) as compared with those on untreated samples. For Cro1, all gene expression levels were upregulated in modified samples as compared with those in untreated samples, with an extreme upregulation seen in the AC20 samples. All treatments showed highly significant differences in gene expression for CyOx as compared with each other, except for the untreated samples as compared with the AC10 samples. The AC15 samples showed the lowest levels of CyOx expression, whereas expression was clearly upregulated in AC20 samples. Upregulation of GOx2 was unambiguous when comparing treated samples with untreated samples. AC15 and AC20 did not differ significantly for this gene, but all other treatments showed highly significant differences. Differences in gene expression of QRD were highly significant between all treatments. A clear upregulation of QRD was not found when comparing treated and untreated samples, although a strong upregulation in samples with high acetylation was observed.

3.2.2. Downregulated Genes

Figure 2 illustrates the results for all genes that were downregulated in treated samples as compared with those in untreated samples. As well, Figure 1 shows the logarithmic values and the *y*-axis scale differs between the genes. Lac1 showed the highest expression levels among untreated samples as compared with all treated samples. The lowest values were observed in AC10, increasing with increased intensity of acetylation. PQT was clearly downregulated in all three acetylation levels as compared with the untreated samples. Results for AlOx1 showed no significant differences between untreated and AC15 samples. AC10 and AC20 showed highly significant lower levels of gene expression as compared the untreated samples. Highly significant differences were also revealed for expression levels of GlyD in all sample treatments. All treated samples had lower expression rates for GlyD as compared with the untreated samples.

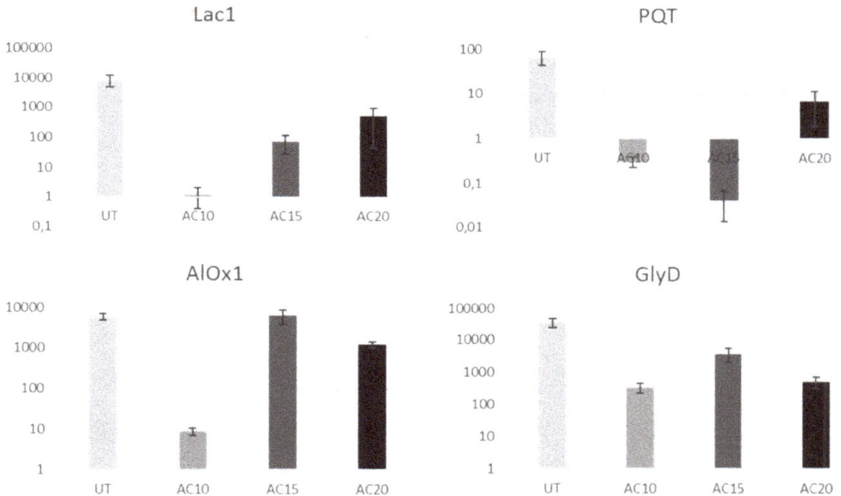

Figure 2. Logarithmic values for gene expression for downregulated genes (UT, untreated; AC10, 10 WPG acetylated; AC15, 15 WPG acetylated; and AC20, 20 WPG acetylated samples; Lac1, laccase 1; PQT, putative quinate transporter; AlOx1, alcohol oxidase 1; GlyD, glyoxylate dehydrogenase).

3.2.3. Effect of Acetylation on Overall Gene Expression

For all genes, significant differences were found between treatments, except in three cases (AlOx1: UT-AC15; CyOx: UT-AC10; and GOx2: AC15-AC20) (Table 4). When UT and AC10 samples were compared, upregulation was observed for only two genes, Cro1 and GOx2, which are likely involved in H_2O_2-production. The comparison of untreated samples with AC15 showned an upregulation of Cro1, GOx2 and, additionally, AlOx2 and AlOx3 were seen. AlOx2 and AlOx3 are also thought to be involved in the production of H_2O_2. The same four genes were also upregulated in the AC20 samples. In addition to those, an upregulation was also seen for CyOx and QRD, which are involved in Fe^{2+} supply for the Fenton reaction.

Table 4. *P*-values for all genes and treatments calculated with t-test. *P*-values are "not significant" for $p > 0.05$; "significant" for $0.05 \geq p < 0.01$, and highly significant for $p \leq 0.01$. Genes in bold are upregulated, others are downregulated.

	AC10	AC15	AC20	Gene
UT	2.08×10^{-9}	7.33×10^{-9}	2.27×10^{-9}	
AC10		7.3×10^{-6}	9.81×10^{-9}	GlyD
AC15			1.14×10^{-5}	
UT	3.60×10^{-5}	5.85×10^{-4}	4.07×10^{-5}	
AC10		3.48×10^{-4}	5.19×10^{-4}	**GOx2**
AC15			1.37×10^{-1}	
UT	1.29×10^{-1}	3.14×10^{-6}	4.03×10^{-4}	
AC10		3.16×10^{-6}	2.21×10^{-4}	**CyOx**
AC15			4.53×10^{-5}	
UT	8.95×10^{-9}	3.26×10^{-6}	1.50×10^{-4}	
AC10		9.69×10^{-9}	1.88×10^{-4}	**Cro1**
AC15			1.60×10^{-4}	
UT	5.64×10^{-7}	6.44×10^{-7}	1.21×10^{-6}	
AC10		7.38×10^{-6}	3.18×10^{-4}	Lac1
AC15			1.41×10^{-3}	
UT	1.32×10^{-6}	1.26×10^{-6}	3.13×10^{-6}	
AC10		2.25×10^{-7}	6.08×10^{-4}	PQT
AC15			4.23×10^{-4}	
UT	7.86×10^{-3}	7.71×10^{-5}	1.19×10^{-12}	
AC10		5.48×10^{-5}	3.75×10^{-14}	**AlOx3**
AC15			5.32×10^{-4}	
UT	4.70×10^{-11}	1.59×10^{-10}	3.72×10^{-10}	
AC10		1.24×10^{-10}	3.51×10^{-10}	**AlOx2**
AC15			7.04×10^{-9}	
UT	1.83×10^{-12}	3.5×10^{-1}	1.71×10^{-11}	
AC10		1.39×10^{-8}	6.56×10^{-12}	AlOx1
AC15			1.63×10^{-7}	
UT	1.26×10^{-11}	4.99×10^{-6}	3.48×10^{-8}	
AC10		1.79×10^{-8}	1.52×10^{-10}	**QRD**
AC15			2.27×10^{-9}	
Highly significant		Significant	Not significant	

4. Discussion

4.1. Mass Loss and Growth Rates

The negative value for the mass loss in AC20 samples could be due to the weight gain through fungal hyphae. This phenomenon has been seen in previous studies [71,77,81].

4.2. Expression of Target Genes

4.2.1. Upregulated Genes

The results of this study show that *R. placenta* upregulates parts of the genes involved in non-enzymatic oxidative degradation when growing on acetylated samples as compared with untreated wood, especially on higher levels of acetylation. An upregulation of GOx2 on acetylated samples was clearly shown as compared with untreated wood (Figure 1). GOx2 belongs to the GMC oxidoreductases, just as AlOx3. Ringman et al. [76] did not find any significant differences between untreated and modified samples for GOx2 expression which differs from our findings. This might be explained by the different sample design and that not only the non-enzymatic degradation phase was captured in the samples from Ringman et al. [76]. An upregulation in treated samples may be of importance for production of higher amounts of H_2O_2 for the Fenton reaction.

The highest levels of Cro1 were found in AC20 samples, but no clear upward trend was observed with increasing acetylation level. An upregulation of Cro1 and GOx2 appeared already in the AC10 samples. This differs from the upregulation of AlOx2 and AlOx3 which were lower in AC10 samples as compared with untreated samples. The reason for the upregulation of Cro1 could be that the fungus increases the H_2O_2 production. Maybe the fungus can upregulate Cro1 and GOx2 faster and with lower effort than the two AlOx genes. Beck et al. [13] found a significant upregulation in samples with a WPG of 21 during initial decay as compared with other treatments, which supports our findings, but the values were higher in the present study. In the study by Ringman et al. [76], Cro1 levels were significantly lower in acetylated samples which is contrary to our findings for all acetylation levels. It should, however, be noted that the sample and experiment design differ between these studies. In Beck et al. [13] and Ringman et al. [76] the samples may have contained mycelia that were in both the non-enzymatic and in the enzymatic phase.

AlOx2 and AlOx3 are both significantly upregulated in AC15 and AC20 samples as compared with untreated samples. Alfredsen et al. [11] compared three different acetylation levels at the same mass loss level with untreated samples and found a significant upregulation in AlOx3 between high-level acetylated and untreated samples. This supports our findings, even though it cannot be decided with certainty that only the oxidative degradation phase was detected due to the sample design in Alfredsen et al. [11]. Beck et al. [13] also reported upregulation of AlOx2 and AlOx3 in acetylated samples as compared with untreated samples, at initial decay. They also found a significantly higher upregulation in the highest level of acetylated samples when looking at samples in early stages of decay, which is similar to our results [13], assuming that Beck et al. [13] only looked at mycelia in the non-enzymatic degradation stage. Ringman et al. [70] reported a small, but not significant, upregulation of AlOx3 in acetylated samples, which differs from our findings. In another study from Ringman et al. [76], an upregulation of AlOx3 in modified samples as compared with untreated samples was seen, as well as in Schmöllerl et al. [69]. This indicates a higher investment in the production of H_2O_2 on modified samples as compared with untreated samples.

AC15 showed the lowest values of CyOx gene expression while the highest values were seen in AC20 [13]. Beck et al. [13] did not find an upregulation in CyOx in acetylated samples, except for the AC17 samples. The reason for these differences might be the different sample design. They also found mass loss in the AC21 samples, showing that the fungus had started the enzymatic degradation phase, which may be another reason for the differences in gene expression. CyOx functions as a catalysator to form acetate and oxalate in filamentous fungi. Enhanced oxalate production forms soluble Fe^{2+}-oxalate complexes needed for the Fenton reaction [82]. This could explain the high expression, especially in samples with high levels of acetylation [83].

R. placenta appears to downregulate the formation of QRD in samples with lower acetylation levels as compared with untreated samples and upregulates it to a level high than that of the untreated samples when growing on higher-level acetylated samples. These results indicate that only higher levels of acetylation induce upregulation of QRD as compared with untreated samples. This suggests that the *R. placenta* QRD could be involved in stress defense rather than in the reduction of chelators. The lower levels of gene expression in AC10 and AC15 could be explained by less stress, since the fungus is still able to degrade the wood. Ringman et al. [70] showed a significant upregulation of QRD in acetylated samples after two and 14 days as compared with untreated samples. The results for QRD gene expression during the initial decay phase differ from the results found in the present study [70], clearly demonstrating the importance of studying the exact function of the QRD found in *R. placenta* in comparison to the two QRD genes found in *G. trabeum*. Beck et al. [13] found no significant upregulation of QRD (BqR) between different harvesting points. Alfredsen et al. [11] observed an increase of QRD expression on treated samples with increasing incubation time which most likely is a result of the enzymatic degradation.

4.2.2. Downregulated Genes

The downregulation of the genes Lac1, PQT, GlyD, and AlOx1 clearly shows that not all genes presumably involved in the non-enzymatic oxidative degradation are upregulated when *R. placenta* is growing on acetylated wood as compared with untreated wood. This leads to the assumption that the downregulated genes may not be as important for the non-enzymatic degradation as assumed or that the acetylation affects the regulation of the genes.

Significantly lower levels of Lac1 expression were shown, in this study, for acetylated samples as compared with untreated, which differs from previous studies. Ringman et al. [77] studied Lac1 on thermally modified wood as compared with untreated wood and no significant differences were shown between treatments and different mass loss levels. In addition, previously, no significant differences have been shown by Zhang et al. [47] for gene expression of Lac1 between oxidative and enzymatic degradation which could indicate that Lac1 does not play an important role during the non-enzymatic oxidative degradation phase or the expression is inhibited through acetylation. However, they found the highest activity for laccase in the hyphal front.

Clearly higher amounts of PQT were found in untreated samples; downregulation in AC10 and further downregulation in AC15 seem to show a tendency. However, levels of PQT expression were higher in AC20, which does not follow any trendline. Since the differences in expression levels between the treatments in this study do not show any clear trends, they are difficult to interpret. The downregulation of PQT in acetylated samples seen in our study, may be due to the fact that this gene may not be involved in oxidative wood degradation after all. On the basisof its sequence, it was proposed to be a quinate transporter [58], however, it is also possible that it is a sugar transporter [59]. Ringman et al. [76] did not find significant differences in expression of PQT between treatments or time points during the non-enzymatic degradation phase but an induction of PQT was seen at 3% mass loss. The results from Ringman et al. [76] indicate that PQT is not induced before the enzymatic degradation phase.

Expression levels of GlyD were significantly lower in all treated samples as compared with untreated. AC10 showed the lowest expression levels and gene expression in AC15 samples was upregulated as compared with the other two treatment levels. Since GlyD levels were lower in AC20 samples as compared with AC15 samples, a clear trend cannot be observed. No significant differences were shown for GlyD in Beck et al. [13], except at the first harvesting point, where untreated samples showed higher expression of GlyD as compared with all levels of acetylation. However, Munir et al. [84] suggested that the major enzyme involved in oxalate production in wood rotting basidiomycetes could be oxaloacetase, which makes the interpretation of our GlyD results difficult. Determination of the total oxalic acid production for comparison of the amounts of different treatments and treatment levels would be a useful goal for further research [85].

In this study, a downregulation of AlOx1 was seen for AC20, and for AC10 markedly lower rates as compared with untreated samples were seen. However, AC15 behaved differently, since there was no significant difference between expression levels of AC15 as compared with untreated samples in this study. Beck et al. [13] reported a downregulation of AlOx1 in AC17 and AC21 samples as compared with AC10 and untreated samples. The results imply that AlOx1 does not serve the same purpose as AlOx2 and AlOx3.

4.2.3. Effect of Acetylation on Overall Gene Expression

In this study, not all genes previously proposed to be involved in oxidative processes were upregulated in acetylated samples as compared with untreated samples. Instead, some were associated with much lower expression levels in acetylated samples as compared with untreated samples. Because of a high number of replicates ($n \geq 11$) and the high significance of the results, these findings can be considered reliable.

Zhang et al. [86] suggested that the shift between non-enzymatic and enzymatic wood degradation is mediated by cellobiose, which is formed through degradation of cellulose. In wood not experiencing mass loss, the levels of cellobiose are assumed to be low, keeping the expression of the genes in the enzymatic degradation phase at a low level. However, the absence of glucose, upregulates the expression of non-enzymatic genes [86]. Therefore, it is assumed, in modified wood, that there is a general upregulation of the genes involved in the non-enzymatic degradation phase, since the fungus is struggling to degrade it. Previous studies have shown that *R. placenta* both up- and downregulates the expression of some genes assumed to be involved in oxidative degradation processes when growing on modified wood, which is in accordance with our results [68–70]. These results, together with previously published gene expression studies on modified wood [11,69,70,75,76,78,79], imply that there might be mechanisms in the regulation of the non-enzymatic degradation in addition to the cellobiose switch proposed by Zhang et al. [86] that are important also in untreated wood. However, it should be noted that in modified wood, degradation products, such as cellobiose, presumably will be modified as well during the wood modification process, which may affect their function as gene expression regulators. Hence, it should be noted that in acetylated wood, the degradation products might look different from the degradation products in untreated wood since they come from wood constituents that were acetylated.

The expression of genes involved in the non-enzymatic degradation phase appeared to be enhanced group-wise according to an increasing level of acetylation. An upregulation of GOx2 and Cro1 (involved in the H_2O_2- production) was seen in AC10 as compared with untreated wood. In AC15, the same two genes were upregulated including two more, also involved in the production of H_2O_2, AlOx2, and AlOx3. In AC20, in addition to GOx2, Cro1, AlOx2, and AlOx3, also CyOx and QRD (both involved in the oxalic acid production) were upregulated. Perhaps the fungus is trying to increase the output of the Fenton reaction by increasing the input of H_2O_2. Perhaps this works for AC10 and AC15, and more sugars are diffusing out into the lumen, but it does not seem to be the case for AC20. Here the fungus must do something more, so it additionally upregulates the production of oxalic acid, trying to set more iron free, and thereby increasing the output of the Fenton reaction. It is possible that upregulating the genes involved in H_2O_2 production presents the lowest "cost" for the fungus and that it is a more direct way to control the Fenton reaction. However, when that does not work, other measures have to be taken.

A clear downregulation of at least some genes, presumably involved in the non-enzymatic oxidative degradation mechanism, implies that the downregulated genes either do not have the function they were assumed to have or that they do not play an important role during the first degradation step. Comparisons with previous studies proved difficult, due to the differences in sample designs. A complete picture of all transcribed genes (transcriptome sequencing) during the early stages of decay and a comparison between untreated samples and treated samples would be of great interest. Useful data on gene expression levels during enzymatic degradation from the same wood wafers for comparison would also be valuable.

5. Conclusions

In this study the gene expression of 10 *R. placenta* genes, presumed to be important in the non-enzymatic degradation phase, were studied in untreated and wood acetylated to three different levels. The expression of the genes involved in the non-enzymatic degradation phase appeared to be enhanced group-wise according to an increasing level of acetylation. *R. placenta* seemed to first increase the output of the Fenton reaction by increasing the input of H_2O_2, trying to establish a satisfying level of degradation, and when this was not possible, the production of oxalic acid was enhanced. Highly interesting is the fact that even though growth and degradation were not inhibited in AC10 samples, results on gene expression showed a relatively strong reaction of the fungus. Furthermore, the results of this study show that *R. placenta* upregulates parts of the genes involved in non-enzymatic oxidative degradation when growing on acetylated samples as compared with untreated wood, especially on higher levels of acetylation. However, only six of the 10 genes are as important for the non-enzymatic degradation phase as previously assumed. The results imply, for example, that AlOx1 does not serve the same purpose as AlOx2 and AlOx3. This confirms previous findings by Beck et al. [13]. Furthermore, the results indicate that only higher levels of acetylation induce upregulation of QRD as compared with untreated samples. This suggests that the *R. placenta* QRD may be involved in stress defense rather than in the reduction of chelators. Highlighting the importance of finding the true function of the QRD gene in *R. placenta*. The results from this study, together with previously published gene expression studies on modified wood [11,69,70,75,76,78,79], imply that there might be mechanisms in the regulation of the non-enzymatic degradation in addition to the cellobiose switch proposed by Zhang et al. [86] in both untreated and modified wood. It would be interesting to apply this method to other modification methods, such as furfurylated and thermally-modified wood. Further research on the whole transcriptome including both the non-enzymatic and the enzymatic degradation phase would be of great interest for an increased understanding of the wood degrading capacities of *R. placenta*.

Author Contributions: Conceptualization, M.K., A.P., and R.R.; methodology, M.K., A.P., and R.R.; measurements, M.K.; writing—original draft preparation, M.K., A.P., and R.R.; writing—review and editing, M.K., A.P., and R.R.; visualization, M.K.

Funding: This research was funded by the Swedish Research Council FORMAS, 942-2015-530.

Acknowledgments: The authors gratefully acknowledge financial support from The Swedish Research Council Formas 942-2015-530.

Conflicts of Interest: The authors declare no conflict of interest. The funders had no role in the design of the study; in the collection, analyses, or interpretation of data; in the writing of the manuscript, or in the decision to publish the results.

References

1. Rowell, R.M. *Wood Chemistry and Wood Composites*; Rowell, R.M., Ed.; Taylor & Francis: Boca Raton, FL, USA, 2005; p. 703. [CrossRef]
2. Huß, W.; Krötsch, S. Energieeffizientes Sanieren und Bauen mit Holz. *LWF Aktuell* **2013**, *97*, 4.
3. Jakes, J.E.; Arzola, X.; Bergman, R.; Ciesielski, P.; Hunt, C.G.; Rahbar, N.; Tshabalala, M.; Wiedenhoeft, A.C.; Zelinka, S.L. Not Just Lumber—Using Wood in the Sustainable Future of Materials, Chemicals, and Fuels. *JOM* **2016**, *68*, 10. [CrossRef]
4. Bergman, R.; Puettmann, M.; Taylor, A.; Skog, K. The Carbon Impacts of Wood Products. *For. Prod. J.* **2014**, *64*, 13. [CrossRef]
5. Zabel, R.A.; Morrell, J.J. *Wood Microbiology: Decay and Its Prevention*; Academic Press: San Diego, CA, USA, 1992.
6. Eaton, R.; Hale, M. *Decay, Pests and Protection*; Chapman & Hall: Cambridge, UK, 1993.
7. Lebow, S. Alternatives to Chromated Copper Arsenate (CCA) for Residential Construction. In Proceedings of the Environmental Impacts of Preservative-Treated Wood, Orlando, FL, USA, 8–10 February 2004; p. 12.
8. Hill, C. *Wood Modification: Chemical, Thermal and other Processes*; Stevens, C.V., Ed.; John Wiley and Sons, Ltd.: Hoboken, NJ, USA, 2006.
9. Townsend, T.; Dubey, B.; Tolaymat, T.; Solo-Gabriele, H. Preservative leaching from weathered CCA-treated wood. *J. Environ. Manag.* **2005**, *75*, 105–113. [CrossRef]
10. Mantanis, G.I. Chemical modification of wood by acetylation or furfurylation: A review of the present scaled-up technologies. *Bioresources* **2017**, *12*, 12. [CrossRef]
11. Alfredsen, G.; Pilgård, A.; Fossdal, C.G. Characterisation of Postia placenta colonisation during 36 weeks in acetylated southern yellow pine sapwood at three acetylation levels including genomic DNA and gene expression quantification of the fungus. *Holzforschung* **2016**, *70*, 11. [CrossRef]
12. Hosseinpourpia, R.; Mai, C. Mode of action of brown rot decay resistance of acetylated wood: Resistance to Fenton's reagent. *Wood Sci. Technol.* **2016**, *50*, 413–426. [CrossRef]
13. Beck, G.; Hegnar, O.A.; Fossdal, C.G.; Alfredsen, G. Acetylation of Pinus radiata delays hydrolytic depolymerisation by the brown-rot fungus Rhondonia placenta. *Int. Biodeterior. Biodegrad.* **2018**, *135*, 39–52. [CrossRef]

14. Engelund, E.; Thygesen, L.G.; Svensson, S.; Hill, C. A critical discussion of the physics of wood-water interactions. *Wood Sci. Technol.* **2013**, *47*, 21. [CrossRef]
15. Popescu, C.-M.; Hill, C.A.S.; Curling, S.; Ormondroyd, G.; Xie, Y. The water vapour sorption behaviour of acetylated birch wood: How acetylation affects the sorption isotherm and accessible hydroxyl content. *J. Mater. Sci.* **2014**, *49*, 2362–2371. [CrossRef]
16. Beck, G.; Thybring, E.; Thygesen, L.; Hill, C. Characterization of moisture in acetylated and propionylated radiata pine using low-field nuclear magnetic resonance (LFNMR) relaxometry. *Holzforschung* **2017**. [CrossRef]
17. Stamm, A.J.; Baechler, R.H. Decay resistance and dimensional stability of five modified woods. *For. Prod. J.* **1960**, *10*, 22–26.
18. Ibach, R.E.; Rowell, R.M. Improvements in Decay Resistance Based on Moisture Exclusion. *Mol. Cryst. Liq. Cryst. Sci. Technol. Sect. A Mol. Cryst. Liq. Cryst.* **2000**, *353*, 23–33. [CrossRef]
19. Larsson Brelid, P.; Simonson, R.; Bergman, Ö.; Nilsson, T. Resistance of acetylated wood to biological degradation. *Holz Als Roh- Und Werkst.* **2000**, *58*, 331–337. [CrossRef]
20. Hill, C. Why does acetylation protect wood from microbiological attack? *Wood Mater. Sci. Eng.* **2009**, *4*, 37–45. [CrossRef]
21. Papadopoulos, A.N.; Hill, C.A.S. The biological effectiveness of wood modified with linear chain carboxylic acid anhydrides against Coniophora puteana. *Holz Als Roh- Und Werkst.* **2002**, *60*, 329–332. [CrossRef]
22. Jakes, J.; Plaza, N.; Stone, D.S.; Hunt, C.; Glass, S.; Zelinka, S. Mechanism of Transport Through Wood Cell Wall Polymers. *J. For. Prod. Ind.* **2013**, *2*, 10–13.
23. Xie, Y.; Xiao, Z.; Mai, C. Degradation of chemically modified Scots pine (Pinus sylvestris L.) with Fenton reagent. *Holzforschung* **2015**, *69*, 153. [CrossRef]
24. Zelinka, S.L.; Ringman, R.; Pilgård, A.; Thybring, E.E.; Jakes, J.E.; Richter, K. The role of chemical transport in the brown-rot decay resistance of modified wood AU - Zelinka, S.L. *Int. Wood Prod. J.* **2016**, *7*, 66–70. [CrossRef]
25. Ringman, R.; Pilgård, A.; Brischke, C.; Richter, K. Mode of action of brown rot decay resistance in modified wood: A review. *Holzforschung* **2014**, *68*, 239. [CrossRef]
26. Hunt, C.; Zelinka, S.; Frihart, C.; Lorenz, L.; Yelle, D.; Gleber, S.-C.; Vogt, S.; Jakes, J.E. Acetylation increases relative humidity threshold for ion transport in wood cell walls—A means to understanding decay resistance. *Int. Biodeterior. Biodegrad.* **2018**, *133*, 230–237. [CrossRef]
27. Zelinka, S.; Glass, S.; Stone, D. A percolation model for electrical conduction in wood with implications for wood-water relations. *Wood Fiber Sci. J. Soc. Wood Sci. Technol.* **2008**, *40*, 544–552.
28. Zelinka, S.L.; Gleber, S.C.; Vogt, S.; López, G.M.R.; Jakes, J.E. Threshold for ion movements in wood cell walls below fiber saturation observed by X-ray fluorescence microscopy (XFM). *Holzforschung* **2015**, *69*, 441–448. [CrossRef]
29. Kirk, P.M.; Cannon, P.F.; Minter, D.W.; Stalpers, J.A. *Dictionary of the Funghi*, 10th ed.; CAB International: Wallingford, UK, 2008.
30. Eriksson, K.; Blanchette, R.A.; Ander, P. *Microbial and Enzymatic Degradation of Wood and Wood Components*; Springer: Berlin, Germany, 1990.
31. Blanchette, R. Delignification by Wood-Decay Fungi. *Annu. Rev. Phytopathol.* **1991**, *29*, 18. [CrossRef]
32. Daniel, G. Use of electron microscopy for aiding our understanding of wood biodegradation. *FEMS Microbiol. Rev.* **1994**, *13*, 199–233. [CrossRef]
33. Riley, R.; Salamov, A.A.; Brown, D.W.; Nagy, L.G.; Floudas, D.; Held, B.W.; Levasseur, A.; Lombard, V.; Morin, E.; Otillar, R.; et al. Extensive sampling of basidiomycete genomes demonstrates inadequacy of the white-rot/brown-rot paradigm for wood decay fungi. *Proc. Natl. Acad. Sci. USA* **2014**, *111*, 9923–9928. [CrossRef]
34. Liese, W. Ultrastructural Aspects of Woody Tissue Disintegration. *Annu. Rev. Phytopathol.* **1970**, *8*, 231–258. [CrossRef]
35. Martin, F. Fair Trade in the Underworld: The Ectomycorrhizal Symbiosis. In *Biology of the Fungal Cell*; Howard, R.J., Gow, N.A.R., Eds.; Springer: Berlin/Heidelberg, Germany, 2007; pp. 291–308. [CrossRef]
36. Schmidt, O. Indoor wood-decay basidiomycetes: Damage, causal fungi, physiology, identification and characterization, prevention and control. *Mycol. Prog.* **2007**, *6*, 261. [CrossRef]

37. Alfredsen, G.; Solheim, H.; Mohn Jenssen, K. Evaluation of decay fungi in Norwegian buildings. In Proceedings of the IRG Annual Meeting, IRG/WP 09-10701, Bangalore, India, 24–28 April 2015; p. 11.
38. Vanden Wymelenberg, A.; Gaskell, J.; Mozuch, M.; Sabat, G.; Ralph, J.; Skyba, O.; Mansfield, S.D.; Blanchette, R.A.; Martinez, D.; Grigoriev, I.; et al. Comparative Transcriptome and Secretome Analysis of Wood Decay Fungi *Postia placenta* and *Phanerochaete chrysosporium*. *Appl. Environ. Microbiol.* **2010**, *76*, 3599–3610. [CrossRef]
39. Goodell, B. Brown-Rot Fungal Degradation of Wood: Our Evolving View. In *Wood Deterioration and Preservation*; American Chemical Society: Washington, DC, USA, 2003; Volume 845, pp. 97–118.
40. Filley, T.R.; Cody, G.D.; Goodell, B.; Jellison, J.; Noser, C.; Ostrofsky, A. Lignin demethylation and polysaccharide decomposition in spruce sapwood degraded by brown rot fungi. *Org. Geochem.* **2002**, *33*, 111–124. [CrossRef]
41. Curling, S.F.; Clausen, C.A.; Winandy, J.E. Relationships between mechanical properties, weight loss, and chemical composition of wood during incipient brown-rot decay. *For. Prod. J.* **2002**, *52*, 34–39.
42. Martínez, Á.T.; Speranza, M.; Ruiz-Dueñas, F.J.; Ferreira, P.; Camarero, S.; Guillén, F.; Martínez, M.J.; Gutiérrez, A.; Del Río, J.C. Biodegradation of lignocellulosics: Microbial, chemical, and enzymatic aspects of the fungal attack of lignin. *Int. Microbiol.* **2005**, *8*, 195–204. [PubMed]
43. Niemenmaa, O.; Uusi-Rauva, A.; Hatakka, A. Demethoxylation of [O14CH3]-labelled lignin model compounds by the brown-rot fungi Gloeophyllum trabeum and Poria (Postia) placenta. *Biodegradation* **2007**, *19*, 555. [CrossRef] [PubMed]
44. Arantes, V.; Goodell, B. Current Understanding of Brown-Rot Fungal Biodegradation Mechanisms: A Review. In *Deterioration and Protection of Sustainable Biomaterials*; American Chemical Society: Washington, DC, USA, 2014; Volume 1158, pp. 3–21.
45. Baldrian, P.; Valášková, V. Degradation of cellulose by basidiomycetous fungi. *FEMS Microbiol. Rev.* **2008**, *32*, 501–521. [CrossRef]
46. Arantes, V.; Jellison, J.; Goodell, B. Peculiarities of brown-rot fungi and biochemical Fenton reaction with regard to their potential as a model for bioprocessing biomass. *Appl. Microbiol. Biotechnol.* **2012**, *94*, 323–338. [CrossRef]
47. Zhang, J.; Presley, G.N.; Hammel, K.E.; Ryu, J.-S.; Menke, J.R.; Figueroa, M.; Hu, D.; Orr, G.; Schilling, J.S. Localizing gene regulation reveals a staggered wood decay mechanism for the brown rot fungus Postia placenta. *Proc. Natl. Acad. Sci. USA* **2016**, *113*, 10968–10973. [CrossRef]
48. Eastwood, D.C.; Floudas, D.; Binder, M.; Majcherczyk, A.; Schneider, P.; Aerts, A.; Asiegbu, F.O.; Baker, S.E.; Barry, K.; Bendiksby, M.; et al. The Plant Cell Wall–Decomposing Machinery Underlies the Functional Diversity of Forest Fungi. *Science* **2011**, *333*, 762–765. [CrossRef]
49. Goodell, B.; Jellison, J.; Liu, J.; Daniel, G.; Paszczynski, A.; Fekete, F.; Krishnamurthy, S.; Jun, L.; Xu, G. Low molecular weight chelators and phenolic compounds isolated from wood decay fungi and their role in the fungal biodegradation of wood1This is paper 2084 of the Maine Agricultural and Forest Experiment Station.1. *J. Biotechnol.* **1997**, *53*, 133–162. [CrossRef]
50. Paszczynski, A.; Crawford, R.; Funk, D.; Goodell, B. De novo synthesis of 4,5-dimethoxycatechol and 2, 5-dimethoxyhydroquinone by the brown rot fungus Gloeophyllum trabeum. *Appl. Environ. Microbiol.* **1999**, *65*, 674–679.
51. Jensen, K.A.; Houtman, C.J.; Ryan, Z.C.; Hammel, K.E. Pathways for Extracellular Fenton Chemistry in the Brown Rot Basidiomycete Gloeophyllum trabeum. *Appl. Environ. Microbiol.* **2001**, *67*, 2705–2711. [CrossRef]
52. Fenton, H. Oxidation of tartaric acid in the presence of iron. *J. Chem. Soc. Trans.* **1894**, *65*, 12. [CrossRef]
53. Goodell, B.; Zhu, Y.; Kim, S.; Kafle, K.; Eastwood, D.C.; Daniel, G.; Jellison, J.; Yoshida, M.; Groom, L.; Pingali, S.V.; et al. Modification of the nanostructure of lignocellulose cell walls via a non-enzymatic lignocellulose deconstruction system in brown rot wood-decay fungi. *Biotechnol. Biofuels* **2017**, *10*, 15. [CrossRef] [PubMed]
54. Martinez, D.; Challacombe, J.; Morgenstern, I.; Hibbett, D.S.; Schmoll, M.; Kubicek, C.P.; Ferreira, P.; Ruiz-Dueñas, F.J.; Martinez, A.T.; Kersten, P.; et al. Genome, transcriptome, and secretome analysis of wood decay fungus *Postia placenta* supports unique mechanisms of lignocellulose conversion. *Proc. Natl. Acad. Sci. USA* **2009**, *106*, 6. [CrossRef] [PubMed]

55. Jensen, K.; Ryan, Z.; Marty, A.; Cullen, D.; Hammel, K.E. An NADH: Quinone Oxidoreductase Active during Biodegradation by the Brown-Rot Basidiomycete Gloeophyllum trabeum. *Appl. Environ. Microbiol.* **2002**, *68*, 2699–2703. [CrossRef] [PubMed]
56. Qi, W.; Jellison, J. Characterization of a transplasma membrane redox system of the brown rot fungus Gloeophyllum trabeum. *Int. Biodeterior. Biodegrad.* **2004**, *53*, 37–42. [CrossRef]
57. Cohen, R.; Suzuki, M.R.; Hammel, K.E. Differential stress-induced regulation of two quinone reductases in the brown rot basidiomycete Gloeophyllum trabeum. *Appl. Environ. Microbiol.* **2004**, *70*, 324–331. [CrossRef] [PubMed]
58. Mueckler, M.; Caruso, C.; Baldwin, S.A.; Panico, M.; Blench, I.; Morris, H.R.; Allard, W.J.; Lienhard, G.E.; Lodish, H.F. Sequence and structure of a human glucose transporter. *Science* **1985**, *229*, 941. [CrossRef]
59. Yan, N. Structural Biology of the Major Facilitator Superfamily Transporters. *Annu. Rev. Biophys.* **2015**, *44*, 257–283. [CrossRef]
60. Wardman, P. Reduction Potentials of One-Electron Couples Involving Free Radicals in Aqueous Solution. *J. Phys. Chem. Ref. Data* **1989**, *18*, 1637–1755. [CrossRef]
61. Wei, D.; Houtman, C.J.; Kapich, A.; Hunt, C.; Cullen, D.; Hammel, K.E. Laccase and Its Role in Production of Extracellular Reactive Oxygen Species during Wood Decay by the Brown Rot Basidiomycete Postia placenta. *Appl. Environ. Microbiol.* **2010**, *76*, 2091–2097. [CrossRef]
62. Floudas, D.; Binder, M.; Riley, R.; Barry, K.; Blanchette, R.A.; Henrissat, B.; Martínez, A.T.; Otillar, R.; Spatafora, J.W.; Yadav, J.S.; et al. The Paleozoic Origin of Enzymatic Lignin Decomposition Reconstructed from 31 Fungal Genomes. *Science* **2012**, *336*, 1715–1719. [CrossRef] [PubMed]
63. Levasseur, A.; Drula, E.; Lombard, V.; Coutinho, P.M.; Henrissat, B. Expansion of the enzymatic repertoire of the CAZy database to integrate auxiliary redox enzymes. *Biotechnol. Biofuels* **2013**, *6*, 41. [CrossRef] [PubMed]
64. Takao, S. Organic Acid Production by Basidiomycetes. *Appl. Microbiol.* **1965**, *13*, 732. [PubMed]
65. Akamatsu, Y.; Shimada, M. Partial purification and characterization of glyoxylate oxidase from the brown-rot basidiomycete Tyromyces palustris. *Phytochemistry* **1994**, *37*, 649–653. [CrossRef]
66. Munir, E.; Yoon, J.-J.; Tokimatsu, T.; Hattori, T.; Shimada, M. New role for glyoxylate cycle enzymes in wood-rotting basidiomycetes in relation to biosynthesis of oxalic acid. *J. Wood Sci.* **2001**, *47*, 368–373. [CrossRef]
67. Tokimatsu, T.; Nagai, Y.; Hattori, T.; Shimada, M. Purification and characteristics of a novel cytochrome c dependent glyoxylate dehydrogenase from a wood-destroying fungus Tyromyces palustris 1. *FEBS Lett.* **1998**, *437*, 117–121. [CrossRef]
68. Pilgård, A.; Alfredsen, G.; Fossdal, C.G.; Long, I.C.J. The effects of acetylation on the growth of Postia placenta over 36 weeks. In Proceedings of the IRG Annual Meeting, Kuala Lumpur, Malaysia, 6–10 May 2012. IRG/WP 12-40589.
69. Schmöllerl, B.; Alfredsen, G.; Fossdal, C.G.; Westin, M.; Steitz, A. Molecular investigation of Postia placenta growing in modified wood. In Proceedings of the IRG Annual Meeting, IRG/WP 11-10756, Queenstown, New Zealand, 8–12 May 2018; p. 10.
70. Ringman, R.; Pilgård, A.; Richter, K. Effect of wood modification on gene expression during incipient Postia placenta decay. *Int. Biodeterior. Biodegrad.* **2014**, *86*, 86–91. [CrossRef]
71. Ringman, R.; Pilgård, A.; Brischke, C.; Windeisen, E.; Richter, K. Incipient brown rot decay in modified wood: Patterns of mass loss, structural integrity, moisture and acetyl content in high resolution. *Int. Wood Prod. J.* **2017**. [CrossRef]
72. Bravery, A.F. A miniaturised wood-block test for the rapid evaluation of preservative fungicides. In *Screening Techniques for Potential Wood Preservative Chemicals. Proceedings of the a Special Seminar Held in Association with the 10th Annual Meeting of IRG, Peebles, UK, 18−22 September 1978*; IRG: Washington, DC, USA, 1978.
73. Alfredsen, G.; Pilgård, A. Postia placenta decay of acetic anhydride modified wood—Effect of leaching. *Wood Mater. Sci. Eng.* **2014**. [CrossRef]
74. Schilling, J.S.; Duncan, S.M.; Presley, G.N.; Filley, T.R.; Jurgens, J.A.; Blanchette, R.A. Colocalizing incipient reactions in wood degraded by the brown rot fungus Postia placenta. *Int. Biodeterior. Biodegrad.* **2013**, *83*, 56–62. [CrossRef]

75. Alfredsen, G.; Ringman, R.; Pilgård, A.; Fossdal, C.G. New insight regarding mode of action of brown rot decay of modified wood based on DNA and gene expression studies: A review. *Int. Wood Prod. J.* **2015**, *6*, 5–7. [CrossRef]
76. Ringman, R.; Pilgård, A.; Kölle, M.; Richter, K. Expression patterns of Postia placenta genes involved in the chelator mediated Fenton degradation in modified wood. In Proceedings of the European Conference on Wood Modification, Arnhem, The Netherlands, 17–18 September 2019; p. 3.
77. Ringman, R.; Pilgård, A.; Kölle, M.; Brischke, C.; Richter, K. Effects of thermal modification on Postia placenta wood degradation dynamics: Measurements of mass loss, structural integrity and gene expression. *Wood Sci. Technol.* **2016**, *50*, 385–397. [CrossRef]
78. Alfredsen, G.; Fossdal, C.G.; Nahy, N.E.; Jellison, J.; Goodell, B. Furfurylated wood: Impact on Postia placenta gene expression and oxalate crystal formation. *Holzforschung* **2016**, *70*, 16. [CrossRef]
79. Pilgård, A.; Schmöllerl, B.; Risse, M.; Fossdal, C.G.; Alfredsen, G. Profiling Postia placenta colonisation in modified wood—microscopy, DNA quantification and gene expression. In Proceedings of the Wood Science and Engineering Conference, Copenhagen, Denmark, 17–18 September 2019.
80. Salame, T.M.; Knop, D.; Levinson, D.; Yarden, O.; Hadar, Y. Redundancy among manganese peroxidases in *Pleurotus ostreatus*. *Appl. Environ. Microbiol.* **2013**, *79*, 2405. [PubMed]
81. Ringman, R.; Beck, G.; Pilgård, A. The Importance of Moisture for Brown Rot Degradation of Modified Wood: A Critical Discussion. *Forests* **2019**, *10*, 522. [CrossRef]
82. Arantes, V.; Qian, Y.; Milagres, A.M.F.; Jellison, J.; Goodell, B. Effect of pH and oxalic acid on the reduction of Fe3+ by a biomimetic chelator and on Fe3+ desorption/adsorption onto wood: Implications for brown-rot decay. *Int. Biodeterior. Biodegrad.* **2009**, *63*, 478–483. [CrossRef]
83. Narayanan, B.C.; Niu, W.; Han, Y.; Zou, J.; Mariano, P.S.; Dunaway-Mariano, D.; Herzberg, O. Structure and function of PA4872 from Pseudomonas aeruginosa, a novel class of oxaloacetate decarboxylase from the PEP mutase/isocitrate lyase superfamily. *Biochemistry* **2008**, *47*, 167–182. [CrossRef]
84. Munir, E.; Yoon, J.; Tokimatsu, T.; Hattori, T.; Shimada, M. A physiological role for oxalic acid biosynthesis in the wood-rotting basidiomycete Fomitopsis palutris. *Proc. Natl. Acad. Sci. USA* **2001**, *98*, 11126–11130. [CrossRef]
85. Schilling Jonathan, S.; Jellison, J. High-performance liquid chromatographic analysis of soluble and total oxalate in Ca- and Mg-amended liquid cultures of three wood decay fungi. *Holzforschung* **2004**, *58*, 682. [CrossRef]
86. Zhang, J.; Schilling, J. Role of carbon source in the shift from oxidative to hydrolytic wood decomposition by Postia placenta. *Fungal Genet. Biol.* **2017**, *106*. [CrossRef]

Article

The Biological Durability of Thermally- and Chemically-Modified Black Pine and Poplar Wood Against Basidiomycetes and Mold Action

Vasiliki Kamperidou

Faculty of Forestry and Natural Environment, Department of Harvesting and Technology of Forest Products, Aristotle University of Thessaloniki, 54124 Thessaloniki, Greece; vkamperi@for.auth.gr; Tel.: +30-231-099-8895; Fax: +30-231-099-8947

Received: 18 November 2019; Accepted: 3 December 2019; Published: 5 December 2019

Abstract: Wood of black pine and poplar species were subjected to thermal modification under variant conditions, while subsequently, a number of the thermally-modified black pine specimens were subjected to surface modification with organosilane solutions, and the biological resistances of the different materials were examined using laboratory agar block tests against the action of basidiomycetes and microfungi. Thermally-modified pine specimens were exposed to the brown rot fungi *Coniophora puteana* and *Oligoporus placenta*, whereas poplar wood was exposed to the white rot fungus *Trametes versicolor* and *O. placenta*. Regarding the biological durability of thermally-chemically-treated pine wood with organosilanes, it was tested against the action of *C. puteana*. Additionally, both of the thermally-treated wood species, as well as thermally-chemically-treated pine wood were exposed to a microfungi mixture, so that the wood treatments efficacy would be evaluated through a visual assessment of fungal growth on the specimen's surface The thermal treatments seem to increase the biological resistance of black pine against *C. puteana* by 9.65–36.73% compared to unmodified wood. The most significant increase in biological durability among all the thermally-treated wood categories was recorded by *O. placenta*, with 28.75–68.46% lower mass losses in treated pine specimens and 31.98–64.72% in thermally-treated poplar, respectively, compared to unmodified wood. The resistance of treated poplar against *T. versicolor* was also found increased (13.25–46.08%), compared to control. Thermal modification affected positively the biological resistance of both species, though it did not manage to protect effectively pine and poplar wood from the microfungi action. The combination of thermal and organosilanes treatment revealed a significant improvement of the durability of pine wood compared to? control (45.68–87.83% lower mass losses against *C. puteana*), as well as against the microfungi action, with the presence of benzin to have a positive effect on the silanes solutions performance and protective action.

Keywords: basidiomycetes; heat; molds; pine wood; poplar wood; silanes; thermal treatment

---

1. Introduction

The resistance of wood to rot is often considered to be synonymous with its duration. The action of various micro-organisms may cause damages and deterioration in the appearance, structure and chemical composition of wood. Several methods have been proposed so far for improving the natural durability of wood [1]. Apart from chemical protection measures, one of the most effective ways to prevent or limit the action of microorganisms in wood is to keep the moisture content low enough or to prevent at least one of the necessary growth requirements of the microorganisms to be fulfilled, such as the presence of oxygen in wood, proper pH (3–8), appropriate temperature (15 °C–40 °C), absence of toxic extracts, presence of other growth factors such as vitamins, nitrogen etc. [2].

Thermal modification consists of a non-biocidal, environmentally friendly wood protection method to enhance the behavior of wood and some characteristics critical for its service life expectancy and its utilization perspectives. Heat induces changes to the chemical constituents of wood, and therefore, to the chemical, physical and mechanical properties of wood, as well. Despite the fact that thermal treatment decreases the density and most of the times the mechanical strength of wood, since it increases its brittleness, it seems to limit the hygroscopic nature and enhance the natural durability of wood, providing the chance, (especially to species characterized by low water resistance, dimensional stability and susceptibility to bio-degradation factors), to be adequately utilized, participating in a much wider range of applications [3–6].

Several researches have been implemented to deal with the biological resistance of thermally-modified wood, the most significant of which are summarized hereupon. Li Shi et al. [7] reported that thermally-modified yellow-poplar and jack pine (210 °C for 15 min) were found to be more resistant to the brown rot fungus *Gloeophyllum trabeum* than the unmodified wood; while referring to aspen and Scots pine, the resistance was improved after the thermal modification, but it remained susceptible to this brown-rot fungus decay. Mburu et al. [8] modified the heartwood of silky oak (250 °C for 7 h) and observed that durability against fungi and termites was greatly improved after treatment, highlighting a good correlation between decay resistance and mass loss due to thermal treatment. Vetter De et al. [9] exposed many species, thermally-modified under various conditions in mycological tests of the fungi *C. puteana, O. placenta* and *T. versicolor*. Heat-modified *Picea abies* recorded a reduced resistance to *O. placenta* action with mass losses close to the control level, while the action of *C. puteana* was inhibited by heat treatment with mass losses less than 3%. Mazela et al. [10] examined the resistance of thermally-modified (160 °C–220 °C for 6 h and 24 h) wood of Scots pine against *C. puteana, G. trabeum, O. placenta* and *T. versicolor*, concluding that the brown rot fungi recorded a more intense action (mass loss of 23–46%) compared to white-rot fungi. *T. versicolor* did not prefer coniferous species (low loss of 15.4%), while *C. puteana* caused a mass loss of 39.4%, *G. trabeum* 23.5% and *O. placenta* a loss of 45.8%, respectively. Mohareb et al. [11] examined the biological durability of *Pinus patula* species after its thermal treatment using different treatment intensities, and concluded that the elemental composition of wood could be used as a valuable marker to predict treatment intensity and decay durability. Since the results of the studies published in the literature so far reveal different tendencies, the kind and range of the thermal treatment impact upon the biological resistance of wood against basidiomycetes seem not to have been thoroughly investigated and elucidated so far.

Organosilicon compounds have been applied by researchers in recent years to modify solid wood to increase hydrophobicity, durability or water resistance and dimensional stability, to prevent leaching of active ingredients or to promote adhesion or to improve fire resistance, etc. [12–15]. Although numerous studies have been conducted on silane modification of solid wood referring to its hygroscopic properties and dimensional stability, there is scant research on the efficacy and influencing mechanism of silanes modification on the biological durability of wood. Additionally, aiming to improve the biological durability of wood, a combination of thermal treatment and organosilanes treatment has not been proposed or investigated so far, according to the bibliography.

Specifically, Donath et al. [13] used the coupling agent γ-methacryloxypropyltrimethoxysilane (TEOS), and organo functional alkoxysilanes and found good incorporation into the cell wall, when conditioned wood was impregnated with alcoholic solutions of the two silanes methyltriethoxysilane and propyltriethoxysilane. Durability of the treated wood towards the white rot fungus *T. versicolor* was increased considerably, but especially when the silane penetrated and bulked the cell wall. In a soil block test, decay was delayed, but not prevented. Hill et al. [16] treated black pine sapwood with the two coupling agents γ-methacryloxypropyltrimethoxysilane and vinyltrimethoxysilane and found incorporation of the silicon material into the cell wall, and concerning the fungal decay tests, only little increase of resistance to *C. puteana* was recorded. Incubation with *T. versicolor* and *Phanerochaete crysosporium* displayed decay resistance of the treated wood above weight percent gain (WPG) of 40%

for *T. versicolor* and around 40–50% for *Phanerochaete crysosporium* [16]. Higher decay resistance was found, when amino-functional silanes were applied.

Donath et al. [13] treated wood with an amino-functional oligomeric silane system and found complete decay resistance of Scots pine sapwood to *C. puteana* (16% WPG) even after prolonged incubation (18 weeks). However, European beech wood treated with the same silane system (11% WPG), revealed considerable mass loss after incubation with *T. versicolor*. Wood treated with silicon compounds without biologically-active substance indicated a rather unsatisfactory resistance against wood decay fungi [12,15].

Microfungi develop mainly upon the surface of wood and the presence of a mold colony may make wood more vulnerable to the growth of a fungal colony, as well [17]. Boonstra et al. [2] reported in their research work that thermal treatment did not manage to inhibit the growth of mold in the surface of the specimens, while Kocaefe et al. [18], who modified Jack pine and white poplar wood thermally, recorded an improvement in wood resistance to mold growth, although this improvement did not correspond to statistically significant differences. Analogous results have been found by Gobakken and Westin [18] who observed a small resistance improvement, but very low compared to other treatments and Fojutowski et al. [19] who examined thermally-treated black poplar and black alder wood.

The aim of this research is to examine the biological resistance of black pine (*Pinus nigra* L.) and poplar (*Populus* sp.) wood after short-term thermal modification for three different durations at two temperature levels, as well as, for the first time, the combination of the described thermal modification with a subsequent chemical surface modification with water-borne and benzin-borne organosilane systems of different ratios on black pine wood, against the action of some of the common European basidiomycetes species and a mixture of commonly found microfungi (molds) species. Poplar has been chosen as a species of low quality, physical properties and natural resistance, and the black pine species was chosen, since especially its sapwood is considered to be non-resistant to wood-destroying fungi attacks, and both species seem to need enhancement e.g., [1]. Additionally, the specific species are of great importance for the Mediterranean, as well as the whole Europe area, concerning availability, timber and wood-products trade, etc., and the thermal and chemical treatments could offer them a chance to compete for specific applications with species of higher cost and value.

2. Materials and Methods

Boards of black pine (*Pinus nigra* L.) and poplar (*Populus* sp.) wood of Greek origin (poplar wood was obtained from the Drama region in North Greece and black pine from the Kalampaka region in the central part of Greece) were placed for about eight months into an air-conditioned room at a temperature of 20 ± 2 °C and 60% ± 5% relative humidity, and left there to dry naturally until there existed a constant weight recording an equilibrium moisture content (EMC) of 11.44% for pine and 10.50% for poplar wood (ISO 13061-1) [20]. The mean density (mass/volume, measured with moisture content at the reported levels) of the pine was 0.662 g/cm^3 (presenting dense structure with thin annual rings) and 0.385 g/cm^3 of poplar (ISO 13061-2) [21]. The dimensions of the boards prior to the treatment were 35 mm thick × 70 mm wide × 400 mm in length, with their length to be in a direction parallel to the grain, while they consisted mainly from the part of sapwood, and the sampling method was based on the methodology of ISO 3129 [22].

2.1. Thermal Treatments

The treatment of the boards was carried out in a laboratory drying chamber (800 mm × 500 mm × 600 mm). The treatments were carried out at 180 °C and 200 °C under atmospheric pressure in the presence of air. The moisture content of the boards was 11.44% and 10.5% for black pine and poplar wood, respectively, when placed in the chamber, which was preheated to the final temperature. In each treatment, 10 identical non-sanded boards were placed into the chamber in specific places of equal space from the walls of the chamber and from one another, in order to ensure the uninterrupted movement of the air between the boards. The treatment durations of 3, 5 and 7 h (shorter than those durations industrially used) were applied, counting 15 min more for the recovery of the temperature in the chamber.

The specific durations were selected as they differentiate from the existing literature of thermal treatment processes, providing new data. After thermal treatment, the boards were placed into large desiccators to return gradually to ambient conditions, and stacked in an air-conditioned room at constant conditions (60% ± 5%, 20 ± 2 °C). The mass loss (ML) of the specimens induced by the process of thermal treatment (thermodegradation) was determined according to the following Equation (1):

$$ML = [(M_0 - M_1)/M_1] \times 100 \quad (1)$$

where:

ML: Mass loss (%)

M_0: Initial oven-dry mass of the specimen before thermal treatment (g)

M_1: Oven-dry mass of the same specimen after thermal treatment (g)

The surfaces of the prepared samples were sanded using 80 and 220 grit sandpaper. Only defect-free material and only sapwood blocks (30 mm × 20 mm × 8 mm) of wood were selected to be used in decay tests, which were prepared from all the mass of each of the prepared treated and untreated boards of the species. The present study investigates only the biological durability of the modified materials against basidiomycetes and mold, but since it constitutes a part of an experimental work, other critical properties of the materials, such as their physical (EMC, density, color etc.), hygroscopic (tangential and radial swelling, adsorption) and mechanical properties (static bending strength, modulus of elasticity, impact bending strength, hardness, compression strength, surface roughness, etc.) have already been investigated [23–26].

2.2. Chemical Treatment

Some of the thermally-modified black pine wood specimens in dimensions of 30 mm × 20 mm × 5 mm and in variant EMC and density levels, according to the conditions of thermal treatment they had been exposed to (Table 1), were subjected to an additional chemical surface treatment with the hydrophobic agent of organosilanes, in order to form a thin layer of the solution in the surface of wood, that could potentially improve even more the biological durability, except for other properties associated to its hygroscopic nature.

Table 1. Mean values of equilibrium moisture content (EMC) and density of unmodified and thermally-modified pine and poplar wood after four weeks of conditioning.

Treatment	EMC (%)		Density (g/cm^3)	
	Pine	Poplar	Pine	Poplar
Control	11.448 (0.172) [1]	10.506 (0.521)	0.662 (0.015)	0.385 (0.002)
180 °C—3 h	9.075 (0.622)	8.535 (0.377)	0.657 (0.022)	0.381 (0.006)
180 °C—5 h	8.875 (0.598)	7.972 (0.500)	0.653 (0.026)	0.356 (0.007)
180 °C—7 h	8.441 (0.963)	7.520 (0.871)	0.645 (0.021)	0.346 (0.007)
200 °C—3 h	8.026 (0.347)	6.181 (0.654)	0.600 (0.021)	0.336 (0.005)
200 °C—5 h	7.531 (0.826)	5.691 (0.246)	0.590 (0.037)	0.333 (0.002)
200 °C—7 h	7.049 (0.412)	5.337 (0.402)	0.528 (0.021)	0.291 (0.005)

[1] Standard deviation values inside the parentheses.

Concerning the EMC values of the thermally-treated wood, statistically significant differences were recorded between the EMC value of untreated wood and the respective values of all the different categories of thermally-treated wood, which applied to both of the wood species, revealing the impact of heat upon the hygroscopicity of wood. Referring to treated black pine wood, statistically significant differences were not recorded, except for the following cases: the EMC value of the mildest treatment (180 °C) differed significantly from the respective values of the specimens treated at 200 °C, regardless of the duration. The EMC values of the specimens treated at 180 °C—5 h, 180 °C—7 h and 200 °C—3 h were found to differ significantly only from the most intensive treatment (200 °C—7 h).

Referring to treated poplar wood, only the following cases appeared to have statistically significant differences: The EMC value of the mildest treatment (180 °C) differed significantly from the respective values of the different categories of specimens treated at 200 °C, as well as the EMC values of the specimens treated at 180 °C—5 h. The value of specimens treated at 180 °C—7 h differed significantly from the respective values of the two most intensive treatments (200 °C—5 h and 200 °C—7 h).

Concerning the density values of the materials, the following statistically significant differences were recorded. Referring to the species of pine wood, the density value of the untreated wood, as well as the treatments of 180 °C, were found to differ significantly from all the respective values of the specimens treated at 200 °C, regardless of the duration. The value of the specimens treated at the most intensive treatment (200 °C—7 h) was found to differ significantly from all the other categories of thermally-treated wood, as well as the untreated wood. Referring to the species of poplar wood, statistically significant differences were recorded between the density value of untreated wood and the respective values of all the different categories of thermally-treated specimens. The density value of specimens treated at the mildest treatment (180 °C—3 h) differed significantly from all the values of the poplar specimens. The density value of specimens treated at 180 °C—5 h differed significantly also from the respective values of all the poplar specimens treated at 200 °C, and the density value of specimens treated at 180 °C—7 h differs significantly also from the values of the specimens subjected to the two most intensive treatments (200 °C—5 h and 200 °C—7 h). The value of the specimens treated at the most intensive treatment (200 °C—7 h) was found to differ significantly from all the other categories of thermally-treated wood, as well as the untreated wood.

Prior to the surface modification, the end-grains of the pine samples were sealed with a single-component polyurethane adhesive in order to prevent penetration of the agents along the grain [27], a process which was commonly followed in the preparation of all the specimens (thermally-treated, thermally-chemically-treated), in order to provide comparable results. Three different organosilanes solutions were used, in which silanes were included as the active substance (XIAMETER OFS-6020 SILANE, San Francisco, California, USA) and Alkyd resin (FTALAK S-6575; Plastbud Sp.z.o.o., Pustków, Poland) of 75% solids content, based on different ratios, while distilled water or benzin were used as a solvent. Specifically, in the three prepared solutions, silanes content corresponded to 5%, resin to 22% and the remaining 73% was the solvent in three different ratios of benzin to distilled water, 50:50 (T-C 1), 70:30 (T-C 2) and 100:0 (T-C 3), respectively. Benzin could not be considered as an environmentally friendly substance, but it was chosen because of its low cost, high availability, satisfying performance and examined the different ratios with water for comparative reasons. A stirring machine was used to homogenize the ingredients (YellowLine DI25 basic; IKA Werke GmbH & Co, Staufen im Breisgau, Germany). Each pine specimen was weighted, immersed in solution of 100 mL for 120 sec. and then left to be drained, weighed again, and finally left to be dried for at least one week in a conditioned room (20 ± 2 °C, 60 ± 5% relative humidity), before the decay tests (Figure 1). For each thermally-treated material category, 16–18 specimens were chemically-treated per each of the three different solutions (T-C 1, T-C 2, T-C 3). For each category of T-C material, approximately 10 specimens were used for the determination of EMC and density, while the rest of the 6–8 specimens were used for each of the decay tests. The weight percent gain (WPG) of the modified pine wood specimens was found to range between 8–15%, with the solutions of T-C 3 to present the highest values of WPG. The EMC and density values of all the different categories of chemically-treated pine specimens were examined.

(a) (b)

Figure 1. (**a**) Organosilanes formulations, (**b**) wood specimens left to drain after surface modification

2.3. Mycological Tests

2.3.1. Basidiomycetes

The methodology of the basidiomycetes decay tests was based on the EN 113 standard processes [28]. According to the mycological test process, 50mL of 'medium' (2000 mL water, 40 g agar, 100 g malt) was placed in each Kollev flask, which were closed and placed in the autoclave for sterilization at 121 °C for 60 min.

During the inoculation process, small, round samples were taken from precultures grown on malt agar in Petri dishes, where there were previously prepared freshly grown monoculture fungi colonies, and they were placed in the center of each Kolle flask. The flasks were placed in the incubation room (25 °C and relative humidity of 70%). After about two weeks, that the fungal mycelia appeared to cover the entire surface of the flasks, the wood specimens were aseptically added inside.

Wood samples had been previously dried (103 °C for 24 h), placed into glass desiccators for 30 min, and weighed accurately to record their initial dry weight. Subsequently, they were sterilized in the autoclave (121 °C, 60 min), before being inserted in the flasks. For each category of wood, at least six Kolle flasks were used, as well as a correction value flask. This flask contained only 50 mL of medium and 2 mL of phenol (as a protective substance against fungi action) without colony fungus, as well as two specimens of each modified specimens category. This is so that the potential weight loss of the specimens, (due to their coexistence with the medium), could be subtracted from the final mass loss results, so that corrected data calculated from the dry mass before and after the test, would be provided [29].

Therefore, eight specimens were used for each category of modified wood, and 36 control specimens for each type of wood (poplar or pine) and for each fungus species tested. Pine wood specimens were exposed to the brown-rot fungi *Coniophora puteana* (Schumach) P. Karst and *Oligoporus placenta* (Fr.) Gilb. & Ryvarden, while poplar specimens to *Trametes versicolor* (white rot) and *Oligoporus placenta* (Fr.) Gilb. & Ryvarden. *C. puteana* was chosen since it strongly decays softwood species, especially pine wood, *T. versicolor* decays hardwood species such as poplar to a large extent, while *O. placenta* is a major threat to several species, especially softwoods. Regarding the decay tests of thermally- and chemically-treated pine wood with organosilanes, the specimens were tested against the action of *C. puteana*.

After the addition of the wooden specimens inside the flasks, they were placed again in the incubation room where they remained exposed to fungal activity for a 4-month period.

Subsequently, the specimens were removed from the flasks, their surface was cleaned from the adhering mycelium and weighed to the nearest thousandth of a gram to calculate their moisture content. Oven-drying of all the specimens followed (103 °C for 24 h) and reweighting, to calculate the dry weight at the end of the decay test. The mass loss due to fungal attack was determined based on the dry weight before and after the decay test by the following equation 2:

$$ML = \left(\frac{W - W_1}{W}\right) \times 100 \quad (2)$$

where W: the dry weight of the specimens before the decay test, W_1: the dry weight of the specimens after the decay test.

2.3.2. Microfungi

The medium used to cultivate microfungi was prepared using water, agar, malt and salt in a specific ratio (1000 mL water, 20 g agar, 30 g malt, 4 g mixture of salt (CZAPEK DOX: $NaNO_3$ 2 g, KH_2PO_4 0.7 g, K_2HPO_4 0.3 g, KCl 0.5 g, $MgSOi_7H_2O$ 0.5 g) and was introduced into the autoclave for sterilization, with all the wood specimens and necessary tools. 50 mL of medium was placed into each of the plastic containers (110 mm × 110 mm) used for the test. The microfungi inoculation chamber had already been sterilized with alcohol and UV radiation. Five specimens were used per each modified wood category, and one modified specimen was placed in each container together with one control specimen.

The dimensions of wood specimens were: 8 mm in thickness × 20 mm in width × 30 mm in length. A spore suspension was prepared by mixing three common cultures of wood-attacking microfungi, *Aspergillus niger* var. niger, *Trichoderma viride* Pers. ex Fries and *Penicillium cyclopium* Westling. Microfungi were inoculated through spreading the spores onto the agar medium surface of the containers, as well as the specimens. After the inoculation, all the containers were placed into an incubator for 21 days at 28 ± 2 °C with relative humidity of 75% ± 5%. The evaluation of wood treatment efficacy was performed after 7, 14 and 21 days by means of visual monitoring of fungal growth on the wood surface in accordance with the scale set out in Table 2, according to ASTM D 5590-17 [30].

Table 2. Rating system for fungal growth [30].

Index Rating System	
0z	no growth of fungi on the specimen, inhibition zone on the medium
0	no growth of fungi on the specimen
1	less than 10% of the specimen area covered by fungi
2	less than 30% of the specimen area covered by fungi
3	less than 60% of the specimen area covered by fungi
4	specimen totally overgrown by fungi

2.4. Statistical Analysis

SPSS Statistics PASW 18 statistical package was used for statistical analysis and processing of the results. The Least Significant Difference two way Analysis of Variance (LSD-Two-way ANOVA) method was also used in the thermal treatment results, which examines the effect of two different independent variables (temperature and duration), but also the effect of any interaction between these two factors upon the dependent variable of the biological durability of wood.

3. Results and Discussion

According to the results, thermal treatment of black pine wood under the abovementioned conditions induced mass losses in the range of 10.63–15.25% of which a 11.44% refers to moisture content lost, whereas thermally-treated poplar wood recorded mass losses of 11.24–18.88%, of which 10.50% corresponds to moisture, generated by the thermo-degradation during the process (Table 3). The wood of black pine contains a significant amount of resin, and the mass loss could be partly attributed also to the evaporation of the resin during the thermal treatment, as well as other volatile compounds and unstable components.

Table 3. Mean mass loss values of black pine and poplar wood specimens due to the heat treatment thermodegradation process.

Species	Mass Loss (%)					
	180 °C—3 h	180 °C—5 h	180 °C—7 h	200 °C—3 h	200 °C—5 h	200 °C—7 h
Pine	10.629 (0.270) [1]	11.262 (0.507)	11.444 (0.572)	12.854 (1.035)	13.984 (1.172)	15.248 (1.516)
Poplar	11.237 (2.512)	11.601 (0.645)	11.935 (0.975)	13.461 (2.192)	16.409 (1.974)	18.884 (2.642)

[1] Standard deviation values in parentheses.

3.1. Basidiomycetes

3.1.1. Thermal Treated Black Pine

The results of the biological resistance tests revealed that the thermal treatment improved, in every case, the biological resistance of the black pine wood specimens against the decay action of *Coniophora puteana* and *Oligoporus placenta* (Figure 2), and this improvement seems to be analogous of the treatment intensity.

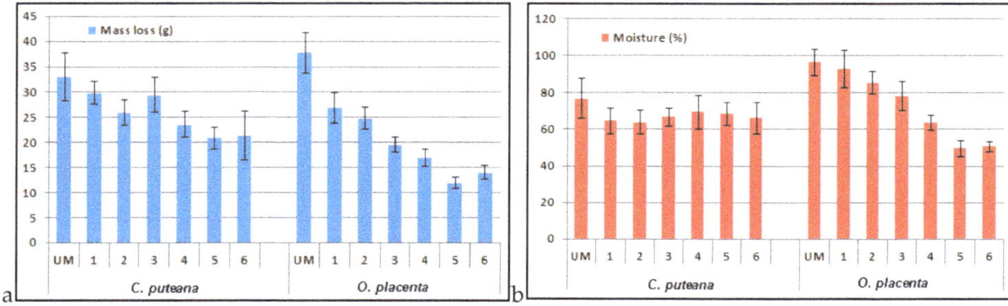

Figure 2. Mean mass loss (**a**) and moisture content rates (**b**) recorded by unmodified (UM) and heat-modified (1: 180 °C—3 h, 2: 180 °C—5 h, 3: 180 °C—7 h, 4: 200 °C—3 h, 5: 200 °C—5 h, 6: 200 °C—7 h) black pine specimens after their exposure to *C. puteana* and *O. placenta*.

Specifically, referring to *C. puteana* test results, the treatments of 200 °C appeared to reduce even more the mass loss rates caused by fungal action and deterioration, compared to treatments of 180 °C. Regardless of the treatment temperature, the treatment duration of 5 h resulted, in each case, in higher resistance of specimens to fungal action, compared to treatment duration of 3 or 7 h. This may indicate that a 3-h duration is not enough to achieve the greatest possible improvement in durability, while the 7-h duration exceeds the point of maximum improvement, and the wood may have suffered enormous mass loss and probably both the critical wood component de-polymerization due to thermal treatment and the density decrease leave, once again, the wood quite vulnerable to fungal biodegradation. The treatments of 180 °C and durations of 3, 5 and 7 h reduced the mean mass loss percentage, compared to unmodified wood, by 9.65%, 21.53% and 10.77%, while the treatments of 200 °C (3, 5 and 7 h) reduced the loss by 28.39%, 36.73% and 35.37%, respectively.

The statistical analysis revealed that the mass loss percentage values of unmodified wood, due to *C. putena* action, statistically differed significantly only from the respective values of the 200 °C treatments. The mass loss of specimens treated at the mildest treatment (180 °C—3 h) as well as the treatment 180 °C—7 h of also differed significantly only from the values of the three 200 °C treatments.

The mass loss of specimens treated at the treatment of 180 °C—5 h differed significantly only from the value of the specimens treated at 200 °C—5 h.

Statistical analysis of Two Way ANOVA showed that the effect of treatment temperature was found statistically significant and influenced its variability by 96.6%, while the treatment duration also presented a statistically significant effect on the mass loss, affecting its variability by 83.5%. The interaction between temperature and duration was significant as well, and affected the mass loss variability by 53.6%. The largest effect of temperature was observed in the 7—h treatments, which corresponded to statistically significant differences.

The moisture content of the thermally-modified pine specimens after the mycological test was found to be, in each case, lower than the respective value of the control specimens. In general, the higher the moisture content value of the specimen found after the mycological test, the more intense the decay by the fungus seems to be, and therefore, the mass loss rate is usually analogous to the moisture content rate of the decayed specimen.

However, treatments of 200 °C showed high moisture content in the specimens after the biological resistance test, which could be translated as an extensive presence of *C. puteana*, but which does not correspond to the continuity of its intense destructive action, since *C. puteana* prefers, as a brown-rot fungus, to degrade holocellulose, e.g., [10], and a significant part of the essential nutrients for this fungi (hemicelluloses, extracts) have been degraded and evaporated, and some by-products thereof have been re-polymerized in new structures not recognizable by the fungal enzymes. This improvement of the thermally-modified wood resistance against fungi action could be possibly attributed to the increased hydrophobicity it acquires after treatment (lower moisture content, EMC), the decomposition of hemicelluloses and amorphous areas of cellulose, which constitute nutrients for fungi especially in the early stages of rot, as well as the production of biocidal extractives during the treatment, which could act as fungicides, e.g., [31,32]. Furthermore, in thermally-modified wood, the enzymes involved in fungal degradation have difficulties to identify the modified polymeric components of wood, and to penetrate in the cell walls due to the decrease in porosity [1].

The results of treated black pine wood resistance against the action of *O. placenta* also seem to be encouraging, since all treatments have greatly improved the bio-resistance of wood, and as shown, this improvement is in line with the intensity of the treatment. As the temperature and duration of treatment increase, lower mean mass loss percentages are being recorded. The only exception is the most intensive treatment (200 °C—7 h), which recorded a slightly higher mass loss compared to the previous treatment of the immediately lighter intensity (200 °C—5 h), possibly attributed to the high mass loss and loss of chemical components carried out during the thermal treatment, critical for the resistance of wood. Specifically, treatments of 180 °C and durations of 3, 5 and 7 h reduced the mass loss compared to the unmodified wood by 28.75%, 34.18% and 48.09% respectively, while the treatments of 200 °C (3, 5 and 7 h) reduced the loss by 55.11%, 68.46% and 62.9%, respectively.

The mean mass loss value recorded by the unmodified pine wood statistically differed significantly from all the respective treated wood categories, revealing the resistance improvement of the materials, which is found higher than that of the same wood species against the fungus *C. puteana*. The mass loss of specimens treated at the mildest treatment (180 °C—3 h) presented similar values to the value of the specimens of next treatment (180 °C—5 h) and both were found to differ significantly from the respective values of all the other treatments. The mass loss of specimens treated at 180 °C—7 h presented similar values to the specimens of the treatment 200 °C—3 h and both were found to differ significantly from the respective values of all the other treatments. Finally, the value of the specimens treated at 200 °C—5 h presented the lowest mass loss value, which differed significantly from the values of all the other categories.

Two-Way ANOVA showed that the effect of temperature on mass loss rate was statistically significant, and influenced its variability by 86.6%, whereas the duration factor did that by 61.4%. The interaction between temperature and duration was also statistically significant, affecting the loss variability by 44.1%. The greatest effect of temperature was observed in 5 h treatments, recording a statistically significant difference.

Finally, the moisture content of the specimens after the mycological test decreased in each case compared to control sample, and its trend was found similar to that of the mass loss. Referring to pine wood, one of the phenolic components arising during thermal treatment (vanillin) may hinder and slow the growth of fungi by inhibiting the production of their enzymes, and it could also be partially responsible for the antifungal activity of pine against *O. placenta* [2]. According to the results of the current research, both fungi species *C. puteana* and *O. placenta* have been proven to be very active and aggressive to black pine wood, with *O. placenta* to cause extended decay to unmodified wood, but the lowest decay in thermally-treated samples.

Mazela et al. [10], who examined the resistance of thermally-treated (160 °C–220 °C, 6 h–24 h) Scots pine, recorded also higher mass losses caused by *C. puteana* (39.4%) and *O. placenta* (45.8%) action, with the thermal treatment of the highest intensity to result in the highest levels of biological resistance recorded.

Tremblay [33] presented an improvement of the biological resistance of *Pinus banksiana* wood after thermal treatment, while Chaouch et al. [34], who subjected Scots pine wood to thermal treatment, did not record differences in resistance levels between modified and unmodified wood to *O. placenta* action. Kortelainen and Viitanen [35] recorded resistance improvement of thermally-modified Scots pine and Norway spruce wood (especially of heartwood) only in the most intense treatment (230 °C) used.

3.1.2. Thermally-Chemically-Treated Black Pine

According to the results (Table 4), all the categories of the specimens that had been subjected to thermal in combination with surface treatment with silanes solutions recorded significantly lower EMC values compared to the control, as well as compared to all the respective values of specimens that had been only thermally treated. This suggests that silanes treatment could limit even more the hygroscopicity of wood, compared to thermal treatment under specific conditions, probably because silane molecules are capable of filling the large pores of wood cell cavities and reacting with the hydroxyl groups. Therefore, the long chains of the silane manage to hinder the degree of moisture uptake [36], improving the dimensional stability and surface hydrophobicity of silane-modified wood. Additionally, by increasing the participation of benzin in the organosilane solutions against the participation of water as a solvent, the EMC seems to be even lower, which could be possibly attributed to better performance and the slightly higher absorption of the silane solution in the mass of wood during the surface modification. This perspective could be supported also by the fact that the density values of pine wood were found almost in all cases higher in the solutions with higher participation of benzin as a solvent (T-C 3).

Table 4. Mean values of equilibrium moisture content (EMC) and density levels of unmodified and thermally-chemically (T-C) modified black pine after two weeks of conditioning.

Thermally and Chemically (T-C) Treatments	EMC (%) Density (g/cm^3)					
	T-C 1	T-C 2	T-C 3	T-C 1	T-C 2	T-C 3
Control	11.45 (0.172) [1]	11.45 (0.172)	11.45 (0.172)	0.662 (0.015)	0.662 (0.015)	0.662 (0.015)
180 °C-3 h	7.194 (0.622)	7.007 (0.541)	7.082 (0.672)	0.665 (0.142)	0.664 (0.115)	0.666 (0.070)
180 °C-5 h	6.433 (0.598)	6.390 (0.738)	6.256 (0.650)	0.667 (0.086)	0.668 (0.042)	0.665 (0.053)
180 °C-7 h	6.003 (0.963)	6.120 (0.570)	6.045 (0.783)	0.656 (0.021)	0.660 (0.120)	0.662 (0.080)
200 °C-3 h	5.724 (0.347)	5.604 (0.732)	5.340 (0.413)	0.624 (0.076)	0.641 (0.063)	0.658 (0.035)
200 °C-5 h	5.231 (0.826)	5.015 (0.280)	4.694 (0.520)	0.601 (0.037)	0.623 (0.048)	0.642 (0.058)
200 °C-7 h	4.845 (0.412)	4.630 (0.687)	4.320 (0.580)	0.557 (0.021)	0.575 (0.084)	0.587 (0.043)

[1] Standard deviation values inside the parentheses.

According to the statistical analysis, statistically significant differences were recorded between the EMC value of untreated wood and the respective values of all the different categories of thermally-chemically-treated wood, which applied to all the different categories of thermally-chemically-treated wood (T-C 1, T-C 2, T-C 3), revealing the great impact of the combination of heat and organosilanes on the hygroscopic nature of wood. Referring to the treated black pine wood, in the cases of T-C 1- and T-C 3-treated specimens, the EMC value of the specimens of the mildest thermal treatment (180 °C—3 h) differed significantly from the respective values of all the categories of specimens thermally-treated at 200 °C, the value of the specimens treated at 180 °C—5 h differed significantly from the respective values of the most intensively thermally-treated specimens at 200 °C—5 h and 200 °C—7 h. There was also recorded a statistically significant difference between the EMC values of the specimens treated at 200 °C—3 h and 200 °C—7 h. In the case of T-C 2-treated specimens, the EMC value of the specimens of the mildest thermal treatment (180 °C—3 h) differed significantly from the respective values of all the categories of specimens thermally-treated at 200 °C, while the value of the specimens treated at 180 °C—5 h differed significantly from the respective values of the most intensively thermally-treated specimens at 200 °C—5 h and 200 °C—7 h.

Generally, the density of thermally-chemically-treated pine wood was found slightly higher in most cases compared to the respective values of specimens treated just thermally. Referring to the statistical analysis of the density values of black pine wood specimens after their thermal and organosilanes treatment, in the case of T-C 1, the value of untreated wood was found to differ significantly only from the respective value of specimens treated at the two most intensive treatments (200 °C—5 h and 200 °C—7 h), and another statistically significant difference was recorded between the value of specimens treated at 180 °C—5 h and the most intensive one (200 °C—7 h). In the case of T-C 2, statistically significant differences were not recorded between the density values, since the values of standard deviation (SD) were found to be higher than the respective values of thermally-treated material. This could be possibly attributed to the fact that the density of these thermally-chemically-treated specimens are strongly influenced, not only by the thermodegradation process during the thermal treatment that decreases the density, but also by the penetration and absorption percentage of the organosilanes solution in the mass of wood. In the case of T-C 3, the only statistically significant difference was recorded between the density value of the untreated wood and the respective value of the specimens subjected to the most intense treatment (200 °C—7 h).

According to the decay tests results (Figure 3), the combination of thermal treatment with silanes solutions treatment improved significantly the biological durability of wood, recording mass losses, caused by the action of fungi, much lower than those of the unmodified wood, as well as lower than the respective mass loss values of thermally-treated wood specimens categories. This improvement of biological resistance of black pine specimens seems to be also in this case of treatments combination analogous of the thermal treatment intensity.

The presence of benzin in silane solutions used in this research work seems to increase the protective effect of the surface modification process on the biological durability of wood. This improved action of silanes solutions in the case of higher benzin participation can be attributed to the hypothesis that benzin contributes to a higher dispersion/mixing of the solution or to a deeper penetration of the solution in wood mass and reaction with its chemical components, and as a result, improved protection by the action of fungi. Specifically, the treatment of T-C 1 resulted in 45.68–79.74% lower mass losses compared to unmodified pine wood, while the treatments of T-C 2 and T-C 3 caused 54.76–83% and 70.21–87.83% lower mass losses, respectively against the action of *C. puteana*.

The statistical analysis of the mass loss values of thermally- and chemically-treated black pine wood revealed that in each case of T-C treatment there were statistically significant differences between the mass loss values of untreated wood and the respective values of all the thermally-chemically-treated wood categories, revealing the positive impact of the methods combination on the biological durability of wood. In the case of T-C 1, the mean mass loss value of the mildest thermal treatment (180 °C—3 h) was found to differ significantly from all the other categories of treated wood, except for the one of 180

°C—5 h. The value of the specimens treated at 180 °C—5 h differed significantly from all the respective values recorded by the specimens of the more intensive thermal treatments. The value of the specimens treated at 180 °C—7 h differed significantly from the respective values of all the other categories of treated or untreated wood.

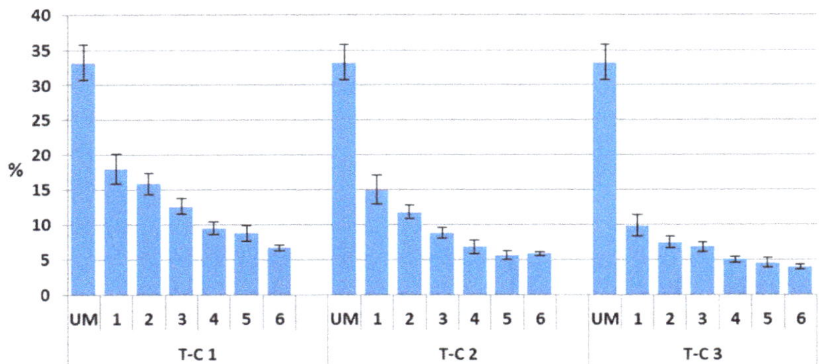

Figure 3. Mean mass loss rates recorded by unmodified (UM) and heat-modified (1: 180 °C—3 h, 2: 180 °C—5 h, 3: 180 °C—7 h, 4: 200 °C—3 h, 5: 200 °C—5 h, 6: 200 °C—7 h) black pine specimens that have been subjected also to chemical modification with organosilanes (T-C 1, T-C 2, T-C 3), after their exposure to *C. puteana*.

Statistically significant differences were presented between the value of the specimens treated at 200 °C—3 h and the respective values of all the other categories of treated or untreated wood, except for the specimens of the treatment 200 °C—5 h. The value of the most intensively treated (200 °C—7 h) specimens differed significantly from all the rest of the categories of specimens. In the case of T-C 2, all the mean mass loss values of different specimens categories differed significantly from one another, except for the three treatments of 200 °C that presented insignificant differences between one another. In the case of T-C 3, the mean mass loss value of the mildest thermal treatment (180 °C—3 h) was found to differ significantly from all the other categories of treated wood. The value of the specimens treated at 180 °C—5 h differs significantly from all the respective values recorded by the specimens of all the treatments, except for the one of 180 °C—7 h. Additionally, the mass loss values recorded by the three specimens categories of 200 °C were revealed to be similar to one another, but differed significantly from all the respective values of the other specimens categories.

3.1.3. Thermally-Treated Poplar

According to the mycological test results of poplar wood against the action of *T. versicolor* and *O. placenta* (Figure 4), the percentage of mean mass loss was found to be, in each case, much lower in the heat-modified poplar specimens than the unmodified wood, which demonstrates an improvement of the biological resistance in thermally-modified wood. Alhough, *T. versicolor*, as well as *O. placenta*, has been proven to be quite active and aggressive against untreated poplar wood species, causing high mass losses.

Referring to the test of *T. versicolor*, the treatments of 200 °C seem to reduce mass loss to a greater extent compared to those of 180 °C. Referring to treatments of 180 °C, the 3 h-duration resulted in a great improvement of resistance, while durations over 3 h (5 or 7 h) contributed to a slight increase in mass loss. This not statistically significant increase may indicate that duration over 3 h at 180 °C contributes to the formation of some components or the loss of some others, which leave the wood vulnerable again to the attack of the specific fungus. Specifically, the treatments of 180 °C and duration of 3, 5 and 7 h reduced the mass loss compared to the control by 34.63%, 30.84% and 26.36%, respectively, while the treatments of 200 °C (3, 5 and 7 h) reduced the loss by 41.02%, 37.29% and 36.11%, respectively.

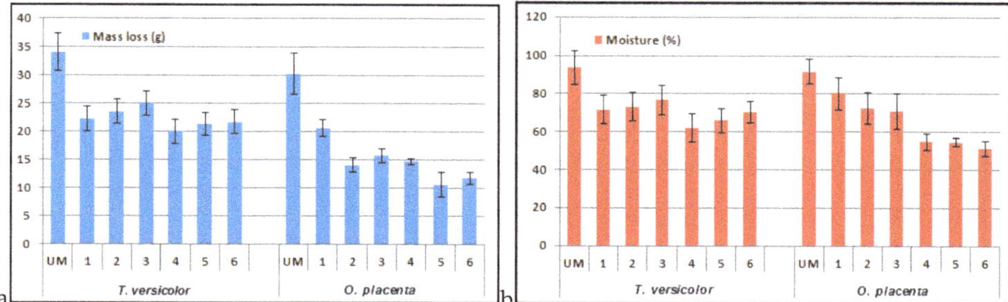

Figure 4. Mean mass loss rates (**a**) and moisture content (**b**) recorded by unmodified (UM) and heat-modified (1: 180 °C—3 h, 2: 180 °C—5 h, 3: 180 °C—7 h, 4: 200 °C—3 h, 5: 200 °C—5 h, 6: 200 °C—7 h) poplar specimens after their exposure to *T. versicolor* and *O. placenta*.

The statistical analysis revealed that the mean mass loss value of unmodified wood statistically differed significantly from the respective values of all the other treatment categories. Between the mass loss values of the thermally-treated poplar of different categories, statistically significant differences were not recorded, except for the case of the specimens treated at 180 °C—7 h that differed significantly from the respective value of the specimens treated at 200 °C—3 h.

The Two-Way ANOVA reported that the effect of temperature appeared to be statistically significant, affecting the variability of resistance by 62.1%, and the duration by 42%. The interaction between temperature–duration was also found statistically significant, affecting the resistance variability by 30%. The greatest effect of temperature was observed in 7-h treatments (statistically significant).

Similarly to the case of pine wood, the moisture content of the modified poplar specimens, after their mycological test process, was found to be in each case lower than the control sample level, presenting a course analogous to that of the mass loss rate.

In the case of testing the resistance of poplar against the *O. placenta* fungus, the results are even more encouraging, since all treatments have increased the bio-resistance of wood. The treatments of 200 °C seem to reduce mass loss to a greater extent, compared to those of 180 °C. Irrespectively of the treatment temperature, the duration of 5 h has caused a greater reduction in the mass loss rate, hence a greater resistance to attack, compared to treatment of 3 or 7 h, indicating probably that the treatment duration of 7 h exceeds the point of maximum improvement and the wood has undergone quite intense mass losses during the treatment, which once again leave it vulnerable to fungal activity. Specifically, the treatments of 180 °C and durations of 3, 5 and 7 h reduced the mass loss percentage compared to unmodified wood by 31.98%, 53.41% and 47.94%, respectively, while the treatments of 200 °C (3, 5 and 7 h), reduced the loss by 51.46%, 64.72% and 60.9%, respectively. Finally, after the mycological test the moisture content of the specimens decreased in each case compared to that of our control, and demonstrated a similar course to that of the mass loss rate. The research results are in line with the claim of Boonstra et al. [2], that the hemicellulose reduction due to treatment favors more the biological resistance of modified wood decayed by brown rot, compared to the white rot. Boonstra et al. [2] have reported that the action of *O. placenta* and *T. versicolor* continues to cause high mass losses in thermally-treated wood, until thermal degradation during treatment becomes so intense to ensure the branching of lignin, whereas the fungus *C. puteana* was inhibited by thermal modification, causing lower mass loss rates. Mazela et al. [37] recorded also an improvement in resistance of thermally-treated poplar wood to all the fungi species tested, with the aggressive fungus *O. placenta* to mark a mass loss of 18.5%.

The mean mass loss rate of the control specimens was found to statistically differ significantly from all the treated wood categories, revealing the improvement of the biological resistance against this fungus species. The mass loss of specimens treated at the mildest treatment (180 °C—3 h) differed significantly from the respective values of all the other treatments. Additionally, among the rest of the mass loss values of the thermally-treated poplar, statistically significant differences were not recorded, except for the case of the respective values of the specimens of the most intensive treatments (200 °C—5 h and 200 °C—7 h).

Two Way ANOVA revealed that the effect of temperature was statistically significant, influencing the variability of the resistance by 75.8%. An equally important factor was the treatment duration, affecting its variability by 76%, while the interaction between temperature against duration was weak. The most remarkable effect of temperature was observed in the treatment duration of 3 h.

3.2. Microfungi

3.2.1. Thermally-Treated Black Pine and Poplar

Results of the microfungi growth test on the surface of thermally-modified and -unmodified specimens (Figure 5) revealed that thermal modification does not effectively protect pine and poplar wood from microfungi growth, since the samples were found totally overgrown in the measurement of the seventh day, probably attributed to the free sugars coming from the hemicellulose decomposition during thermal treatment [2]. Nevertheless, the treatments recorded a delay of small extent in the attack and growth of microfungi on the wood surface. The surface of the unmodified specimens, from the first visual assessment implemented in the fourth day, was almost entirely covered by mold, while the thermally-modified samples were covered to a lower extent (Figure 6). This retardation tendency of the microfungi development is evident in both of the wood species tested.

As the intensity of treatment (temperature and duration) increases, the degree of mold growth (coverage) in poplar and pine specimens is found to decrease, compared to the unmodified wood, only until the fourth day. In this resistance improvement course, which is analogous to treatment intensity course, exceptions include the two more intensive treatments (200 °C—5 and 7 h) in both wood species, the specimens of which presented a small increase of surface coverage by mold, compared to the lighter treatments; nevertheless, these percentages do not approach the respective values of unmodified wood.

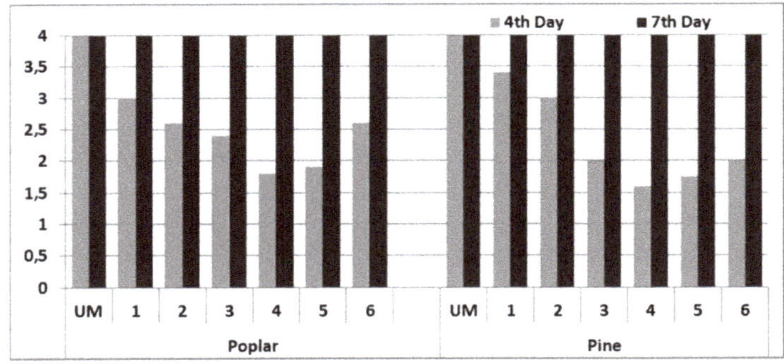

Figure 5. Microfungi growth degree evaluation according to the fungal growth rating system (Table 2) on the surface of unmodified (UM) and thermally-modified (1: 180 °C—3 h, 2: 180 °C—5 h, 3: 180 °C—7 h, 4: 200 °C—3 h, 5: 200 °C—5 h, 6: 200 °C—7 h) poplar and black pine wood.

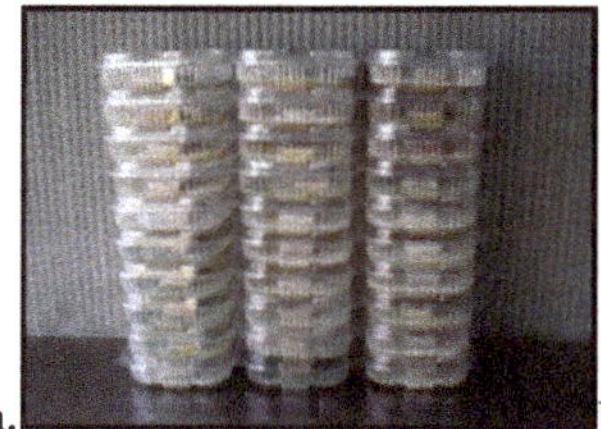

Figure 6. **(a)** Plastic containers, **(b)** mold growth on the surface of a thermally-modified pine sample at 180 °C-7 h (on the top) and the control specimen (on the bottom) recorded in the 4th day of exposure, **(c)** the respective mold growth on the surface of a thermally-modified pine sample at 180 °C–5 h (top) and the control specimen (bottom) (4th day of exposure)

3.2.2. Thermally-Chemically-Treated Black Pine

The combination of thermal treatment and surface modification process seems to increase in great extent the resistance of black pine wood against the action of microfungi (Figure 7). The unmodified wood was found fully covered with mold from the first visual assessment in the fourth day of exposure, while the thermally-chemically-modified pine wood was in a very small extent covered only in the last measurements in the 14th and 21st days of exposure. As the intensity of thermal treatment increases, the growth of mold is lower, except for the most intensive treatment, which reveals the protective effect of thermal treatment on wood. The results of these treatments combination could be considered as an evidence of an adequately protective technique against the action of microfungi.

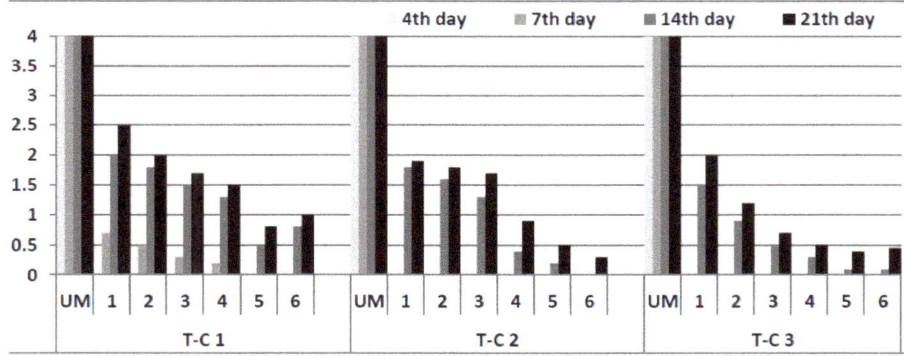

Figure 7. Microfungi growth degree evaluation according to fungal growth rating system (Table 2) on the surface of unmodified (UM) and thermally (1: 180 °C—3 h, 2: 180 °C—5 h, 3: 180 °C—7 h, 4: 200 °C—3 h, 5: 200 °C—5 h, 6: 200 °C—7 h) and silanes modified black pine wood.

4. Conclusions

According to the findings, short-term thermal treatments (even in the presence of air, performed in simple drying laboratory equipment) could enhance the biological resistance of wood, and therefore expand its service life.

Thermally-treated black pine wood presented 9.65–36.73% lower mass loss caused by *C. puteana* compared to the unmodified wood. As the intensity of thermal treatment increases, black pine presents a higher resistance against the *O. placenta* fungus and the mass loss of treated wood was found 28.75–68.46% less than that of the unmodified wood. Both fungi species, *C. puteana* and *O. placenta*, have been proven to be very active and aggressive to black pine wood, with *O. placenta* to cause extended decay to unmodified wood, but the lowest decay action in thermally-treated samples of this research.

Thermally-treated poplar displayed a significant decrease in mass losses (26.36–41.02%) caused by the action of *T. versicolor* compared to control. All thermal treatments improved the biological resistance of poplar wood against the fungus of *O. placenta*, presenting a mass loss reduction of 31.98–64.72%, compared to our control. The treatment duration of 5 h (middle intensity) revealed the highest biological resistance values for almost all fungi and both wood species tested, which suggests that the 3-h duration was not adequate to achieve the desirable chemical reactions in wood mass, necessary to improve durability at the highest possible level, while the 7-h treatment duration seemed to be long enough to induce great weight losses (poplar: 18.88%, black pine: 15.25%), leaving wood once again susceptible to the catastrophic action of basidiomycetes.

The variability of biological resistance improvement referring to thermal treatments was found to be highly influenced by the factor of treatment temperature, to a lower extent by the treatment duration, and even less by the interaction between the temperature-duration factors.

The combination of thermal treatment with silanes components treatment improved significantly the biological durability of black pine wood, recording mass losses, caused by the action of fungi, much lower than those of the unmodified wood, as well as lower than the respective mass loss values of thermally-treated wood specimens, with this biological resistance improvement to be analogous to the thermal treatment intensity. T-C 1, T-C 2 and T-C 3 resulted in 45.68–79.74%, 54.76–83% and 70.21–87.83% lower mass losses, respectively, compared to unmodified wood against the action of *C. puteana*. The presence of benzin in silane solutions increased the protective effect of the surface modification process and the biological durability of wood, probably attributed to the fact that benzin contributed to a higher dispersion of the solution or a deeper penetration of the solution in wood mass (WPG approximately 15%).

Thermal modification did not manage to effectively protect pine and poplar wood from microfungi growth, but it only contributed to a retardation of the attack and growth process, confirming the tendency and results of previous studies, examining the durability of other thermally-treated species. On the contrary, the combination of thermal treatment and surface modification process increased in great extent the resistance of black pine wood against the action of microfungi, and can be considered as an adequately protective technique against the mold action.

Generally, the biological strength of thermally-treated pine and poplar wood against basidiomycetes was found to be enhanced to a small extent, and they should not be exposed directly above ground without the application of additional protective techniques, whereas the treatment with silane components enhanced significantly the resistance against basidiomycetes and mold action, and could be suitable for numeral interior, under shelter or even exterior applications. Additional research should be carried out on this direction of thermal and silanes treatment methods combination, in order to achieve even more enhanced durability by optimizing the treatments conditions and process through the simultaneous testing of the physical, chemical and mechanical properties of the studied wood species.

Funding: This research received no external funding.

Acknowledgments: The author warmly thanks the team of Institute of Chemical Wood Technology (Faculty of Wood Technology) in Poznan University of Life Sciences, for the provision of decay tests equipment.

Conflicts of Interest: The author declares no conflict of interest.

References

1. Hill, C. *Wood Modification, Chemical, Thermal and other Processes*; John Wiley & Sons Ltd: Hoboken, NJ, USA, 2006.
2. Boonstra, M.J.; Acker, J.V.; Kegel, E.; Stevens, M. Optimization of a two-stage heat treatment process: Durability aspects. *Wood Sci. Technol.* **2007**, *41*, 31–57. [CrossRef]

3. Militz, H. *Thermal Treatment of Wood: European Processes and Their Background*; IRG/WP 02-40241; The International Research Group on Wood Preservation: Stockholm, Sweden, 2002.
4. Brischke, C.; Welzbacher, C.R.; Brandt, K.; Rapp, A.O. Quality control of thermally modified timber: Interrelationship between heat treatment intensities and CIE L*a*b* color data on homogenized wood samples. *Holzforschung* **2007**, *61*, 19–22. [CrossRef]
5. Esteves, B.M.; Pereira, H.M. Wood modification by heat treatment-A review. *Bioresources* **2009**, *4*, 370–404.
6. Salman, S.; Thévenon, M.F.; Pétrissans, A.; Dumarçay, S.; Candelier, K.; Gérardin, P. Improvement of the durability of heat-treated wood against termites. *Maderas. Ciencia y tecnología* **2017**, *19*, 317–328. [CrossRef]
7. Li Shi, J.; Kocaefe, D.; Amburgey, T.; Zhang, J. A comparative study on brown-rot fungus decay and subterranean termite resistance of thermally-modified and ACQ-C-treated wood. *Holz als Roh-und Werkst.* **2007**, *65*, 353–358. [CrossRef]
8. Mburu, F.; Dumarcay, S.; Huber, F.; Petrissans, M.; Gerardin, P. Evaluation of thermally modified Grevillea robusta heartwood as an alternative to shortage of wood resource in Kenya: Characterisation of physicochemical properties and improvement of bio-resistance. *Bioresour. Technol.* **2007**, *98*, 3478–3486. [CrossRef]
9. Vetter De, L.; Depraetere, G.; Janssen, C.; Stevens, M.; Van acker, J. Methodology to assess both the efficacy and ecotoxicology of preservative-treated and modified wood. *Ann. For. Sci.* **2008**, *65*, 1–10. [CrossRef]
10. Mazela, B.; Zakrzewski, R.; Grześkowiak, W.; Cofta, G.; Bartkowiak, M. Resistance of thermally modified wood to basidiomycetes. *Electron. J. Pol. Agric. Univ. Wood Technol.* **2004**, *7*, 253–262.
11. Mohareb, A.; Sirmah, P.; Petrissans, M.; Gerardin, P. Effect of heat treatment intensity on wood chemical composition and decay durability of Pinus patula. *Eur. J. Wood Wood Prod.* **2012**, *70*, 519–524. [CrossRef]
12. Mai, C.; Militz, H. Modification of wood with silicon compounds. Treatment systems based on organic silicon compounds-A review. *Wood Sci. Technol.* **2004**, *37*, 453–461. [CrossRef]
13. Donath, S.; Militz, H.; Mai, C. Creating water repellent effects on wood by treatment with silanes. *Holzforschung* **2006**, *60*, 40–46. [CrossRef]
14. Weigenand, O.; Militz, H.; Tingaut, P.; Sebe, G.; de Jeso, B.; Mai, C. Penetration of amino-silicone micro- and macro-emulsions into Scots pine sapwood and the effect on water-related properties. *Holzforschung* **2007**, *61*, 51–59. [CrossRef]
15. Vetter De, L.; Van den Bulcke, J.; Acker Van, J. Impact of organosilicon treatments on the wood-water relationships of solid wood. *Holzforschung* **2010**, *64*, 463–468. [CrossRef]
16. Hill, C.A.S.; Farahani, M.R.M.; Hale, M.D.C. The use of organo alkoxysilane coupling agents for wood preservation. *Holzforschung* **2004**, *58*, 316–325. [CrossRef]
17. Gobakken, L.R.; Westin, M. Surface mould growth on five modified wood substrates coated with three different coating systems when exposed outdoors. *Int. Biodeterior. Biodegrad.* **2008**, *62*, 397–402. [CrossRef]
18. Kocaefe, D.; Shi, J.L.; Yang, D.; Bouazara, M. Mechanical properties, dimensional stability, and mold resistance of heat-treated jack pine and aspen. *For. Prod. J.* **2008**, *58*, 88–93.
19. Fojutowski, A.; Kropacz, A.; Noskowiak, A. The susceptibility of poplar and alder wood to mould fungi attack. Annals of Warsaw University of Life Sciences–SGGW. *For. Wood Technol.* **2011**, *74*, 56–62.
20. *ISO 13061-1:2014 Physical and Mechanical Properties of Wood—Test Methods for Small Clear Wood Specimens—Part 1: Determination of Moisture Content for Physical and Mechanical Tests*; ISO: Geneva, Switzerland, 2014.
21. *ISO 13061-2:2014 Physical and Mechanical Properties of Wood—Test Methods for Small Clear Wood Specimens—Part 2: Determination of Density For Physical and Mechanical Tests*; ISO: Geneva, Switzerland, 2014.
22. *ISO 3129:2012 Wood—Sampling Methods and General Requirements for Physical And Mechanical Testing of Small Clear Wood Specimens*; ISO: Geneva, Switzerland, 2014.
23. Kamperidou, V.; Barboutis, I. Mechanical Performance of Thermally Modified Black Pine (Pinus nigra L.) Wood. *Electron. J. Pol. Agric. Univ.* **2017**, *20*, 7–13. [CrossRef]
24. Kamperidou, V.; Barboutis, I. Mechanical Strength and Surface Roughness of thermally modified Poplar wood. *PRO Ligno* **2017**, *13*, 107–114.
25. Kamperidou, V.; Barboutis, I. Changes in hygroscopic properties of poplar and black pine induced by thermal treatment. *PRO Ligno* **2018**, *14*, 57–64.
26. Kamperidou, V.; Barboutis, I. Physical and Hygroscopic properties of pine and poplar wood after Heat Treatment. In Proceedings of the 5th International Conference on Processing Technologies for the Forest and Bio-based Products Industries (PTF BPI 2018), Freising/Munich, Germany, 20–21 September 2018; pp. 192–199.

27. Hochmanska, P.; Mazela, B.; Krystofiak, T. Hydrophobicity and weathering resistance of wood treated with silanemodified protective systems. *Drewno* **2014**, *57*. [CrossRef]
28. BS EN 113:1997. *Wood Preservatives. Test. Method for Determining the Protective Effectiveness Against Wood Destroying Basidiomycetes. Determination of the Toxic Values*; British Standards Institution: London, UK, 1997.
29. Broda, M.; Mazela, B.; Frankowski, M. Durability of wood treated with AATMOS and Caffeine—Towards the long-term carbon storage. *Maderas. Ciencia y Tecnología* **2018**, *20*, 455–468. [CrossRef]
30. *ASTM D5590-17 Standard Test Method for Determining the Resistance of Paint Films and Related Coatings to Fungal Defacement by Accelerated Four-Week Agar Plate Assay*; ASTM International: West Conshohocken, PA, USA, 2017.
31. Lekounougou, S.; Petrissans, Z.M.; Jacquot, Z.J.P.; Gelhaye, Z.E.; Gerardin, Z.P. Effect of heat treatment on extracellular enzymatic activities involved in beech wood degradation by *Trametes versicolor*. *Wood Sci. Technol.* **2009**, *43*, 331–341. [CrossRef]
32. Unsal, O.; Kartal, S.N.; Candan, Z.; Arango, R.A.; Clausen, C.A.; Green, F. Decay and termite resistance, water absorption and swelling of thermally compressed wood panels. *Int. Biodeterior. Biodegrad.* **2009**, *63*, 548–552. [CrossRef]
33. Tremblay, C. Physical Properties of Jack Pine Thermally Modified at Three Temperature Levels. In Proceedings of the European Conference on Wood Modification 2007, Cardiff, UK, 15–16 October 2007; pp. 183–186.
34. Chaouch, M.; Pétrissans, M.; Pétrissans, A.; Gérardin, P. Use of wood elemental composition to predict heat treatment intensity and decay resistance of different softwood and hardwood species. *Polym. Degrad. Stab.* **2010**, *95*, 2255–2259. [CrossRef]
35. Kortelainen, S.M.; Viitanen, H. Decay resistance of sapwood and heartwood of untreated and thermally modified Scots pine and Norway spruce compared with some other wood species. *Wood Mater. Sci. Eng.* **2009**, *3–4*, 105–114. [CrossRef]
36. Ouyang, L.; Huang, Y.; Cao, J. Hygroscopicity and Characterization of Wood Fibers Modified by Alkoxysilanes with Different Chain Lengths. *Bioresources* **2014**, *9*, 7222–7233. [CrossRef]
37. Mazela, B.; Plenzler, R.; Olek, W. Effects of Thermal modification on decay resistance and mechanical properties of Poplar and Beech wood. *Wood Struct. Prop.* **2010**, *10*, 141–143.

 *forests* 

Article

The Impact of Paraffin-Thermal Modification of Beech Wood on Its Biological, Physical and Mechanical Properties

Ladislav Reinprecht * and Miroslav Repák

Department of Wood Technology, Faculty of Wood Sciences and Technology, Technical University in Zvolen, T. G. Masaryka 24, SK-96001 Zvolen, Slovakia; xrepak@tuzvo.sk

* Correspondence: reinprecht@tuzvo.sk; Tel.: +421-45-520-6383

Received: 5 November 2019; Accepted: 26 November 2019; Published: 2 December 2019

Abstract: The European beech (*Fagus sylvatica* L.) wood was thermally modified in the presence of paraffin at the temperatures of 190 or 210 °C for 1, 2, 3 or 4 h. A significant increase in its resistance to the brown-rot fungus *Poria placenta* (by 71.4%–98.4%) and the white-rot fungus *Trametes versicolor* (by 50.1%–99.5%) was observed as a result of all modification modes. However, an increase in the resistance of beech wood surfaces to the mold *Aspergillus niger* was achieved only under more severe modification regimes taking 4 h at 190 or 210 °C. Water resistance of paraffin-thermally modified beech wood improved—soaking reduced by 30.2%–35.8% and volume swelling by 26.8%–62.9% after 336 h of exposure in water. On the contrary, its mechanical properties worsened—impact bending strength decreased by 17.8%–48.3% and Brinell hardness by 2.4%–63.9%.

Keywords: beech; paraffin; thermal modification; fungi; swelling; mechanical properties

---

1. Introduction

European beech (*Fagus sylvatica* L.) is one of the most popular commercial broad leaved tree species in Central Europe [1,2]. Beech wood is preferred due to its properties; especially because it is easily workable and impregnable. On the contrary, beech wood suffers from low resistance to fungi and insects, and also from high volume shrinking, i.e., during drying it warps and splits, and under changing weather conditions its dimensions change significantly [3]. Low decay and mold resistance is considered one of the basic disadvantages of beech wood. According to the Standard EN 350 [4], it is non-durable and in exterior cannot be used without convenient treatments, mainly for structural elements.

In comparison to traditional chemical treatment of wood with toxic biocides, the processes of its thermal, biological and chemical modification do not usually affect the environment; therefore, investigation of new wood modification modes is very prospective [5,6].

Thermal modification of wood at high temperatures is connected with changes in its molecular structure. These changes are associated primarily with the hemicelluloses degradation, creation of hemicelluloses–lignin linkages and extinction of some hydroxyl groups [7–9]. An increase in the resistance of wood to water and biotic agents is the main aim of its thermal modification [5,10]. Types and extent of changes in molecular structure and subsequently in properties of thermally modified wood depend not only on the temperature and its duration, but also on the wood species, its initial moisture content, as well as on parameters of air, nitrogen, plant oil or other heating medium [5,6,10,11]. Biological resistance of wood is not affected positively by the thermal modification carried out in the atmosphere at lower temperatures ranging from 130 to 160 °C [10,12,13]. In practice, an increase in wood resistance to wood-decaying fungi and insects results from the application of higher temperatures of air ranging from 160 to 220 °C, or also from temperatures up to 260 °C at the limited quantity of oxygen present [5,12,14,15].

In comparison to standard thermal modification of wood in hot air, higher resistance to decaying-fungi and water is ensured when hot plant oils and other hydrophobic agents are applied [12–14,16–18].

Waxes are due to hydrophobic properties used to protect wood against water, as the water sorption kinetics of wood is reduced and dimensional stability is improved [19]. Waxes also improve resistance of wood to termites [20,21]. Mechanical properties, as compressive strength [22], bending strength [21] and hardness [21,22], of wood treated with waxes do not change or increase slightly. Moreover, waxes are almost non-toxic and environmentally acceptable, so they can also be used in other sectors, e.g., pharmaceutical, beauty and food-processing industries.

Paraffin belongs to the group of waxes. It is a mixture of solid linear aliphatic hydrocarbons, especially straight chain alkanes synthesized from crude oil using Fischer–Tropsch synthesis or from coal tar. Paraffin is cheap, healthy clean and hydrophobic substance used in beauty industry, candle-making business, civil engineering, as well as in preservation of wood materials in order to reduce the hygroscopicity and to improve the dimensional stability. Paraffin and paraffin emulsions have been used to reduce water soaking and to improve dimensional stability of particleboards [23,24] and in hydrophobic treatment of solid wood [19,21,25] over a long period. Combining the technology of paraffin impregnation of wood with subsequent thermal modification in paraffin can have a synergic effect resulting in an effective improvement of selected wood properties [21,26,27].

The aim of the experiment was to determine the effect of thermal modification of European beech wood in the presence of paraffin in order to increase its resistance to rot, mold and water and at the same time to evaluate the effect of modification on selected mechanical properties of wood.

2. Materials and Methods

2.1. Wood

European beech (*Fagus sylvatica* L.) heart-wood specimens of high quality, i.e., without rot, insect gallery, growth defects, tension wood or red-false wood were prepared from the sawn timber naturally seasoned to a moisture content of 13.5% ± 2%. Three types of specimens were used in the experiment—type (a): 25 mm × 25 mm × 5 mm in mycological tests and in testing the Brinell hardness; type (b): 5 mm × 50 mm × 25 mm in testing the soaking and swelling, and type (c): 120 mm × 10 mm × 10 mm in testing the impact bending strength (Figure 1). The top and bottom surfaces of specimens of the type (a) and type (b) were milled.

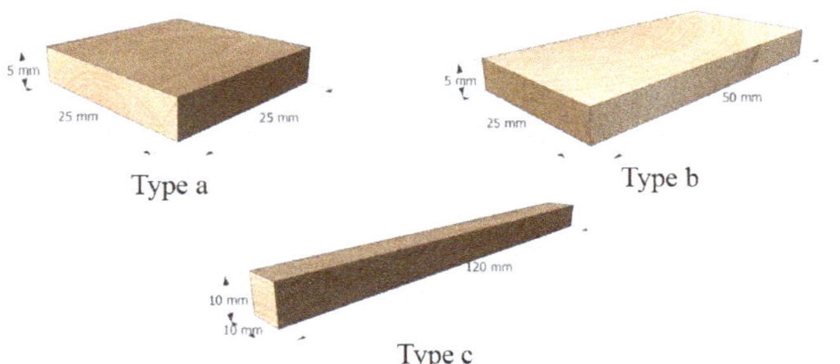

Figure 1. Types of beech wood specimens used in the paraffin-thermal modification and for testing the selected properties. Note: Type (a): 64 modified and 32 reference specimens for attack by wood-decaying fungi; 32 modified and 16 reference specimens for attack by the mold *Aspergillus niger*; 54 modified and six reference specimens for testing the Brinell hardness. Type (b): 54 modified and six reference specimens for testing the soaking and swelling. Type (c): 54 modified and six reference specimens for testing the impact bending strength.

Prior to modification, the beech specimens were dried at 103 ± 1 °C to the oven-dry state in the kiln Memmert UNB 100 (Memmert, Schwabach, Germany), and subsequently cooled in desiccators to a temperature of 20 ± 2 °C and weighed with an accuracy of 0.001 g.

2.2. Paraffin

Clear paraffin wax (MOL, Hungary) with the melting point ranging from 60 to 62 °C was used to modify the beech specimens.

2.3. Paraffin-Thermal Modification

The beech specimens were thermally modified with hot paraffin in the kiln Memmert UNB 100. In the first phase, paraffin melted in stainless steel containers at the temperatures from 80 to 100 °C/1 h. In the second phase, beech wood specimens were impregnated in the melt of paraffin wax at atmospheric pressure and at a temperature of 100 °C/1 h. In the third phase, the temperature of paraffin increased continuously to 190 °C (or 210 °C) during 1 h. In the fourth phase, the temperature in the containers with paraffin and beech specimens remained at 190 °C (or 210 °C) for 1, 2, 3 or 4 h. In the last fifth phase, non-absorbed hot paraffin ran off surfaces of beech specimens. The process of modification is shown in Figure 2a.

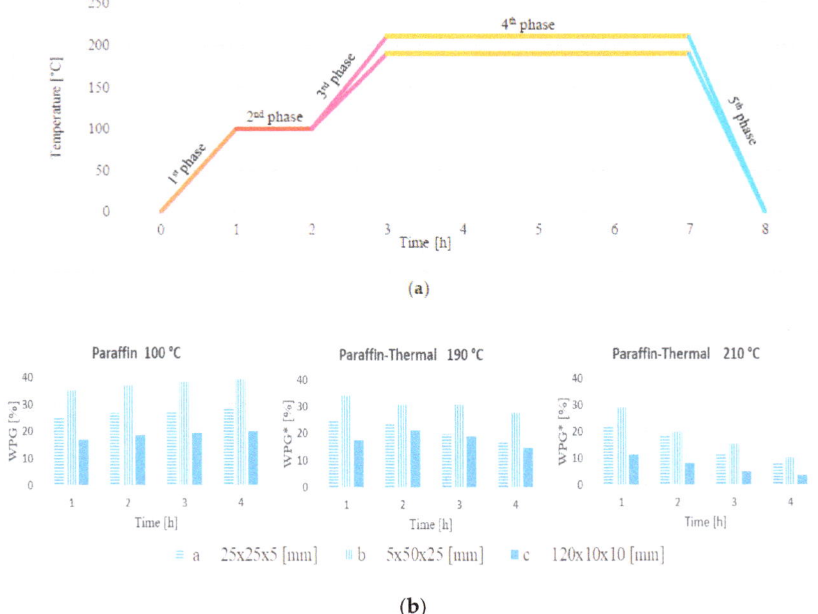

(b)

Figure 2. Phases of the paraffin-thermal modification of beech specimens (**a**) and weight percent gain (WPG) values of the paraffin modified specimens and also so-called real WPG* values of the paraffin-thermally modified specimens—within the 4th phase lasting from 1 to 4 h (**b**). Note: WPG values of paraffin at the end of the 2nd modification phase: Type (a) 23.91%; Type (b) 33.37% and Type (c) 13.98%. Calculation of the theoretical weight losses ($\Delta m_{Thermal}$ = WPG − WPG*) of paraffin-thermally modified specimens caused by thermal effects in the 4th phase could be performed as well, for example, for specimens of Type (a): $\Delta m_{Thermal}$ at 190 °C/1 h = 0.10%; 2 h = 3.14%; 3 h = 7.82%; 4 h = 12.02%; and $\Delta m_{Thermal}$ at 210 °C/1 h = 2.72%; 2 h = 8.68%; 3 h = 15.78%; 4 h = 20.81%.

Modified specimens of beech wood were cooled in desiccators to a temperature of 20 ± 2 °C, weighed with an accuracy of 0.001 g, their dimensions were determined with an accuracy of 0.01 mm, and transferred were to the desiccators again.

The weight percent gain (WPG, and also WPG*) values of paraffin into beech wood specimens were affected by their dimension, at which the lowest WPG (or WPG*) values had specimens of Type (c) 120 mm × 10 mm × 10 mm (L × T × R) with the smallest portion of axial surfaces (Figure 2b). The so-called WPG* (WPG − $\Delta m_{Thermal}$) values were simultaneously affected by weight losses of specimens caused by degradation of hemicelluloses and other components of wood at the high temperatures of 190 or 210 °C acting from 1 to 4 h (Figure 2b).

Specimens used in mycological tests (i.e., resistance to wood-decaying fungi and molds) and in testing the mechanical properties (i.e., impact bending strength and hardness) were air-conditioned at a temperature of 20 ± 2 °C and a relative air humidity of 60% ± 5% for 14 days. Oven-dry specimens with 0% moisture content were used in soaking and swelling tests.

2.4. Attack by Wood-Decaying Fungi

The specimens (25 mm × 25 mm × 5 mm) were subjected to attack by the brown-rot fungus *Poria placenta* (Fries) Cooke sensu J. Eriksson, strain FPRL 280 (Building Research Establishment, Garston-Watford-Herst, UK) or to attack by the white-rot fungus *Trametes versicolor* (Linnaeus ex Fries) Pilat, strain BAM 116 (Bundesanstalt für Materialforschung und -prüfung, Berlin, Germany).

Fungal attack of specimens was performed in Petri dishes with a diameter of 100 mm according to modified Standard EN 113 [28], i.e., specimens with another shape and another method of sterilization were used, and their exposition in fungal mycelia at a temperature of 24 ± 2 °C and a relative air humidity of 90% ± 5% lasted for 6 weeks instead of 16 weeks, according to the rapid screening test by Van Acker et al. [29].

Two replicates of the equally modified beech specimen and one replicate of the reference beech unmodified specimen were placed into each Petri dish in the vaccination box (Merci Ferrera, Italy) (Figure 3). Specimens were deposited on plastic mats under which a fungal mycelium had already been grown on a sterilized 4.5 wt% malt agar medium (HiMedia, Ltd., Mumbai, India) with a thickness from 3 to 4 mm. After the fungal attacks, mycelium was carefully cleaned from the surface of specimens. Specimens were then air-conditioned for 14 days at a temperature of 20 ± 2 °C and a relative air humidity of 60% ± 5%, and weighed to determine the weight loss (Δm), using Equation (1):

$$\Delta m = \frac{m_0 - m_{0deg}}{m_0} \times 100\ (\%) \tag{1}$$

where: m_0—mass of specimen in the conditioned state before mycological test (g) and m_{0deg}—mass of specimen in the conditioned state after mycological test (g).

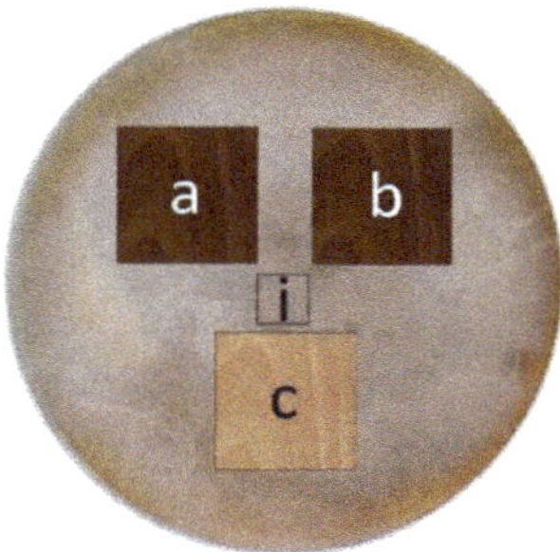

Figure 3. Petri dish with three beech wood specimens (a, b—modified by the same mode, c—reference) and fungal inoculum (i).

2.5. Attack by the Mold Aspergillus niger

The mold resistance test of the specimens (25 mm × 25 mm × 5 mm) was performed with the microscopic fungus *Aspergillus niger* Tiegh. according to modified Standard EN 15457 [30], i.e., specimens with other shape and other methods of sterilization were used. Firstly, all surfaces of specimens were sterilized with the 30 W germicidal lamp (Chirana, Medical, a. s., Stará Turá, Slovakia) from a distance of 1 m at a temperature of 22 ± 2 °C/0.5 h. Sterilized specimens were placed into Petri dishes with a diameter of 100 mm on a sterilized 4.9 wt% Czapek-Dox agar medium (HiMedia, Ltd., Mumbai, India) with a thickness from 3 to 4 mm and inoculated with a spore suspension of *A. niger* in a sterile water (10^6–10^7 spores/mL). Incubation of inoculated specimens in the thermostat Nahita 636 Plus (Nahita, France) took 21 days at a temperature of 24 ± 2 °C and a relative air humidity of 90%–95%. Growth activity of the mold *A. niger* on the top surfaces of specimens was evaluated on the 7th, 14th and 21th day by these criteria: 0 → growth 0% = no growth of the mould on the top surface of specimen; 1 → growth ≤10%; 2 → growth >10% and ≤30%; 3 → growth >30% and ≤50% and 4 → growth >50%.

2.6. Soaking and Swelling

In order to determine the ability of the reference and modified beech wood specimens (5 mm × 50 mm × 25 mm) to absorb distilled water (S_i), the soaking test according to the Standard STN 49 0104 [31] was performed, using Equation (2):

$$S_i = \frac{m_i - m_0}{m_0} \times 100\ (\%) \tag{2}$$

where: m_i—mass of the moist specimen at the defined time of soaking (g) and m_0—mass of the oven-dry specimen (g).

At the same time, volume swelling of wood (β_{Vi}) was evaluated, using Equation (3):

$$\beta_{Vi} = \frac{V_i - V_0}{V_0} \times 100\ (\%) \tag{3}$$

where: V_i—volume of the moist specimen at the defined time of soaking (mm^3) and V_0—volume of the oven-dry specimen (mm^3).

2.7. Impact Bending Strength and Brinell Hardness

Impact bending strength of specimens in tangential direction (I) was determined according to the Standard ISO 3348 [32], using Equation (4):

$$I = \frac{W}{b \times h} \left(J.cm^{-2}\right) \tag{4}$$

where: W—work done for cutting the specimen (J) and b and h—specimen cross section dimensions (cm).

Brinell hardness of specimens in radial direction (H_B) was evaluated according to the Standard EN 1534 [33] using a steel ball with a diameter of 11.284 mm impressed into the wood surface with a force of 500 N. It was calculated by Equation (5):

$$H_B = \frac{F}{S} = \frac{2 \times F}{\pi \times D \times \left(D - \sqrt{D^2 - d^2}\right)}\ (MPa) \tag{5}$$

where: F—force on the ball (N), D—ball diameter (mm) and D—diameter of the impressed area (mm).

3. Results and Discussion

3.1. The Rot Resistance

All paraffin-thermal modifications of beech wood specimens resulted in an increase in their rot resistance (Table 1). The highest weight losses (Δm) caused by wood-decaying fungi were observed in the case of the reference beech specimens (22.14% with *Poria placenta* and 29.65% with *Trametes versicolor*). On the contrary, the lowest weight losses <1% were observed in the specimens modified with the most severe modification regimes, i.e., at 210 °C for 3 h or 4 h. The modified specimens were usually more intensively attacked by the white-rot fungus *T. versicolor* (Δm from 0.15% to 14.8%) than by the brown-rot fungus *P. placenta* (Δm from 0.35% to 6.34%). However, in the case of their exposure to the most severe regimes (210 °C/3 or 4 h) the result obtained was opposite. This disproportion arose probably due to a greater inhibitory effect of specific carbonized wood components created with prolonged exposure to 210 °C on the growth and enzymatic activities of *T. versicolor*.

Table 1. Weight losses (Δm) of the reference and modified beech specimens after action of the wood-decaying fungi *Poria placenta* and *Trametes versicolor*.

Paraffin-Thermal Modification	*P. placenta* Δm (%)	*T. versicolor* Δm (%)
Reference	22.14 *(4.50)*	29.65 *(6.53)*
190 °C/1 h	6.34 *(2.91)* a	14.80 *(6.19)* a
190 °C/2 h	5.32 *(2.17)* a	10.05 *(3.86)* a
190 °C/3 h	5.33 *(2.04)* a	9.68 *(3.70)* a
190 °C/4 h	3.76 *(1.66)* a	7.83 *(3.02)* a
210 °C/1 h	5,25 *(1.95)* a	7.74 *(2.98)* a
210 °C/2 h	2.67 *(1.07)* a	4.87 *(1.93)* a
210 °C/3 h	0.92 *(0.42)* a	0.25 *(0.19)* a
210 °C/4 h	0.35 *(0.28)* a	0.15 *(0.11)* a

Note: mean values are from four replicates, and from 16 reference replicates. Standard deviations are in italics and parantheses. The Duncan test, with significance levels a = 99.9%, b = 99%, c = 95% and d < 95%, was performed in relation to reference specimens—at which differences always occurred at the 99.9% significance level (a).

Lesar and Humar [34] found out that wood attacked with wood-decaying fungi could degrade slower when impregnation with waxes is applied. The wax-barrier located in the cell lumens of wood slows down the diffusion, both of fungal enzymes and products resulting from the cell wall degradation, between hyphae of wood-decaying fungi and wood structural components. Specific changes in the molecular structure of wood, causing an increase in its resistance to decay, occur as a result of its parallel wax and thermal modification. The fungal growth inhibition is here also caused due to polysaccharide dihydroxylation and resulting in a lower moisture content of wood [35]. Boonstra et al. [36] mentioned that the thermal modification of wood could result in chemical transformation of components like minerals, vitamins and lower molecular weight carbohydrates necessary for the activity of wood-decaying fungi. At the higher temperatures of 210 °C, hemicelluloses degrade to substances with lower hygroscopicity and higher toxicity to fungi, e.g., like furfural polymers [37,38]. According to Lacić et al. [39], an increase in the rot resistance of alder wood, markedly to the brown-rot fungus *P. placenta* and less significantly to the white-rot fungus *T. versicolor*, resulted from its thermal modification at the temperatures of 180 °C and 200 °C for 6 h and 10 h in proportion to an increase in temperature and time. Similarly, several researchers found out that the rot resistance of spruce, pine and other wood species to wood-decaying fungi *Coniophora puteana* and *T. versicolor* is increased by their thermal modification in atmosphere [40] or in plant oils [14,41]. For example, Rapp and Sailer [14] determined that the rot resistance of thermally modified spruce wood and sapwood of pine to *C. puteana* increased at the plant oil temperature ranging from 190 °C to 220 °C—since the weight losses of wood during the decay processes decreased from 40% to 5.5% in the case of spruce wood or from 48% to 11% in the case of pine sapwood. Yılgör and Kartal [15] found out that the rot resistance of the thermally modified sugi (*Cryptomeria japonica* D.) sapwood, exposed to a temperature of 180 °C for 2 and 4 h or to a temperature of 220 °C for 2 h, increased significantly against the white-rot fungus *T. versicolor*—since its weight losses decreased evidently from 41.4% to 4.1% in dependence to an increase in the temperature and time. According to Kartal [42], thermal modification of sugi sapwood at a temperature of 180 °C for 2 and 4 h did not result in an increase in its rot resistance to the brown-rot fungus *Fomitopsis palustris*, but there was a slight increase in its resistance to the white-rot fungus *T. versicolor*.

3.2. The Mold Resistance

The growth intensities of the mold *Aspergillus niger* on the surface of beech wood specimens subjected to different paraffin-thermal modifications were markedly reduced only at the beginning of testing, after the 7th and 14th day (Table 2). Only beech specimens modified for the longest time, for 4 h at the temperatures of 190 °C or 210 °C, showed lower growth intensity of this mold (GIM ≤ 2) on the final 21st day of testing (Table 2).

Table 2. Growth intensities of the mold *Aspergillus niger* (GIM from 0 to 4) on the top surfaces of beech wood specimens subjected to different paraffin-thermal modifications.

Paraffin-Thermal Modification	*A. niger* *GIM (0–4)*	*A. niger* *GIM (0–4)*	*A. niger* *GIM (0–4)*
	7th day	**14th day**	**21st day**
Reference	3	3	4
190 °C/1 h	1	2	2.75
190 °C/2 h	1	2	3
190 °C/3 h	0.5	1.5	2.5
190 °C/4 h	0.25	1	2
210 °C/1 h	1	1.75	3
210 °C/2 h	1	2.25	2.5
210 °C/3 h	0	1.5	2.25
210 °C/4 h	0	1	1.25

Note: mean values are from four replicates, and from 16 reference replicates..

Results of the experiment can be compared to the results obtained by [17] who searched mold resistance of beech and pine woods thermally modified by by the oil heat treatment (OHT) technology in rapeseed oil—when both wood species showed markedly increased resistance to the mold A. niger only in the case of modification performed at the highest temperature of 220 °C for 3 h or 6 h. Similarly, Yılgör and Kartal [15] found out that the resistance of sugi wood, modified at a temperature of 180 °C for 2 and 4 h and at a temperature of 220 °C for 2 h, to the molds *Rhizopus javanicus* and *Trichoderma virens* increased slightly, whereby to the mold *A. niger* increased only negligibly. Impregnation of beech and spruce woods with various wax emulsions had no significant impact on an increase in the resistance of mentioned wood species to molds [34].

3.3. The Soaking and Swelling Resistance

Paraffin is a hydrophobic substance. In the impregnated wood it is located in the cell lumens and on the S_3 layer of cell walls, due to what it reduces its soaking and swelling [43]. However, paraffin does not penetrate the wood cell walls, which allows water vapor diffusion in its structure, and thus the effect of paraffin on improving the dimensional stability of wood is only temporary [26]. Lesar and Humar [34] mentioned the effect of waxes in spruce wood on a decrease in the moisture absorption and soaking kinetics of modified wood.

Similar inhibition effects of paraffin in wood against its soaking and swelling in distilled water were investigated and founded in this experiment (Table 3, Figures 4 and 5). The alone paraffin reduced the capacity of water penetration into wood during the soaking test (Figure 4). Paraffin up to 8 h evidently suppressed kinetics of wood swelling, however, after 24 h its anti-swelling effect was already smaller (Table 3, Figure 5).

Table 3. Soaking and volume swelling in distilled water of beech wood specimens subjected to different paraffin-thermal modifications—results after 24 h and 336 h.

Paraffin-Thermal Modification	Soaking—S_i (%)		Swelling—β_{Vi} (%)	
	24 h	336 h	24 h	336 h
Reference	74.34 *(10.82)*	96.92 *(12.52)*	19.43 *(3.46)*	20.55 *(3.83)*
Paraffin only	27.67 *(0.78)* a	78.68 *(1.45) b*	16.56 *(3.19)* c	19.19 *(1.16) d*
190 °C/1 h	38.15 *(4.67)* a	66.46 *(2.88)* a	14.46 *(0.99)* b	14.67 *(1.20)* a
190 °C/2 h	39.49 *(2.90)* a	64.46 *(2.37)* a	14.68 *(0.59)* b	15.04 *(0.82)* a
190 °C/3 h	34.03 *(3.41)* a	63.26 *(2.16)* a	13.58 *(0.71)* a	14.25 *(0.81)* a
190 °C/4 h	31.24 *(3.18)* a	65.50 *(1.46)* a	12.21 *(0.98)* a	12.81 *(1.08)* a
210 °C/1 h	36.06 *(4.46)* a	66.13 *(1.38)* a	12.36 *(0.61)* a	12.16 *(1.04)* a
210 °C/2 h	31.95 *(3.55)* a	67.61 *(1.33)* a	10.14 *(0.96)* a	10.86 *(0.82)* a
210 °C/3 h	22.75 *(1.08)* a	67.22 *(1.79)* a	7.81 *(0.45)* a	8.37 *(0.24)* a
210 °C/4 h	20.26 *(2.29)* a	62.20 *(1.62)* a	6.99 *(0.53)* a	7.62 *(0.52)* a

Note: mean values are from six replicates. Standard deviations are in italics and parantheses. The Duncan test, with significance levels a = 99.9%, b = 99%, c = 95% and d < 95%, was performed in relation to reference specimens—at which differences usually occurred at the 99.9% significance level (a), or at swelling also at lower significance levels, the 99% (b) and 95% (c), respectively none for specimens modified only with paraffin (d)

In accordance with several research studies [44–46] the fact that swelling of wood was reduced especially due to its thermal modification while the role of paraffin was only to slow down kinetics of swelling was confirmed (Table 3, Figure 5). Thermal modification of wood results in changes in its molecular structure, i.e., the polysaccharides (especially hemicelluloses) depolymerized, the microcrystalline cellulose portion increased, the lignin linkage occurred and therefore, the presence of free hydroxyl groups in wood decreased [47].

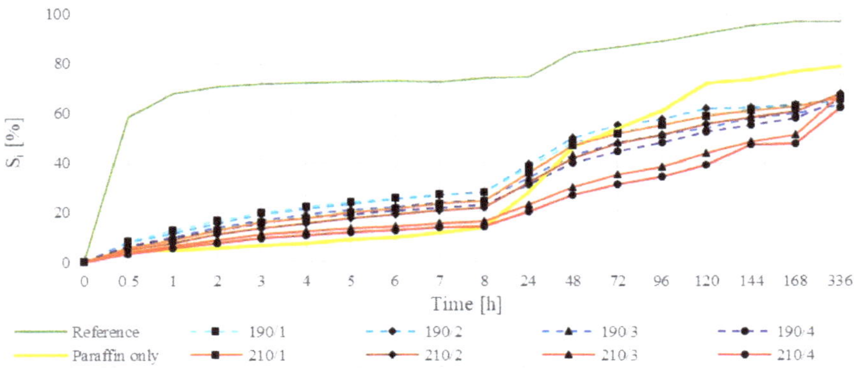

Figure 4. Soaking kinetics of reference and paraffin-thermally modified beech wood specimens.

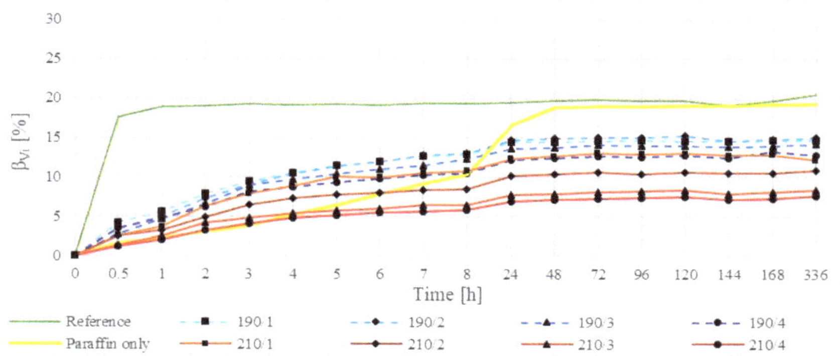

Figure 5. Volume swelling kinetics of reference and paraffin-thermally modified beech wood specimens.

The values of soaking and swelling of paraffin-thermally modified beech wood were in all cases lower in comparison to the reference beech wood specimens (Table 3, Figures 4 and 5). The swelling of beech wood reduced proportionally to an increase in temperature and time of modification (Table 3, Figure 5). Significantly reduced swelling of beech wood, by approximately 60%, was observed in the case of specimens modified under the regimes at higher temperature of 210 °C for 3 h and 4 h. However, significantly reduced soaking, by approximately 30%, already occurred when modifying the wood under the mildest regime at a temperature of 190 °C for 1 h, especially due to a high presence of paraffin in cell lumens (Table 3, Figure 4).

Similarly by Reinprecht and Vidholdová [17], the soaking and swelling of beech and spruce woods modified by the OHT process in rapeseed oil were reduced especially in the case of the highest temperature of 220 °C acting for the longest time of 6 h. Bal [48] determined physical properties, including volume swelling, of beech thermowood modified in the hot oil medium or hot air at the temperatures of 160 °C, 190 °C and 220 °C for 2 h. In accordance with our experiment, the volume swelling decreased in the case of the most severe modification regime from 17.6% to 4.2%—i.e., in percentage it reduced proportionally to higher temperature and prolonged time of thermal modification from about 8.5% up to 76%. The swelling of beech wood thermally modified in hot oil was reduced more in comparison to the modification process performed in hot air.

In general, the fact that the effect of thermal modification of wood in hot oils or waxes on reducing the swelling is more significant than the effect of thermal modification by hot air can be stated. It can be due to the presence of hydrophobic media located in cell lumens also after thermal modification, therefore, the barrier reducing the process of water absorption by cell walls is made [16,34,49,50]. The correlation between the weight loss of wood during the process of thermal modification and soaking and swelling was mentioned in several previous research works [51–53]. It is observed mainly in the case of wood modification at higher temperatures when an increase in hydrophobicity is affected more by degrading and linking its structural components.

3.4. The Impact Bending Strength and Brinell Hardness

Lots of researchers found out that mechanical properties of wood like compressive strength [22] bending strength [21] and hardness [21,22] can be improved by impregnating the wood with waxes. Mechanical properties of wood like hardness [17,54], bending strength [54], compressive strength [54–56] and modulus of elasticity [57] were improved also under milder conditions of its thermal modification. An increase in the compressive strength of thermowood can be explained with a relative increase in the lignin content and its condensation confirmed by the near infrared spectroscopy (NIR) [55]. It is supposed that an increase in the modulus of elasticity of thermowood relates to forming new chemical bonds with higher bond energy in comparison to the energy of absent hydrogen bonds [57].

However, the mechanical properties get worse when modified temperatures are higher, usually above 220 °C, and the time of modification is longer [10,17,54,58–61]. It is especially due to significant thermal degradation of hemicelluloses situated between cellulose microfibrils in wood cell walls and also due to organic acids resulting from the hemicellulose decomposition catalyzing the cleavage of lignin–carbohydrate complex of wood [62]. Mechanical properties of commercial thermowood get worse by 10%–30% and such wood is not recommended to be used in load-bearing structural elements [63].

The impact bending strength of beech wood decreased slightly about 20.63% due to the presence of paraffin, however, more rapidly after paraffin-thermal modifications in the range from 17.84% to 48.33%—proportionally to an increase in temperature and time of modification (Table 4). The impact bending strength of poplar wood heated at a temperature of 210 °C for 3 h decreased even by about 61% [64]. The fact that this mechanical property decreases significantly at higher temperatures especially in the case of hardwoods is generally known. It is owing to pentosans decomposition and loss of their elastic-mechanical function in cell walls. Beech and other hardwoods contain 2 or 2.5 times more of pentosans in comparison to coniferous wood [65].

Table 4. Impact bending strength and Brinell hardness of beech specimens subjected to different paraffin-thermal modifications.

Paraffin-Thermal Modification	Impact Bending Strength I (J.cm^{-2})	Brinell Hardness H_B (MPa)
Reference	5.38 *(0.83)*	31.56 *(5.29)*
Paraffin only	4.27 *(0.40)* b	32.22 *(4.86)* d
190 °C/1 h	3.76 *(0.65)* a	30.81 *(5.02)* d
190 °C/2 h	4.42 *(0.47)* c	29.09 *(1.94)* d
190 °C/3 h	3.87 *(0.23)* a	25.46 *(4.96)* c
190 °C/4 h	3.36 *(0.77)* a	27.86 *(3.47)* d
210 °C/1 h	3.90 *(0.31)* a	18.87 *(1.88)* a
210 °C/2 h	3.66 *(0.31)* a	18.46 *(0.94)* a
210 °C/3 h	3.40 *(0.22)* a	13.15 *(0.70)* a
210 °C/4 h	2.78 *(0.55)* a	11.38 *(1.37)* a

Note: mean values are from six replicates. Standard deviations are in italics and parantheses. The Duncan test, with significance levels a = 99.9%, b = 99%, c = 95% and d < 95%, was performed in relation to reference specimens—at which mechanical properties of specimens modified at 210 °C always decreased on the 99.9% significance level (a).

The Brinell hardness of paraffin-thermally modified beech wood decreased in the range from 2.38% to 63.94%, especially at a higher temperature of 210 °C. However, harness was not affected by alone paraffin (Table 4). Significant decrease in hardness was investigated especially in the case of specimens modified at 210 °C. Reinprecht and Vidholdová [17] determined a decrease in the Brinell hardness by 1.5%–33% when modifying beech wood in rapeseed oil at a temperature of 180 °C and 220 °C for 3 h to 6 h. Bakar et al. [63] investigated a decrease in hardness of oak wood by 33.3% resulting from thermal modification at a temperature of 190 °C for 8 h. Borůvka et al. [66] determined different changes in the Brinell hardness of thermally modified beech and birch woods (210 °C/3 h)—the hardness of beech wood decreased by 37% and the hardness of birch wood increased by 9%. An increase in hardness of birch wood was explained due to a higher content of mannans in hemicelluloses. By Bonstra et al. [36], the finding for pine and spruce woods that the Brinell hardness parallel to grains increased by 48% or perpendicular to grains by 5%—as a result of their primary hydrothermal modification and subsequent thermal modification—can be considered interesting.

4. Conclusions

- Paraffin-thermal modification resulted in a significant increase in the rot resistance of beech wood to decaying fungi—the brown-rot fungus *Poria placenta* by 71.4%–98.4% and the white-rot fungus *Trametes versicolor* by 50.1%–99.5%. The lowest weight loss, less than 1%, was observed in the case of beech wood specimens modified in paraffin at a temperature of 210 °C for 3 h or 4 h.
- The mold resistance of paraffin-thermally modified beech wood to the microscopic fungus *Aspergillus niger* increased significantly in the first days of testing. However, on the final 21st day of the mold test, the growth intensity of *A. niger* reduced only in the case of specimens modified under the most severe modification regimes.
- The soaking and volume swelling of beech wood reduced markedly as a result of paraffin-thermal modification—the soaking was reduced in all cases by more than 30% and the volume swelling was reduced by 26.8%–62.9%. The specimens of beech wood modified at a temperature of 210 °C for 3 h or 4 h were the most resistant to swelling. On the contrary, soaking was not affected by the temperature and the time of modification.
- Mechanical properties of beech wood got worse as a result of an increase in temperature and time of modification. There was a decrease in the impact bending strength in the range from 17.8% to 48.3% and in the Brinell hardness in the range from 2.4% to 63.9%.
- Generally, beech wood modified with hot paraffin showed significantly better resistance to wood-decaying fungi, slightly better resistance to mold growth, and reduced soaking and swelling. Wood modified this way can be used as a material for making products especially in the demanding interior projects, e.g., sauna, bathroom, kitchen paneling, as well as in exterior projects, e.g., facade or swimming pool panels. However, due to lower values of the impact bending strength and Brinell hardness, the wood modified this way is not convenient for load-bearing structural elements and in projects where good mechanical properties are necessary.

Author Contributions: Conceptualization—L.R. and M.R.; Methodology—L.R.; Software—M.R.; Validation—L.R. and M.R.; Formal Analysis—L.R. and M.R.; Investigation—L.R. and M.R.; Resources—L.R. and M.R.; Data Curation—L.R. and M.R.; Writing—Original Draft Preparation—L.R. and M.R.; Writing—Review and Editing—L.R.; Visualization—L.R. and M.R.

Funding: The article was funded from the project APVV-17-0583.

Acknowledgments: The authors would like to thank the Slovak Research and Development Agency under the contract No. APVV-17-0583 and the VEGA project 1/0729/18 for funding and financial support.

Conflicts of Interest: The authors declare no conflict of interest.

References

1. Kúdela, J.; Čunderlík, I. *Bukové Drevo—Štruktúra, Vlastnosti, Použitie (Beech Wood—Structure, Properties and Usage)*; Technical University in Zvolen: Zvolen, Slovakia, 2012; p. 152. ISBN 978-80-228-2318-0.
2. Klement, I.; Vilkovská, T. Color characteristic of red false heartwood and mature wood of beech (*Fagus sylvatica* L.) determining by different chromacity coordinates. *Sustainability* **2019**, *11*, 690. [CrossRef]
3. Klement, I.; Vilkovská, T.; Uhrín, M.; Barański, J.; Konopka, A. Impact of high temperature drying process on beech wood containing tension wood. *Open Eng.* **2019**, *9*, 2391–5439. [CrossRef]
4. EN 350. *Durability of Wood and Wood-Based Products—Testing and Classification of the Durability to Biological Agents of Wood and Wood-Based Materials*; European Committee for Standardization: Brussels, Belgium, 2016.
5. Hill, C.A.S. *Wood Modification—Chemical, Thermal and Other Processes*; John Wiley & Sons, Ltd.: Chichester, UK, 2006; p. 260. [CrossRef]
6. Reinprecht, L. *Wood Deterioration, Protection and Maintenance*; John Wiley & Sons, Ltd.: Chichester, UK, 2016; p. 357. ISBN 978-1-119-10653-1.
7. Tjeerdsma, B.F.; Boonstra, M.; Pizzi, A.; Tekely, P.; Militz, H. Characterisation of thermally modified wood: Molecular reasons for wood performance improvement. *Holz als Roh-und Werkstoff* **1998**, *56*, 149–153. [CrossRef]

8. Reinprecht, L.; Vidholdová, Z. *TermoDřevo-ThermoWood*; Šmíra print s.r.o.: Ostrava, Czech Republic, 2011; p. 89. ISBN 978-80-87427-05-7.
9. Srinivas, K.; Pandey, K.K. Effect of heat treatment on color changes, dimensional stability, and mechanical properties of wood. *J. Wood Chem. Technol.* **2012**, *32*, 30416. [CrossRef]
10. Esteves, B.; Pereira, H. Wood modification by heat treatment: A review. *BioResources* **2009**, *4*, 370–404.
11. Kocaefe, D.; Huang, X.; Kocaefe, Y. Dimensional Stabilization of Wood. *Curr. For. Rep.* **2015**, *1*, 151–161. [CrossRef]
12. Rapp, A.O.; Sailer, M. Oil heat treatment of wood in Germany—State of the art. In *Review on Heat Treatments of Wood*; Rapp, A.P., Ed.; BFH: Hamburg, Germany, 2001; pp. 45–62. ISBN 3-926301-02-3.
13. Hasan, M.; Despot, R.; Šafran, B.; Lacić, R.; Peršinović, M. Oil heat treatment of alder wood for increasing biological durability of wood. *Drv. Ind.* **2008**, *65*, 143–150. [CrossRef]
14. Rapp, A.O.; Sailer, M. Oil-heat-treatment of wood—Process and properties. *Drv. Ind.* **2001**, *52*, 63–70.
15. Yilgör, N.; Kartal, S.N. Heat Modification of wood: Chemical properties and resistance to mold and decay fungi. *For. Prod. J.* **2010**, *60*, 357–661. [CrossRef]
16. Wang, J.Y.; Cooper, P.A. Effect of oil type, temperature and time on moisture properties of hot oil-treated wood. *Holz als Roh-und Werkstoff* **2005**, *63*, 417–422. [CrossRef]
17. Reinprecht, L.; Vidholdová, Z. Mould resistance, water resistance and mechanical properties of OHT-thermowoods. In Proceedings of the Final Conference of COST Action E37, Bordeaux, France, 29–30 September 2008; UGent—Faculty of Bioscience Engineering—Laboratory of Wood Technology: UGent, Belgium; pp. 159–165.
18. Dubey, M.K.; Pang, S.; Chauhan, S.; Walker, J. Dimensional stability, fungal resistance and mechanical propertiesof radiate pine after combined thermos-mechanical compression and oil heat-treatment. *Holzforschung* **2016**, *70*, 793–800. [CrossRef]
19. Scholz, G.; Krause, A.; Militz, H. Capillary water uptake and mechanical properties of wax soaked Scots pine. In Proceedings of the European Conference on Wood Modification, Stockholm, Sweden, 27–29 April 2009; SP Technical Research Institute of Sweden: Stockholm, Sweden.
20. Scholz, G.; Militz, H.; Gascón, P.; Ibiza-Palacios, M.S.; Oliver-Villanueva, J.V.; Peters, B.C.; Fitzgerald, C.J. Improved termite resistance of wood by wax impregnation. *Int. Biodeterior. Biodegrad.* **2010**, *64*, 688–693. [CrossRef]
21. Esteves, B.; Nunes, L.; Domingos, I.; Pereira, H. Improvement of termite resistance, dimensional stability and mechanical properties of pine wood by paraffin impregnation. *Eur. J. Wood Wood Prod.* **2014**, *72*, 609–615. [CrossRef]
22. Rapp, A.O.; Beringhausen, C.; Bollmus, S.; Brischke, C.; Frick, T.; Haas, T.; Sailer, M.; Welzbacher, C.R. Hydrophobierung von Holz-Erfahrungen nach 7 Jahren Freilandtest. In Proceedings of the 24th Holzschutztagung der DGFH, Leipzig Germany, 2005; pp. 157–170.
23. Amthor, J. Paraffin dispersions for waterproofing of particle board. *Holz als Roh-und Werkstoff* **1972**, *30*, 422–425. [CrossRef]
24. Deppe, H.J.; Ernst, K. *MDF–Mitteldichte Faserplatten*; DRW-Verlag: Leinfelden-Echterdingen, Germany, 1996; p. 200. ISBN 978-3871813290.
25. Neimsuwan, T.; Wang, S.; Via, B.K. Effect of processing parameters, resin, and wax loading on water vapor sorption of wood strands. *Wood Fiber Sci.* **2008**, *40*, 495–504.
26. Wang, W.; Zhu, Y.; Jinzhen, C.; Guo, X. Thermal modification of Southern pine combined with wax emulsion preimpregnation: Effect on hygroscopicity and dimensional stability. *Holzforschung* **2015**, *69*, 405–413. [CrossRef]
27. Humar, M.; Kržišnik, D.; Lesar, B.; Thaler, N.; Ugovšek, A.; Zupanič, K.; Žlahtič, M. Thermal modification of wax-impregnated wood to enhance its physical, mechanical, and biological properties. *Holzforschung* **2017**, *71*, 57–64. [CrossRef]
28. EN 113. *Wood Preservatives. Test Method for Determining the Protective Effectiveness against Wood Destroying Basidiomycetes. Determination of the Toxic Values*; European Committee for Standardization: Brussels, Belgium, 1996.
29. Van Acker, J.; Stevens, M.; Carey, J.; Sierra-Alvarez, R.; Militz, H.; Le Bayon, I.; Kleist, G.; Peek, R.D. Biological durability of wood in relation to end-use. *Holz als Roh-und Werkstoff* **2003**, *61*, 35–45. [CrossRef]
30. EN 15457. *Paints and Varnishes. Laboratory Method for Testing the Efficacy of Film Preservatives in a Coating against Fungi*; European Committee for Standardization: Brussels, Belgium, 2014.

31. STN 490104. *Skúšky Vlastností Rasteného dreva. Metóda Zist'ovania Nasiakavosti a Navlhavosti. (Tests of Native Wood Properties. Method of Water Absorptivity and Hygroscopicity Determining)*; Slovak Standards Institute: Bratislava, Slovakia, 1987.
32. ISO 3348. *Wood—Determination of Impact Bending Strength*; International Organization for Standardization: Geneva, Switzerland, 1975.
33. EN 1534. *Wood Flooring. Determination of Resistance to Indentation. Test Method*; European Committee for Standardization: Brussels, Belgium, 2010.
34. Lesar, B.; Humar, M. Use of wax emulsion for improvement of wood durability and sorption properties. *Eur. J. Wood Wood Prod.* **2011**, *69*, 231–238. [CrossRef]
35. Archer, K.; Leebow, S. *Primary Wood Processing: Principles and Practice*; Springer: Dordrecht, The Netherlands, 2006; Chapter 9–J.C.F. Walker, Wood Preservation; pp. 297–338. ISBN 978-1-4020-4393-2.
36. Boonstra, M.J.; Van Acker, J.; Kegel, E.; Stevens, M. Optimisation of a two-stage heat treatment process: Durability aspects. *Wood Sci. Technol.* **2007**, *41*, 31–57. [CrossRef]
37. Stamm, A.J. *Wood and Cellulose Science*; Ronald Press Co.: New York, NY, USA, 1964; p. 549. ISBN 978-0826084958.
38. Kamdem, D.P.; Pizzi, A.; Jermannaud, A. Durability of heat treated wood. *Holz als Roh-und Werkstoff* **2002**, *60*, 1–6. [CrossRef]
39. Lacić, H.; Hasan, M.; Trajković, J.; Šefc, B.; Šafran, B.; Despot, R. Biological durability of oil-heat treated alder wood. *Eur. J. Wood Wood Prod.* **2014**, *69*, 231–238. [CrossRef]
40. Tjeerdsma, B.F.; Stevens, M.; Militz, H.; Van Acker, J. Effect of process conditions on moisture content and decay resistance of hydrothermally treated wood. *Holzforsch. Holzverwert.* **2002**, *54*, 94–99.
41. Welzbacher, C.; Rapp, O. Comparison of thermally modified wood originating from four industrial scale processes—durability. In *International Research Group on Wood Preservation*; Cardiff, Wales, UK, Document No. IRG/WP 02-40229; IRG-WP: Stockholm, Sweden, 2002.
42. Kartal, S.N. Combimed effect of boron compounds and heat treatments on wood properties. Boron release and decay and termite resistance. *Holzforschung* **2006**, *60*, 455–458. [CrossRef]
43. Scholz, G.; Krause, A.; Militz, H. Exploratory study on the impregnation of Scots pine sapwood (*Pinus sylvestris* L.) and European beech (*Fagus sylvatica* L.) with different hot melting waxes. *Wood Sci. Technol.* **2010**, *44*, 379–388. [CrossRef]
44. Kocaefe, D.; Poncsak, S.; Doré, G.; Younsi, R. Effect of heat treatment on the wettability of white ash and soft maple by water. *Holz als Roh-und Werkstoff* **2008**, *66*, 355–361. [CrossRef]
45. Cao, Y.; Lu, J.; Huang, R.; Jiang, J. Increased dimensional stability of Chinese fir through steam-heat treatment. *Eur. J. Wood Wood Prod.* **2012**, *70*, 441–444. [CrossRef]
46. Sinković, T.; Govorčin, S.; Sedlar, T. Comparison of physical properties of heat treated and untreated hornbeam wood, beech wood, ash wood and oak wood. In Proceedings of the Hardwood Science and Technology—the 5th Conference on Hardwood Research and Utilisation in Europe, Sopron, Hungary, 10–11 September 2012; pp. 63–70, ISBN 978-963-9883-97-0.
47. Boonstra, M.J.; Van Acker, J.; Tjeerdsma, B.F.; Kegel, E. Strength properties of thermally modified softwoods and its relation to polymeric structural wood constituents. *Ann. For. Sci.* **2007**, *64*, 679–690. [CrossRef]
48. Bal, B.C. Physical properties of beech wood thermally modified in hot oil and in hot air at various temperatures. *Cienc. Tecnol.* **2015**, *17*, 789–798. [CrossRef]
49. Awoyemi, L.; Cooper, P.A.; Ung, T.Y. In-treatment cooling during thermal modification of wood in soy oil medium: Soy oil uptake, wettability, water uptake and swelling properties. *Eur. J. Wood Wood Prod.* **2009**, *67*, 465–470. [CrossRef]
50. Dubey, M.K.; Pang, S.; Walker, J. Changes in chemistry, color, dimensional stability and fungal resistance of Pinus radiata D. Don wood with oil heat-treatment. *Holzforschung* **2012**, *66*, 49–57. [CrossRef]
51. Almeida, G.; Brito, J.O.; Perre, P. Changes in wood-water relationship due to heat treatment assessed on micro-samples of three *Eucalyptus* species. *Holzforschung* **2009**, *63*, 80–88. [CrossRef]
52. Bal, B.C.; Bektaş, I. The effects of heat treatment on the physical properties of juvenile wood and mature wood of *E. grandis*. *BioResources* **2012**, *7*, 5117–5127. [CrossRef]
53. Bal, B.C. A comparative study of the physical properties of thermally treated poplar wood and plane wood. *BioResources* **2013**, *8*, 6493–6500. [CrossRef]

54. Percin, O.; Peker, H.; Atilgan, A. The effect of heat treatment on the some physical and mechanical properties of beech (*Fagus orientalis* Lipsky) wood. *Wood Res.* **2016**, *61*, 443–456.
55. Windeisen, E.; Strobel, C.; Wegener, G. Chemical changes during the production of thermo-treated beech wood. *Wood Sci. Technol.* **2007**, *41*, 523–536. [CrossRef]
56. Taghiyari, H.R.; Enayati, A.; Gholamiyan, H. Effects of nano-silver impregnation on brittleness, physical and mechanical properties of heat-treated hardwoods. *Wood Sci. Technol.* **2013**, *47*, 467–480. [CrossRef]
57. Straže, A.; Fajdiga, G.; Pervan, S.; Gorišek, Ž. Hygro-mechanical behaviour of thermally treated beech subjected to compression loads. *Constr. Build. Mater.* **2016**, *113*, 28–33. [CrossRef]
58. Bekhta, P.; Niemz, P. Effect of high temperature on the change in color, dimensional stability and mechanical properties of spruce wood. *Holzforschung* **2003**, *57*, 539–546. [CrossRef]
59. Unsal, O.; Ayrilmis, N. Variations in compression strength and surface roughness of heat-treated Turkish river red gum (*Eucalyptus camaldulensis*) wood. *J. Wood Sci.* **2005**, *51*, 405–409. [CrossRef]
60. Yildiz, S.; Gezerb, D.; Yildiz, C. Mechanical and chemical behaviour of spruce wood modified by heat. *Build. Environ.* **2006**, *41*, 1762–1766. [CrossRef]
61. Korkut, D.S.; Guller, B. The effects of heat treatment on physical properties and surface roughness of red-bud maple (*Acer trautvetteri* Medw) wood. *Bioresour. Technol.* **2008**, *99*, 2846–2851. [CrossRef]
62. Zaman, A.; Alen, R.; Kotilainen, R. Thermal behaviour of Scots pine (*Pinus sylvestris*) and silver birch (*Betula pendula*) at 200–230 degrees C. *Wood Fiber Sci.* **2000**, *32*, 138–143.
63. Bakar, B.F.A.; Hiziroglu, S.; Tahir, P.M. Properties of some thermally modified wood species. *Mater. Des.* **2013**, *43*, 348–355. [CrossRef]
64. Reinprecht, L. Thermal degradation of chemically pretreated wood. Part 1. Course of thermal degradation of wood valued on the basis of loss of mass, impact strength in bending and cutting strength. In *Wood Burning 92, Proceedings of the 2nd International Scientific Conference, High Tatras, Hotel Patria, 1–5 June 1992*; Technical University in Zvolen: Zvolen, Slovakia, 1992; pp. 57–67. ISBN 80-228-0178-X.
65. Fengel, D.; Wegener, G. *Wood—Chemistry, Ultrastructure, Reactions*; Verlag Kessel: Remagen, Germany, 2003; p. 613. ISBN 3935638-39-6.
66. Borůvka, V.; Zeidler, A.; Holeček, T.; Dudík, R. Elastic and strength properties of heat-treated beech and birch wood. *Forests* **2018**, *9*, 197. [CrossRef]

Article

On the Effect of Heat Treatments on the Adhesion, Finishing and Decay Resistance of Japanese cedar (*Cryptomeria japonica* D. Don) and Formosa acacia (*Acacia confuse* Merr.(Leguminosae))

Chia-Wei Chang [1], Wei-Ling Kuo [2] and Kun-Tsung Lu [1,*]

[1] Department of Forestry, National Chung Hsing University, Taichung 402, Taiwan
[2] Forestry Bureau, Council of Agriculture, Executive Yuan, Taichung 402, Taiwan
* Correspondence: lukt@dragon.nchu.edu.com; Tel.: +886-4-22840345 (ext. 122)

Received: 14 June 2019; Accepted: 11 July 2019; Published: 13 July 2019

Abstract: In Taiwan, it is important to maintain sustainable development of the forestry industry in order to raise the self-sufficiency of domestic timber. Japanese cedar (*Cryptomeria* D. Don and Formosa acacia (*Acacia confusa* Merr.(Leguminosae)) have abundant storage options and are the potential candidates for this purpose. Heat treatment is a new environment-friendly method used to enhance the dimensional stability and durability of wood. On treatment, a surface with new characteristics is produced because of wood component changes. Consequently, an inactivated surface and a weak boundary layer are generated, and the wettability for adhesives and coatings is reduced. Furthermore, it decreases the pH value of the wood surface, and results in delay or acceleration during the curing of adhesives. This phenomenon must be paid attention to for practical applications of heat-treated wood. Ideal heat-treated conditions of *C. japonica* and *A. confusa* woods with productive parameters such as temperature, holding time, heating rate, and thicknesses of wood were identified in our previous study. In this research work, we focus on the normal shear strength of heat-treated wood with adhesives such as urea-formaldehyde resin (UF) and polyvinyl acetate (PVAc), and the finishing performances of heat-treated wood with polyurethane (PU) and nitrocellulose lacquer (NC) coatings as well as assessing the decay-resistance of heat-treated wood. The results show that heat-treated wood had a better decay resistance than untreated wood. The mass decrease of heat-treated wood was only 1/3 or even less than the untreated wood. The normal shear strength of heat-treated wood with UF and PVAc decreased from 99% to 72% compared to the untreated wood, but the wood failure of heat-treated wood was higher than that of the untreated one. Furthermore, the adhesion and impact resistance of wood finished by PU and NC coatings showed no difference between the heat-treated wood and untreated wood. The finished heat-treated wood had a superior durability and better gloss retention and lightfastness than that of the untreated wood.

Keywords: heat-treated wood; decay resistance; adhesion; finishing

---

1. Introduction

Wood is an excellent and environment-friendly renewable material, which possesses good mechanical strength, low thermal expansion, aesthetic appeal and is easy to process for a variety of applications. Owing to these characteristics and its abundance, wood has been used for building, structure, furniture, and other objects that meet the demands of people in the modern world. However, being a biological material, it is susceptible to warping, splitting, photodegradation and biodeterioration, highlighting natural anisotropic and hygroscopic characteristics. These defects restrict its use in certain applications. It is thus necessary to enhance its dimensional stability and durability to expand applications of the material. Wood treatments include chemical modification methods such as esterification and etherification [1,2] and using chromated copper arsenate (CCA) or alkaline copper quaternary ammonium compounds (ACQ) preservatives [3–5], which, however, cause environment concerns.

A new environment-friendly method used to enhance the dimensional stability and durability of wood is heat treatment. This technique is suitable for wood due to its non-toxic characteristics. In heat-treated wood, physical and chemical properties such as density, equilibrium moisture content, roughness, and strength are altered by submitting wood to high temperatures (150–230 °C) in an inert or restricted air environment. The dimensional characteristics and decay resistance of wood can be improved and the treatment results in a uniform dark color. However, its strength is decreased as compared to those of untreated control samples. The use of heat treatment as a wood modification method and its products have increased significantly over the last decade and have been well studied [6–15]. However, in practical utilization, the adhesion, finishing, and decay resistance performances of heat-treated wood have not been examined enough.

After heat treatment, new surface characteristics are produced because of wood component changes. The results affect the wettability of heat-treated wood and further affect its bonding strength [11,12,16]. In addition, the hydrophobic extractives and volatile organic compounds that are derived from the depolymerization of wood components, especially from hemi-cellulose [17], moved to or deposited on the wood surface. As a result, an inactivated wood surface and a weak boundary layer were produced. Consequently, the wettability of wood for adhesives and coatings [18–21] was reduced. Furthermore, it decreases the pH value of wood surface and results in delay or acceleration during the curing of adhesives [18,22]. This phenomenon must be paid attention to for practical applications of heat-treated wood.

In Taiwan, the domestic timber yield accounts for merely 0.5% of its timber consumption, which implicates that raw timber materials are almost completely imported. In view of this, relevant policies—such as enlarging planting area and increasing the utilization of domestic woods—should be formulated to raise the self-sufficiency of domestic timber, which is vital for Taiwan to maintain sustainable development of its forestry industry. Japanese cedar (*Cryptomeria japonica* D. Don), a softwood, and Formosa acacia (*Acacia confuse* Merr.(Leguminosae)), a hardwood, have abundant storage options and are potential candidates for this purpose. The total cultivation area of *C. japonica* in Taiwan is 39,100 ha (or 2.68%) and of *A. confusa* is 21,200 ha (or 1.43%) [23]. *C. japonica* and *A. confusa* woods be used for making furniture, building, structures and other objects.

In our previous study, the heat treatments of *C. japonica* and *A. confusa* woods with moisture content of 12% were performed using a steel kiln under an N_2 atmosphere. The parameters of heat-treated conditions such as temperature (130, 160,190, 220, 250 and 280 °C), holding times (1, 2 and 3 h), thicknesses (30, 50 and 70 mm) and heating rates (10, 20 and 30 °C /h) were examined for the best performances of heat-treated woods. The results indicated that the best heat-treated conditions for the superior of dimensional stability, hardness and static bending strength of woods included heating rates of 10 °C/h, temperature of 190 °C, wood thickness of 50 mm, and holding time of 2 h for *C. japonica* and 3 h for *A. confusa* [24]. Furthermore, for practical utilization, in this report, we focus on the normal shear strength of heated *C. japonica* and *A. confusa* woods with urea-formaldehyde resin (UF) and polyvinyl acetate (PVAc) wood adhesives, the performances of finishing properties of woods with polyurethane (PU) and nitrocellulose lacquer (NC) wood coatings, and decay-resistance of woods by soil block test.

2. Materials and Methods

20–30-year-old *A. confusa* and *C. japonica* woods were obtained from Hui-Sun Forest Station of the Experimental Forest of National Chung Hsing University in Nan-Tou County, Taiwan. The wood was cut into the desired dimensions after air-drying and then conditioned for one week at 20 °C and 65% relative humidity. The adhesives, including urea-formaldehyde resin (UF) with a solid content of 65%, was purchased from Wood Glue Industrial Co., Ltd., Tainan, Taiwan. Polyvinyl acetate (PVAc) was supplied by Nan Pao Resins Chemical Co., Ltd., Tainan, Taiwan. The coatings, including two-package polyurethane resin (PU) and nitrocellulose lacquer (NC), were purchased from Yan Chang Trading Co., Ltd., Taichung, Taiwan.

2.1. Heated Treatment of Wood

A. confusa and *C. japonica* with dimensions of 50 mm (T) × 50 mm (R) × 330 mm (L) and moisture content of 12% were placed in a stainless steel cylindrical equipment (390 mm diameter and 800 mm height), as shown in Figure 1. The autoclave body was flushed with a continuous N_2 gas flow first with a flowing rate of 500 mL/min for 30 min and then the flowing rate was adjusted to 50 mL/min for the heated procedure. The temperature was first heated from room temperature to 100 °C at a heating rate of 10 °C/min and holding time of 6 h and then heated to 190 °C under the same heating rate and holding time of 2 h for *C. japonica* wood and 3 h for *A. confusa* wood. After the heated treatment, the wood samples were cooled to room temperature and conditioned for one week at 20 °C and 65% relative humidity. The moisture content, density, environmental moisture content (at 40 °C, a relative humidity of 65%), contact angle of radial section, modulus of rupture were measured in our previous study [24], and the results were 2.6%, 387 kg/m^3, 2.7%, 23.4%, 57.2 MPa for *C. japonica* heat-treated wood, and 2.6%, 684 kg/m^3, 5.4%, 37.8%, 44.4 MPa for *A. confusa* heat-treated wood.

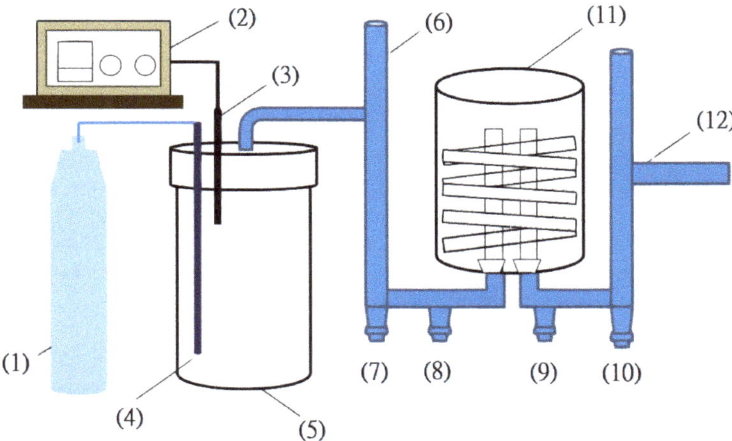

Figure 1. Schematic diagram of the heat-treated equipment. (1) N_2 cylinder, (2) Controller, (3) Thermocouple, (4) Air supply tube, (5) Heat-treated body, (6) Condenser, (7~10) Liquid-collecting vessels, (11) Water cooling tower, (12) Exhauster.

2.2. Compression Shear Bonding Strength of Heat-Treated Wood

The shear bonding strength of heat-treated wood specimens were measured by compression loading in accordance with Chinese National Standards (CNS) 5809. The heat-treated and untreated woods were cut into thickness of 10 mm specimens and the out-layer with radial section of the heat-treated wood was as a glue surface. The UF resin (with adding 1% NH_4Cl as a catalyst) and PVAc were applied on both sides of the interface (200 g/m^2 for each layer), respectively. The specimens were pressed by compression molding machine at room temperature and pressure of 1.47 MPa for 48 h. After conditioning at room temperature for 48 h, the glued samples were cut into 30 mm (L) × 25 mm (R) × 20 mm (T) as shown in Figure 2. Normal shear strength test of the specimens was conducted with a Universal Testing Machine (Shimadzu UEH-10, SHIMADZU Co., Kyoto, Japan) at a crosshead speed of 1 mm/min with duplicates of 10, and wood failure was noted.

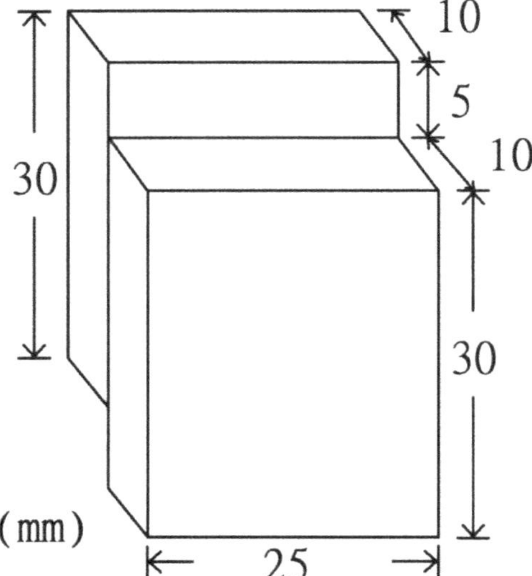

Figure 2. Dimension of testing sample for the compression shear bonding strength test.

2.3. Heat-Treated Wood Finishing and Characterization of Finishing Performances

The specimens for wood finishing were quarter-sawed, and had a dimension around 15 cm (L) × 10 cm (R) × 1.5 cm (T). The viscosity of 2-package PU coating (polyol mixed with aliphatic isocyanate hardener by wt. of 1:1) and NC lacquer were adjusted to 20 s with thinner using No. 4 Ford cup at room temperature. The two coatings were finished on the heat-treated and untreated woods, respectively, by using an air spray gun with the proper ratio of compressed air and fluid coatings to obtain a smooth pattern of spay flow, and thickness of films. The thickness of the dry film on the wood specimens surface were 100 μm and 60 μm for the PU coating and NC lacquer, respectively. The finished specimens were then reconditioned at room temperature for 1 week prior to examination of the properties of the films.

The adhesion of tested films was determined by the cross-cut method according to CNS K 6800 with duplicates of 3; the best adhesion was 10 grade followed by 8, 6, 4, 2 and 0 grade. The impact resistance of the films was determined based on the striking height of the finished wood panels that do not crack on a knowing falling weight of 300 g and impact hammer diameter of 1/2 inch by using a Dupont Impact Tester IM-601 (DU PONT Co., Wilmington, DE, USA) according to American Society for Testing and Materials (ASTM) D5420. The gloss of films finished on wood and parallel to grain was detected by using Dr. Lange Reflectometer 60° Gloss Meter (DR LANGE Co., Dusseldorf, Germany) according to Deutsches Institut für Normung & International Organization for Standardization (DIN ISO) 4630 with duplicates of 10. The durability of films was evaluated using hot and cold cycling tests, in which one tested cycle of the specimens were first placed them into −20 °C refrigerator for 2 h, and then transferred to 50 °C oven for another 2 h according to CNS15200-7-4. The cycle number was recorded if the sample film cracked. If the films were intact after the cycling test, gloss retention of film was measured.

The lightfastness of films finished on wood was carried out with a Paint Coating Fade Meter (SUGA TEST INSTRUMENTS Co., Tokyo, Japan), the light source was mercury light (H400-F) and chamber temperature was at 32 ± 4 °C. After 100 h exposure, the changes in color of the specimens were measured with nine points on the radial section using a spectrophotometer (CM-3600d, Minolta. Osaka, Japan) fitted with a D65 light source with a measuring angle of 10° and a test-window diameter of 8 mm. The tristimulus values X, Y, and Z of all specimens were obtained directly from the colormeter.

The brightness (L^*), redness-greenness (a^*), and yellowness-blueness (b^*) color parameters were then computed according to the Commission internationale de l'éclairage (CIE) 1976 L*a*b* color system, followed by calculating the brightness difference (ΔL^*), color difference (ΔE^*) and yellowness difference (ΔYI, calculated by ASTM E313) directly from the Minolta Managed Content Services (MCS) software system (version of 3.1, Newport Beach, CA, USA).

2.4. Soil Block Test for Decay Resistance of Heat-Treated Woods

The soil block test was performed, with partial reference to the AWPA E14-94 standard. The samples were cut into a size of 200 mm (L) × 20 mm (R) × 20 mm (T) with duplicates of 5. The test samples were buried upright with 200 mm length of the samples in the ground and each sample was installed 50 mm apart. The testing site was located in National Chung Hsing University, Taichung, Taiwan. The surface layer ca. 100 mm of the soil is humus and underlayer is yellow-brown color soil. The experiment was performed for 50 days and the weight loss, brightness difference (ΔL^*) and color difference (ΔE^*) of the test samples were assessed.

3. Results and Discussion

3.1. The Normal Shear Strength

The normal shear strength of heat-treated and untreated *C. japonica* and *A. confusa* wood with UF and PVAc adhesives are listed in Table 1. For UF adhesive, the shear strength of heat-treated *C. japonica* and *A. confusa* woods were 7.66 and 13.98 MPa, respectively, slightly lower than those of untreated woods at 7.73 and 19.50 MPa, respectively. The results may be due to preferential degradation of the less-ordered molecules of amorphous cellulose and hemi-cellulose with heat-treatment, and the increase of the crystallinity of cellulose as well as the affecting of wettability of UF adhesive on the heat-treated wood surface [13,25]. It also may be caused by the extractives of wood migrating to the surface with higher temperature heat treatment, which affected the wettability of the UF adhesive [20,26,27]. In addition, the decomposition products during the heat-treated processing (such as acidic components including acetic acid and formic acid) decrease the pH value on the wood surface, which accelerate the curing of the UF adhesive and further reduce the gluing ability of wood [20,28]. The wood failure of heat-treated *C. japonica* was 90% and was higher than the untreated one at 40%. The appearance of wood failure occurred in the wood itself and not in the interface of the glue layer, as shown in Figure 3. This result was due to the degradation of hemi-cellulose and cellulose of wood during the heat-treated processing and the decrease of the hardness as well as the brittle surface of heat-treated wood [29,30]. Similar results were also obtained for heat-treated *A. confusa* wood with wood failure of 100% and untreated wood of 80%.

Table 1. Normal shear strength of untreated and treated Japanese cedar (*Cryptomeria japonica* D. Don) and Formosa acacia (*Acacia confuse* Merr.(Leguminosae)) wood with urea-formaldehyde resin (UF) and polyvinyl acetate (PVAc) adhesives.

Wood		Shear Strength (MPa)	
		UF	PVAc
C. japonica	Untreated	7.73 ± 0.35 (40) [2]	8.25 ± 0.30 (90)
	Heat-treated [1]	7.66 ± 0.33 (90)	7.43 ± 0.33 (100)
A. confusa	Untreated	19.50 ± 0.22 (80)	15.39 ± 0.31 (20)
	Heat-treated	13.98 ± 0.30 (100)	11.39 ± 1.05 (40)

[1] Heat-treated conditions are heating rates of 10 °C /h, temperature of 190 °C, wood thickness of 50 mm, and holding time of 2 h for *C. japonica* and 3 h for *A. confusa*. [2] represents wood failure (%).

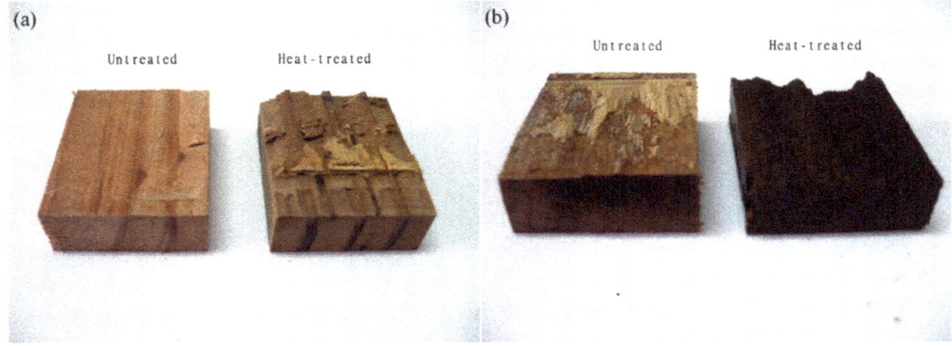

Figure 3. Appearance of untreated and heat-treated Japanese cedar (*Cryptomeria japonica* D. Don) (**a**) and Formosa acacia (*Acacia confuse* Merr.(Leguminosae)) (**b**) wood with urea-formaldehyde resin (UF) adhesive after the shear strength test.

For the PVAc adhesive, the similar results of adhesion performances were as the UF adhesive. The shear strength of heat-treated *C. japonica* and *A. confusa* wood was 7.43 and 11.39 MPa, which was lower than that of untreated wood at 8.25 and 15.39 MPa, respectively. However, the wood failure of heat-treated *C. japonica* and *A. confusa* wood was 100 and 40% and was higher than that of untreated ones at 90 and 20%, respectively. The appearances of untreated and heat-treated *C. japonica* (a) and *A. confusa* (b) wood with PVAc adhesives after normal shear strength test are as shown in Figure 4.

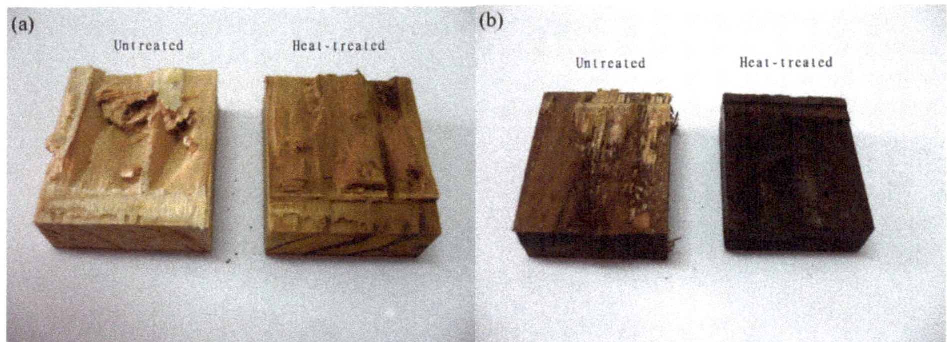

Figure 4. Appearances of untreated and heat-treated Japanese cedar (*Cryptomeria japonica*) (**a**) and Formosa acacia (*Acacia confusa*) (**b**) wood with polyvinyl acetate (PVAc) adhesive after the normal shear strength test.

3.2. The Finishing Performances

The appearance, adhesion and impact resistance of untreated and heat-treated *C. japonica* and *A. confusa* wood after PU and NC lacquer finishing are shown in Table 2, respectively. The result showed that the heated wood had a more even and darker color than the untreated wood. The adhesion of both PU and NC lacquer films of heat-treated and untreated *C. japonica* and *A. confusa* was grade 10, showing excellent adhesion. The striking height was the same of 10 cm for PU finished untreated and heat-treated *C. japonica*. However, it was 40 and 35 cm for untreated and heat-treated *A. confusa*, respectively, highlighting that the impact resistance of heat-treated *A. confusa* was slightly inferior to that of untreated wood. This may be due to the generation of a hydrophobic surface after heat treatment, resulting in a decrease of wettability for coatings. In addition, the NC lacquer finished on untreated or heat-treated *C. japonica* and *A. confusa* wood had the same values of 5 cm. In summary, the adhesion and impact resistance of heat-treated wood is similar to those of untreated wood for PU and NC lacquer finishing.

Table 2. Adhesion and impact resistance of untreated and heat-treated Japanese cedar (*Cryptomeria japonica*) (**a**) and Formosa acacia (*Acacia confusa*) wood finished with polyurethane (PU) and nitrocellulose lacquer (NC) coatings.

Wood		PU		NC	
		Adhesion (Grade)	Impact Resistance (cm)	Adhesion (Grade)	Impact Resistance (cm)
C. japonica	Untreated	10	10	10	5
	Heat-treated [1]	10	10	10	5
A. confusa	Untreated	10	40	10	30
	Heat-treated	10	35	10	30

[1] Heat-treated conditions are heating rates of 10 °C /h, temperature of 190 °C, wood thickness of 50 mm, and the holding time of 2 h for *C. japonica* and 3 h for *A. confusa*.

The hot and cold cycling test and gloss retention of untreated and heat-treated *C. japonica* and *A. confusa* woods finished with PU and NC lacquer coatings are shown in Table 3. The results indicated that all samples were intact after 20 cycles of the hot and cold cycling test. However, the gloss retention of both heat-treated woods was higher than that of the untreated woods for both PU and NC lacquer finishing. Because the dimension stability of wood could be improved after heat-treatment, the gloss lowering of film derived from wood swelling and shrinkage during the hot and cold cycling test could be reduced.

Table 3. Cycling test and gloss retention of untreated and heat-treated Japanese cedar (*Cryptomeria japonica*) and Formosa acacia (*Acacia confusa*) wood finished with polyurethane (PU) and nitrocellulose lacquer (NC) coatings.

Wood		PU		NC	
		Cycling Test (Cycles)	Gloss Retention (%)	Cycling Test (cycles)	Gloss Retention (%)
C. japonica	Untreated	>20	93	>20	93
	Heat-treated [1]	>20	96	>20	100
A. confusa	Untreated	>20	90	>20	95
	Heat-treated	>20	100	>20	98

[1] Heat-treated conditions are heating rates of 10 °C /h, temperature of 190 °C, wood thickness of 50 mm, and holding time of 2 h for *C. japonica* and 3 h for *A. confusa*.

The lightfastness of films finished on wood was carried out on a Paint Coating Fade Meter with 100 h exposure; the brightness difference (ΔL^*), color difference (ΔE^*) and yellowness difference (ΔYI are listed in Tables 4 and 5. The light induced degradation of wood is mainly due to the photochemica radical reactions occurring in lignin and the generation of chromophoric groups. These structures a responsible for the wood color changes. For both PU and NC lacquer finishing, the ΔE^* and ΔYI valu of heat-treated woods after UV-light irradiation were lower than those of untreated woods. During heat-treatment of wood, the mainly thermal degradation occurs in hemicellulose and cellulose and relative content of lignin increased in the heat-treated wood. The best photostability of heat-treat *japonica* and *A. confusa* woods' color could be partially explained by the increase of lignin stabil condensation during the heat-treatment [31]. Furthermore, the ΔL^* values of heat-treated woo higher than those of untreated woods, meaning the heat-treated woods possessed a brighter

This may be due to the producing of more Stab lignin and less chromophoric groups of heat-treated wood after UV-light exposure. In addition, the ΔE^* and ΔYI values of both heat-treated and untreated woods finished with PU were lower than those with NC lacquer, showing that the PU finished wood had superior lightfastness.

Table 4. Lightfastness of untreated and heat-treated Japanese cedar (*Cryptomeria japonica*) and Formosa acacia (*Acacia confusa*) wood finished with polyurethane (PU) coating.

Wood		Brightness Difference (ΔL^*)	Color Difference (ΔE^*)	Yellowness Difference (ΔYI)
C. japonica	Untreated	−6.17 ± 0.19	32.61 ± 0.57	15.25 ± 0.24
	Heat-treated [1]	−0.47 ± 0.97	18.68 ± 2.63	3.28 ± 0.87
A. confusa	Untreated	−5.27 ± 2.24	11.22 ± 2.26	0.89 ± 0.09
	Heat-treated	−0.25 ± 0.23	1.70 ± 0.66	0.07 ± 0.15

[1] Heat-treated conditions are heating rates of 10 °C /h, temperature of 190 °C, wood thickness of 50 mm, and the holding time of 2 h for *C. japonica* and 3 h for *A. confusa*.

Table 5. Lightfastness of untreated and heat-treated Japanese cedar (*Cryptomeria japonica*) and Formosa acacia (*Acacia confusa*) wood finished with nitrocellulose lacquer (NC) coating.

Wood		Brightness Difference (ΔL^*)	Color Difference (ΔE^*)	Yellowness Difference (ΔYI)
C. japonica	Untreated	−8.28 ± 0.87	63.63 ± 1.67	23.42 ± 1.00
	Heat-treated [1]	4.56 ± 0.63	28.22 ± 4.14	4.23 ± 0.15
A. confusa	Untreated	−8.10 ± 0.81	11.04 ± 0.62	5.46 ± 1.15
	Heat-treated	1.69 ± 0.69	7.98 ± 1.29	3.02 ± 2.81

[1] Heat-treated conditions are heating rates of 10 °C /h, temperature of 190 °C, wood thickness of 50 mm, and the holding time of 2 h for *C. japonica* and 3 h for *A. confusa*.

3.3. *Decay Resistance Performances*

The appearances of untreated and heat-treated wood after a 50-day soil block decay resistance
's shown in Figure 5. The untreated woods showed an uneven color on the surface; there were
l dark spots in the untreated *A. confusa*. However, the heat-treated wood had a uniform color
decay resistance test. The weight loss and color changes of untreated and heat-treated *C.*
d *A. confusa* wood after the resistance test are drawn in Figure 6. The weight loss of untreated
and *A. confusa* were 9.3 and 1.6%, while that of the heat-treated woods were 3.0 and
ively. In a study by Vidholdová and Reinprecht [32], 21 tropical woods were attacked
r six weeks by brown-rot fungus, resulting in 0.08% to 6.48% weight loss of the wood.
sults showed higher values (1.6% and 9.3%). This pheromone can be explained by
our soil block test—moisture vibration changes, water flow, temperature cycles,
s. Nonetheless, the medium difference shows that the brown-rot fungus is a key
The results showed that the heat-treated wood had superior decay resistance
crease of hydrophobicity and decrease of water absorbency [17,25]. There also
oup content, which affected enzyme recognition blocking of the fungi [33],
lose content of the wood, therefore cutting down on the main nutritive
]. In addition, the extractives of wood and pyrolysis products such as
s, which have fungicidal efficiency, are deposited on the wood surface,
i [20,27,33,35]. The color changes on ΔL^* of untreated *C. japonica* and
after the 50-day block decay test, meaning a darkening color was
urface; the ΔL^* was 4.0 and 3.5, respectively for the heat-treated
r obtained after the decay test. The ΔE^* of the untreated woods

were 11.5 and 20.9, which was greater than the values of 5.9–0.5 from a previous study [32]. The ΔE^* values of heat-treated wood were also lower than that of the untreated wood. The best photostability of heat-treated wood color could be partially explained by the increase of lignin stability by condensation during the heat-treatment, restricting light from attacking the lignin. In addition, the phenol content increased after heating and the antioxidant compounds can limit degradation caused by oxygen and radicals [31].

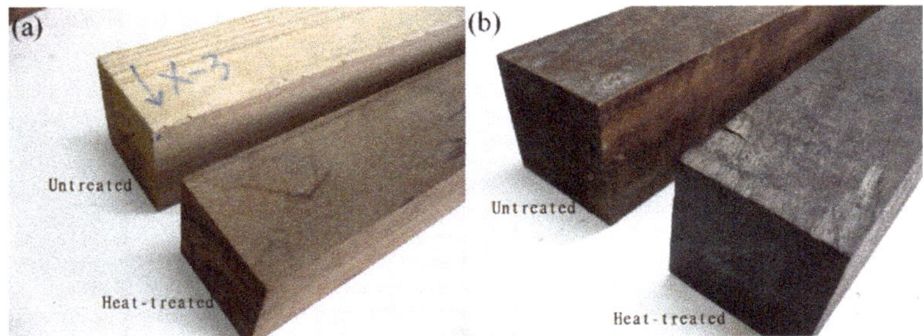

Figure 5. Appearances of untreated and heat-treated Japanese cedar (*Cryptomeria japonica*) (**a**) and Formosa acacia (*Acacia confusa*) (**b**) wood after the 50-day decay resistance test.

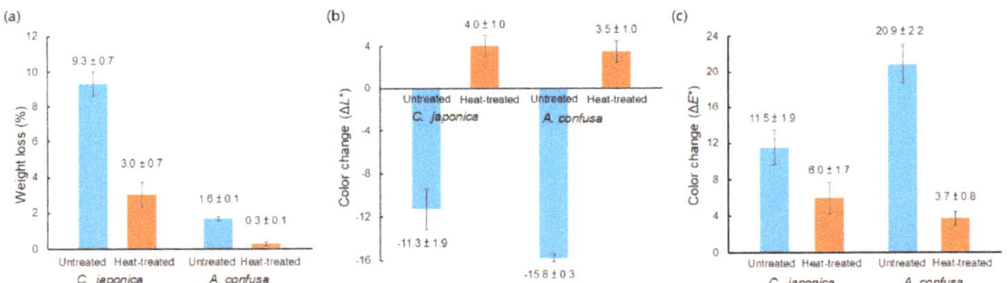

Figure 6. Weight loss (**a**) and color changes (**b**,**c**) of untreated and heat-treated Japanese cedar (*Cryptomeria japonica*) and Formosa acacia (*Acacia confusa*) wood after the 50-day decay resistance test.

4. Conclusions

The results showed that the normal shear strength of heat-treated *C. japonica* and *A. confusa* with UF and PVAc adhesives was lower than that of untreated wood (the decrease is around 99% to 72%), but the wood failure of heat-treated wood was higher than that of the untreated one. Furthermore, the adhesion and impact resistance of heat-treated *C. japonica* and *A. confusa* finished with PU and NC coatings showed no difference with untreated wood. The finished heat-treated wood also showed superior durability and had better gloss retention and lightfastness than those of the untreated wood. The heat-treated wood also had better decay-resistance than untreated wood, ascertained by a soil block test. The weight loss of heat-treated *C. japonica* and *A. confusa* were 3.0% and 0.3%, respectively. Both values of mass decrease were only 32% and 19% compared to the untreated wood.

Author Contributions: Conceptualization, C.-W.C. and K.-T.L.; methodology, W.-L.K. and C.-W.C.; formal analysis, W.-L.K., C.-W.C. and K.-T.L.; investigation, W.-L.K.; data curation, C.-W.C. and W.-L.K.; writing—original draft preparation, K.-T.L.; writing—review and editing, C.-W.C. and K.T. Lu; supervision, K.-T.L.; project administration, K.-T.L.

Funding: This research was funded by Forestry Bureau, Council of Agriculture, Executive Yuan, grant number tfbp-990506.

Conflicts of Interest: The authors declare no conflict of interest.

References

1. Antonios, N.P.; Georgia, P. Mechanical Behaviour of Pine Wood Chemically Modified with A Homologous Series of Linear Chain Carboxylic Acid Anhydrides. *Bioresour. Technol.* **2010**, *101*, 6147–6150. [CrossRef]
2. Chang, H.T.; Chang, S.T. Modification of Wood with Isopropyl Glycidyl Ether and Its Effects on Decay Resistance and Light Stability. *Bioresour. Technol.* **2006**, *97*, 1265–1271. [CrossRef] [PubMed]
3. Shi, J.L.; Kocaefe, D.; Amburgey, T.; Zhang, J. Acomparative Study on Brown-Rot Fungus Decay and Subterranean Termite Resistance of Thermally-Modified and ACQ-C-Treatedwood. *Holz. Roh. Werkst.* **2007**, *65*, 353–358. [CrossRef]
4. Kartal, S.N.; Kakitani, T.; Imamura, Y. Bioremediation of CCA-C Treated Wood by *Aspergillus Niger* Fermentation. *Holz. Roh. Werkst.* **2004**, *62*, 64–68. [CrossRef]
5. Kartal, S.N.; Clausen, C.A. Leachability and Decay Resistance of Particleboard Made from Acid Extracted and Bioremediated CCA-Treated Wood. *Int. Biodeterior. Biodegrad.* **2001**, *47*, 183–191. [CrossRef]
6. Srinivas, K.; Panda, K.K. Photodegradaion of Thermally Modified Wood. *J. Photochem. Photobiol. B Biol.* **2012**, *117*, 140–145. [CrossRef]
7. Dilik, T.; Hiziroglu, S. Bonding Strength of Heated Compressed Eastern Redcedar Wood. *Mater. Des.* **2012**, *42*, 317–320. [CrossRef]
8. Korkut, S. Performance of Three Thermally Treated Tropical Wood Species Commonly Used in Turkey. *Ind. Corp. Prod.* **2012**, *36*, 355–362. [CrossRef]
9. Akyildiz, M.H.; Ates, S. Effect of Heat-Treatment on Equilibrium Moisture Content (EMC) of Some Wood Species in Turkey. *Res. J. Agric. Biol. Sci.* **2008**, *4*, 660–665. [CrossRef]
10. Del Menezzi, C.H.S.; De Souza, R.Q.; Thompson, R.M.; Teixeria, D.E.; Okino, E.Y.A.; Da Costa, A.F. Properties after Weathering and Decay Resistance of a Thermally Modified Wood Structural Board. *Int. Biodeterior. Biodegrad.* **2008**, *62*, 448–454. [CrossRef]
11. Gobakken, L.R.; Westin, M. Surface Mould Growth on Five Modified Wood Substrates Coated with Three Different Coating Systems when Exposed Outdoors. *Int. Biodeterior. Biodegrad.* **2008**, *62*, 397–402. [CrossRef]
12. Metsä-Kortelainen, S.; Antikainen, T.; Viitaniemi, P. The Water Absorption of Sapwood and Heartwood of Scots Pine and Norway Spruce Heat-Treated at 170 °C, 190 °C, 210 °C and 230 °C. *Holz. Roh. Werkst.* **2006**, *64*, 192–197. [CrossRef]
13. Hakkou, M.; Pétrissans, M.; Zoulalian, A.; Gérardin, P. Investigaation of Wood Wettability Changes During Heat-Treatment on The Basis of Chemical Analysis. *Polym. Degrad. Stab.* **2005**, *89*, 1–5. [CrossRef]
14. Hakkou, M.; Pétrissans, M.; Gérardin, P.; Zoulalian, A. Investigation of the Reasons for Fungal Durability of Heat-Treated Beech Wood. *Polym. Degrad. Stab.* **2006**, *91*, 393–397. [CrossRef]
15. Manninen, A.M.; Pasanen, P.; Holopainen, J.K. Comparing the VOC Emissions Between Air-Dried and Heat-Treated Scots Pine Wood. *Atmos. Environ.* **2002**, *36*, 1763–1768. [CrossRef]
16. Günduz, G.; Aydemir, D. Some Physical Properties of Heated Hornbeam (*Carpinus betulus* L.) Wood. *Dry. Technol.* **2009**, *27*, 714–720. [CrossRef]
17. Boonstra, M.J.; Tjeerdsma, B. Chemical Analysis of Heat Treated Softwoods. *Holz. Roh. Werkst.* **2006**, *64*, 204–211. [CrossRef]
18. Šernek, M.; Boonstra, M.; Pizzi, A.; Despres, A.; Gérardin, P. Bonding Performance of Heat-Treated Wood with Structural Adhesives. *Holz. Roh. Werkst.* **2008**, *66*, 173–180. [CrossRef]
19. Follrich, J.; Müller, U.; Gindl, W. Effects of Thermal Modification on the Adhesion Between Spruce Wood (*Picea abies Karst.*) and a Thermoplastic Polymer. *Holz. Roh. Werkst.* **2006**, *64*, 373–376. [CrossRef]
20. Šernek, M.; Kamke, F.A.; Glasser, W.G. Comparative Analysis of Inactivated Wood Surface. *Holzforschung* **2004**, *58*, 22–31. [CrossRef]
21. Vick, C.B. Adhesive Bonding of Wood Materials. In *Wood Handbook–Wood as an Engineering Material*; USDA Forest Service, Forest Products Laboratory: Madison, WI, USA, 1999; pp. 9-1–9-23. [CrossRef]
22. Shi, S.Q.; Gardner, D.J. Dynamic Adhesive Wettability of Wood. *Wood Fiber Sci.* **2001**, *33*, 58–68. [CrossRef]
23. Forestry Statistics of Taiwan Region. Available online: http://www.forest.gov.tw (accessed on 6 January 2005).
24. Kuo, W.L.; Lu, K.T. Characteristics of *Cryptomeria japonica* and *Acacia confusa* Heat-Treated Woods with Different Temperatures. *Q. J. For. Res.* **2012**, *34*, 269–286. [CrossRef]
25. Bhuiyan, M.T.R.; Sobue, N.H.N. Changes of Crystallinity in Wood Cellulose by Heat-Treatment under Dried and Moist Conditions. *J. Wood Sci.* **2000**, *46*, 431–436. [CrossRef]

26. Follrich, J.; Teischinger, A.; Gindl, W.; Müller, U. Adhesive Bond Strength of End Grain Joints in Softwood with Varying Density. *Holzforschung* **2008**, *62*, 237–242. [CrossRef]
27. Nuopponen, M.; Vuorinen, T.; Jämsä, S.; Viitaniemi, P. The Effects of a Heat-Treatment on the Behavior of Extractives in Softwood Studied by FTIR Spectroscopic Methods. *Wood Sci. Technol.* **2003**, *37*, 109–115. [CrossRef]
28. Pizzi, A. *Wood Adhesives: Chemistry and Technology*; Marcel Dekker Inc.: New York, NY, USA, 1983; pp. 307–387. [CrossRef]
29. Kocaefe, D.; Poncsak, S.; Tang, J.; Bouazara, M. Effect of Heat-Treatment on the Mechanical Properties of North American Jack Pine: Thermogravimetric Study. *J. Mater. Sci.* **2009**, *45*, 681–687. [CrossRef]
30. Poncsák, S.; Shi, S.Q.; Kocaefe, D.; Miller, G. Effect of Thermal Treatment of Wood Lumbers on Their Adhesive Bond Strength and Durability. *J. Adhesion Sci. Technol.* **2007**, *21*, 745–754. [CrossRef]
31. Ayadi, N.; Lejeune, F.; Charrier, F.; Charrier, B.; Merlin, A. Color Stability of Heat-Treated Wood During Artificial Weathering. *Holz. Roh. Werkst.* **2003**, *61*, 221–226. [CrossRef]
32. Vidholdová, Z.; Reinprecht, L. The Colour of Tropical Woods Influenced by Brown Rot. *Forests* **2019**, *10*, 322. [CrossRef]
33. Hill, C.A.S. *Wood Modification: Chemical, Thermal and Other Processes*; John Wiley & Sons: Chichester, UK, 2006; pp. 100–127. [CrossRef]
34. Chaouch, M.; Pétrissans, M.; Pétrissans, A.; Gérardin, P. Use of Wood Elemental Composition to Predict Heat-Treatment Intensity and Decay Resistance of Different Softwood and Hardwood Species. *Polym. Degrad. Stab.* **2010**, *95*, 2255–2259. [CrossRef]
35. Weiland, J.J.; Guyonnet, R. Study of Chemical Modifications and Fungi Degradation of Thermally Modified Wood Using DRIFT Spectroscopy. *Holz. Roh. Werkst.* **2003**, *61*, 216–220. [CrossRef]

Article

Effects of Wood Moisture Content and the Level of Acetylation on Brown Rot Decay

Samuel L. Zelinka *, Grant T. Kirker, Amy B. Bishell and Samuel V. Glass

US Forest Service Forest Products Laboratory, Madison, WI 53726, USA; grant.kirker@usda.gov (G.T.K.); amy.b.bishell@usda.gov (A.B.B.); samuel.v.glass@usda.gov (S.V.G.)

* Correspondence: samuel.l.zelinka@usda.gov; Tel.: +1-608-231-9277

Received: 16 January 2020; Accepted: 5 March 2020; Published: 7 March 2020

Abstract: Acetylation is one of the most common types of wood modification and is commercially available throughout the world. Many studies have shown that acetylated wood is decay resistant at high levels of acetylation. Despite its widespread use, the mechanism by which acetylation prevents decay is still not fully understood. It is well known that at a given water activity, acetylation reduces the equilibrium moisture content of the wood cell wall. Furthermore, linear relationships have been found between the acetylation weight percent gain (WPG), wood moisture content, and the amount of mass loss in decay tests. This paper examines the relationships between wood moisture content and fungal growth in wood, with various levels of acetylation, by modifying the soil moisture content of standard soil block tests. The goal of the research is to determine if the reduction in fungal decay of acetylated wood is solely due to the reduction in moisture content or if there are additional antifungal effects of this chemical treatment. While a linear trend was observed between moisture content and mass loss caused by decay, it was not possible to separate out the effect of acetylation from fungal moisture generation. The data show significant deviations from previously proposed models for fungal moisture generation and suggest that these models cannot account for active moisture transport by the fungus. The study helps to advance our understanding of the role of moisture in the brown rot decay of modified wood.

Keywords: acetylated wood; wood modification; brown rot decay; fungal decay; soil block test; wood moisture content; soil bottle assays

---

1. Introduction

When wood is used in exterior applications it is subject to biodeterioration from wood decay fungi. Wood modifications, such as acetylation, are often used to protect wood from decay [1]. Unlike preservative treatments where the chemical treatments are targeted to repel or kill wood decay fungi, acetylation modifies many wood properties and a complete understanding of how acetylation prevents decay is a topic of active research. Ringman et al. reviewed the literature and concluded that diffusion inhibition through the reduction of void volume was the only mechanism that was consistent with all observed behaviors in modified wood [2]. Researchers have further proposed that diffusion likely occurs through regions of softened hemicelluloses and that acetylation may prevent decay by arresting diffusion by increasing the moisture-induced glass transition temperature of the hemicelluloses [3–5]. Recent work by Hunt et al. has confirmed that acetylated wood indeed has a higher relative humidity threshold for the diffusion of potassium ions [6]. Furthermore, Ringman et al. have suggested several possible alternative mechanisms for the inhibition of diffusion beyond cell wall plasticization [7].

Much of what is known about how treatments and wood modifications affect fungal decay comes from laboratory soil bottle tests, such as the American Wood Protection Association (AWPA) E-10 soil block assay, and the corresponding European tests, such as EN-113 [8,9]. In these tests, a fungal monoculture is cultivated in a sterile bottle with soil and a wood feeder strip and then a wood test block is added once the fungus is established. The efficacy of the treatment is typically examined by measuring the mass loss (ML) of the wood test block, which is calculated using the following equation:

$$ML = \frac{m_{od,\ pre} - m_{od,\ post}}{m_{od,\ pre}} \quad (1)$$

where m is the mass of the specimen and the subscripts "pre" and "post" refer to the mass before and after the soil block test, respectively, and the subscript "od" stands for oven dried. Note that in this paper, the mass, m, includes the added mass of the acetylation chemical, where applicable. In the standard soil block tests, the wood moisture content is not controlled, but instead the soil is preconditioned to a certain moisture content (specified as a percent of the water holding capacity of the soil), and the moisture content of the wood test block is affected by capillary uptake from the soil and feeder strip, water vapor sorption from the humidity within the soil bottle, the moisture produced from the breakdown of wood polymers, and the moisture translocated from the fungus.

Soil block tests have shown that increasing the level of acetylation results in lower mass loss [10–14]. Typically, the mass loss of the test block after the test is plotted as a function of the weight percent gain (WPG, a measure of the intensity of the wood modification). An example of such a plot is shown in Figure 1a from the thesis of Forster [10]. Note that the mass used in all calculations includes the mass of the treatment chemical. These plots are helpful for determining the level of modification at which decay is negligible; however, such plots cannot give us insight as to the mechanism by which the wood modifications provide protection. In addition to measuring the weight loss caused by decay, Forster [10] also measured the moisture content at the end of the decay test and normalized to the dry mass after degradation, which we show in Figure 1b. We can see that there is a strong correlation between the moisture content and the mass loss, and, in fact, when the wood moisture content at the end of the test is plotted against the mass loss (with the WPG plotted implicitly, Figure 1c), we see that the mass loss increases with the moisture content at the end of the test.

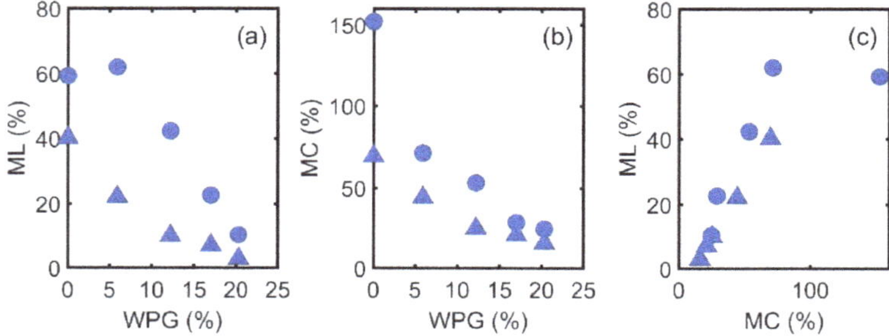

Figure 1. The interrelationship between the amount of acetylation, mass loss caused by decay, and wood moisture content at the end of the test. (**a**), mass loss as a function of weight percent gain. (**b**) final moisture content as a function of weight percent gain. (**c**) mass loss as a function of the final moisture content. The data replotted from Forster [10], ▲ = *Gloeophyllum trabeum* and •= *Coniophora puteana*.

The fact that the wood moisture content in soil bottle tests varies with acetylation level has important implications for understanding the mechanism of protection as well as understanding whether wood that appears to be decay resistant in soil bottle tests may be subject to decay under different environmental conditions [7]. Decay fungi are most active when the wood moisture content is between 40% and 80% moisture content (MC) [15]. In the absence of fungi, wood moisture content would depend on the water activity of the environment and the kinetics of water absorption [16]. In the soil block tests, the moisture content of the wood specimen would also depend on moisture generation from the breakdown of wood polymers by the fungus and active water translocation by the fungus. It has been shown that for untreated wood in the standard soil block tests, the wood moisture content increases during the first 3 weeks of the decay test and then levels off in an optimal range for fungal growth [17]. Measurements of this type are lacking for acetylated wood. Given the differences in moisture properties between acetylated and unmodified wood and the various phenomena that influence wood moisture content during decay tests, from the available literature data, it is unclear whether acetylation would provide protection against decay under higher wood moisture contents.

Thybring reviewed literature data and developed a model based upon assumed cell wall chemistry and structure that suggested that it is not surprising that wood moisture content and mass loss are correlated [18]. The model includes the theoretical amount of water produced by degradation of cellulose and hemicellulose, which is 0.56 g of water per 1 g of polymer, and is similar to an earlier model developed by Peterson and Cowling in the 1970s [19]. Thybring suggested that instead of plotting conventional moisture content as a function of mass loss, a "reduced moisture content" should instead be used where the MC is calculated with the original dry mass [18]. Since there are several definitions associated with "reduced moisture content", we use the nomenclature "MC_O", where the subscript stands for the moisture content calculated from the original dry mass, i.e.,

$$MC_o = \frac{m_{post} - m_{od,\ post}}{m_{od,\ pre}} \tag{2}$$

in this paper to describe the "reduced moisture content" described in the work of Thybring [18]. Thybring stated that MC_o should be directly proportional to the mass loss as the consumption of lignocellulosic biomass results in the production of water [18]. However, Thybring did not provide additional experimental tests to confirm whether this model can fully explain the effects of moisture dynamics on decay in acetylated wood.

In this paper, we attempt to separate the moisture exclusion effects of acetylation from other mechanisms of wood protection provided by acetylation. To do this, we examine the relationship between the soil moisture content and the wood moisture content at different levels of acetylation in the presence and absence of fungus. By adjusting the soil moisture content, we are able to test different levels of acetylation at the same wood moisture content in the absence of fungus. Furthermore, performing parallel experiments with and without fungus allow us to investigate the moisture generated by the fungus.

2. Materials and Methods

2.1. Materials

Southern pine (*Pinus spp.*) sapwood was examined. The southern pine was cut into 10 mm cubes ("mini-blocks"). Four different levels of acetylation were examined: 0 WPG (control, untreated), 8 WPG, 11 WPG, and 18 WPG. The acetylation was performed in a similar fashion to our previous work [6,20]. Wood was vacuum-impregnated with acetic anhydride and cured in an oven at 140 °C for various times to provide a range of WPGs. The specimens were then rinsed to remove unbound acetic acid. For each different test and treatment combination.

2.2. Soil Block Tests

With the exception of the soil moisture content, the soil block tests followed the AWPA E-10 standard [9]. The soil bottles were made by adding soil, water, and a feeder strip of southern pine sapwood to a flint glass jar with a metal lid and autoclaving it. In this study, *Rhodonia placenta* (Fr.) Niemelä, K.H. Larss. and Schigel 2005 (Mad-698-R) was examined. Following the autoclave, two 5-mm plugs from the leading edge of a fungus grown on 2% malt extract plus a 0.1% yeast extract agar plate incubated in a 30 °C, 70% RH room were placed in each bottle, one at each end of the feeder, and this was allowed to colonize the bottle for 3 weeks before introducing an autoclaved test block to the jar. The test was allowed to continue for 8 weeks, after which the blocks were removed. Prior to autoclaving the test block, an initial oven-dry weight was determined. After removal from the jar, the mycelia were removed, the test block was weighed, oven dried at 105 °C overnight, and reweighed. The mass loss caused by decay was determined by Equation (1) and the moisture content after the test (MC) was determined by the following equation:

$$MC = \frac{m_{post} - m_{od,\ post}}{m_{od,\ post}} \tag{3}$$

In some experiments, the wood moisture content without the fungus was examined. In these tests, the soil, water, feeder strip, and test block were all autoclaved at the same time and the bottle was disassembled after eight weeks and the block was examined.

2.3. Soil Water Holding Capacity

The E-10 standard specifies that the testing should utilize a soil moisture content that is 130% of the soil water holding capacity (SWHC). For our soil, collected from an agricultural site, the SWHC was found to be 31% soil moisture content (by mass, dry basis). SWHC was varied in the tests as one of the main variables.

In the control experiments, which were intended to follow the AWPA standard, the soil moisture content prior to soil bottle preparation was not independently verified and was assumed to be the same as the previous experiment. Therefore, it was inadvertently lower than expected and, as a result, the "standard" measurements were conducted with a soil moisture content of 38.3% MC or 124% SWHC instead of the desired 130%.

In addition to the tests at 124% SWHC, additional experiments were run at 120% SWHC and 140% SHWC on all levels of the acetylated wood.

2.4. Wood Moisture Content Preconditioning

In preliminary experiments, the starting moisture content of the wood block was examined as another variable. It is expected that the moisture content of the blocks will change throughout the test as the specimen comes into equilibrium with the soil. If the moisture uptake of the acetylated wood has different kinetics than untreated wood, preconditioning the wood may affect the final moisture content or the amount of decay exhibited by the block. Figure 2 shows data collected at 124% SWHC for blocks that had been autoclaved in the presence of liquid water to bring their starting moisture contents above 40% MC; the actual MCs varied between 40% and 55% MC. The starting moisture content was shown to have little effect on the amount of mass loss; a t-test with unequal variance (Welch test) showed no statistical differences between the mass loss for dry and pre-wetted blocks 0%, 8%, and 11% WPG ($p = 0.05$). The final moisture content of the blocks was shown to be statistically different ($p = 0.05$) for the 8% WPG and 18% WPG but not the 0% and 11% WPG conditions. Because preconditioning had such a small effect on the mass loss, except at the highest level of acetylation, it was not investigated further.

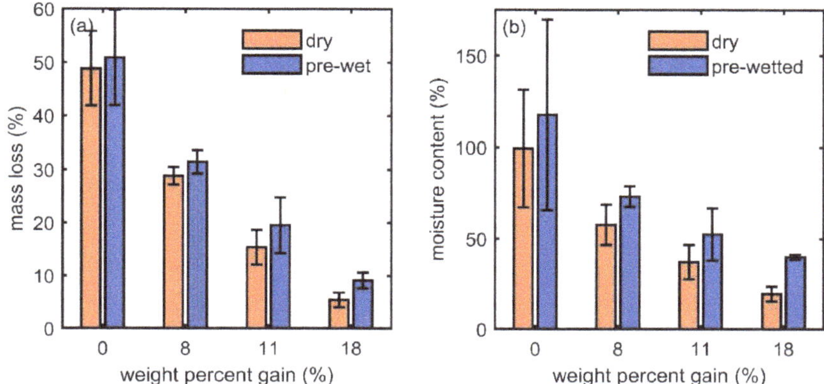

Figure 2. (**a**) mass loss and (**b**) moisture content as a function of percent weight loss from preliminary experiments where specimen pre-wetting was examined. The dry blocks had starting moisture contents (MCs) of less than 10%, and the pre-wetted blocks had starting moisture contents between 40% and 55%. Error bars represent the standard deviation from 5 replicates.

3. Results

3.1. Effect of Acetylation and Soil Moisture Content

In order to assess the effects of soil moisture on the final block MC, a preliminary trial was conducted using the standard E10 soil bottle setup only without fungi. We refer to the moisture content from this test as MC^{ster} and calculate it as:

$$MC^{ster} = \frac{m^{ster}_{post} - m^{ster}_{od,\ post}}{m^{ster}_{od,\ pre}} \tag{4}$$

where the superscript "*ster*" indicates sterile, uninoculated controls. Soil water holding capacities were varied at 110%, 120%, 140%, and 150% (Figure 3). A Tukey Honest Significance Difference (HSD) test was performed on the data. For each SWHC, the MC^{ster} of acetylated wood was significantly different (at p = 0.05) from the untreated controls, however, the amount of acetylation did not cause significant difference in MC^{ster}.

These results indicate that the standard soil bottles at 130% SWHC should achieve an MC that is well suited for fungal decay activity.

The effect of SWHC and acetylation on mass loss are presented in Figure 4. The amount of decay decreases with increasing levels of acetylation. This is not surprising as many others have demonstrated similar results [10]. Furthermore, the wood acetylated to 18% WPG showed almost no mass loss; this is in line with previous works suggesting that wood acetylated to 20% WPG is very decay resistant [11,21]. The most interesting results for this investigation, however, are the results of the 8% and 11% WPG. These two levels of acetylation show measurable amounts of decay but less decay than the untreated wood. These groups will later be used to examine the interactions between acetylation, wood moisture content, and susceptibility to decay.

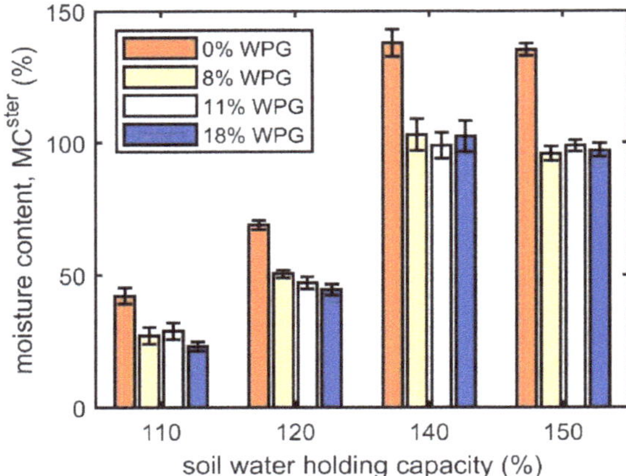

Figure 3. Effect of the soil water holding capacity on the wood moisture content after 8 weeks with no fungus present (MC^{ster}). The error bars represent the standard deviation from five replicates.

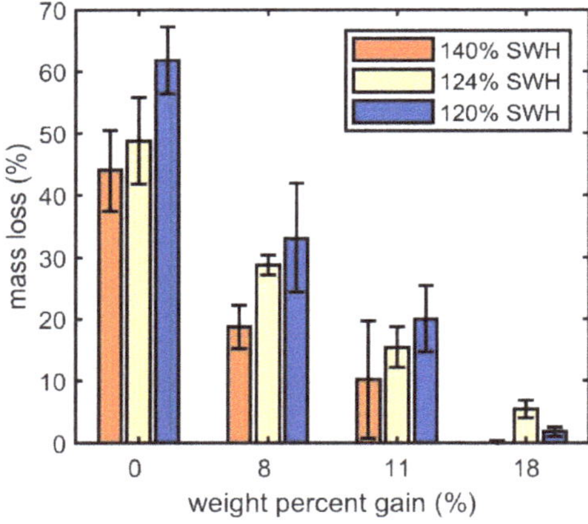

Figure 4. Percent weight loss caused by decay as a function of the soil water holding capacity and the level of acetylation. The error bars represent the standard deviation from five replicates.

The data in Figure 4 are also useful for understanding the effect of the soil moisture content on the decay fungi; for reference, the AWPA standard soil bottle test specifies 130% SWHC. For all treatments, the soil water holding capacity affected the mass loss caused by the fungus. It appears that for the three soil moisture contents examined, *Rhodonia placenta* is most active at 120% SWHC. For the 0%, 8%, and 11% WPG samples, the mass loss increases with the decreasing SWHC. For the samples acetylated to 18% WPG, the 124% SWHC exhibited a higher mass loss than the 120% SWHC group ($p = 0.01$), however, all the SWHCs had a mass loss of less than 10%.

3.2. Relationships Between Moisture Content and Mass Loss

A major motivation of this work was to understand the relationship between acetylation, wood moisture content, and decay. This is illustrated in Figure 5, which plots the mass loss against the post-test moisture content calculated three different ways. In all subfigures, the level of acetylation is plotted implicitly; the four data points correspond with the average moisture content and mass loss of the four levels of acetylation with the 18% WPG specimen being the closest to the origin and the control specimen having the highest moisture content and mass loss.

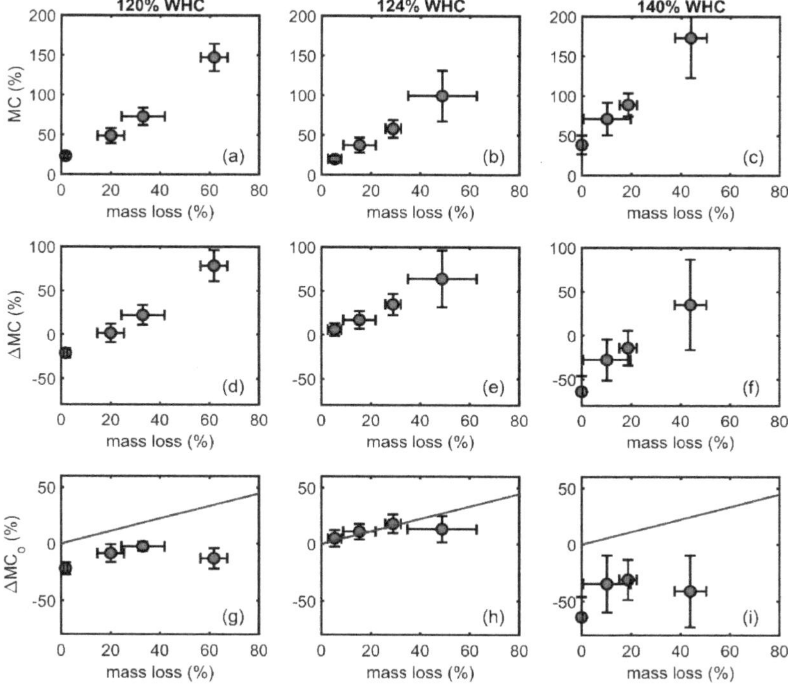

Figure 5. Relationship between mass loss and the moisture content at the end of the experiment calculated three different ways from Equations (3), (5), and (6) in rows 1, 2, and 3. The columns represent different soil water holding capacities. The error bars represent the standard deviation from five replicates.

In the first row, the data are plotted as a function of the moisture content at the end of the test calculated in the traditional manner (Equation (3)). A clear linear trend can be seen with higher mass losses having higher moisture content, similar to literature data (Figure 1c). Because the fungus that causes the mass loss itself produces water, it is not possible to determine from the data the first row whether the lower mass loss is a result of the lower wood moisture contents or whether the lower moisture contents are a result of the lower mass loss.

To examine the extra effect of fungal growth on the final moisture content, we plot "ΔMC" as a function of mass loss in the middle row of Figure 5. We define ΔMC using the following equation:

$$\Delta MC = MC - MC^{ster} = \frac{m_{post}^{inoc} - m_{od,post}^{inoc}}{m_{od,post}^{inoc}} - \frac{m_{post}^{ster} - m_{od,post}^{ster}}{m_{od,post}^{ster}} \tag{5}$$

where "*inoc*" indicates inoculated specimens and "*ster*" indicates sterile uninoculated controls. We are able to calculate ΔMC because we ran the identical experiment two times; once where wood was exposed to the soil bottle for 8 weeks in the absence of fungus and the second time where it was placed in the soil bottle for 8 weeks with the culture of *R. placenta*. In theory, ΔMC should represent the extra amount of moisture generated by the fungus beyond any changes in the wood moisture content that occur from being in contact with moist soil.

Two observations can be made from the ΔMC. The first is that all samples show a linearly increasing trend with mass loss. That is, higher amounts of fungal metabolism are associated with higher final moisture contents. This is not surprising as water is a byproduct of fungal metabolism. The slope of the line in the middle column of Figure 5 cannot, however, be directly related to the stoichiometry of the metabolic reaction since the slope is in units of water per gram of post-test dry mass and the metabolic reaction can only be understood in terms of grams of water per grams of polymer consumed (mass loss and moisture content both need to be expressed with respect to pre-test dry mass).

The second, and more surprising observation from the middle row of Figure 5 is that some of the data from the 120% and 140% SWHC experiments are negative. That is, in the experiments conducted with fungus, the moisture content was lower than in the experiments without the fungus. The reason for this behavior is not clear. One potential explanation of this result is that the block was potentially inhibited from reaching the higher MC due to physiological envelopment by the fungus, thereby reducing the overall block MC. The blocks in the soil bottle begin at less than 10% MC and get to 100% MC without fungus after 8 weeks of exposure in the test. This suggests that the fungus is either inhibiting the block from taking up moisture or physically excluding moisture from entering the block. Similar trends were also reported by Peterson and Cowling, where a durable species (Sitka spruce) exposed to *T. versicolor* exhibited a lower final moisture content (68%) after 8 weeks in comparison to an uninoculated block (98%) [19]. They concluded that *T. versicolor* can adjust the MC of the wood by the transfer of water through its own mycelium.

In short, both the plots of ΔMC and show linear trends with increasing fungal mass loss. While not surprising, it unfortunately not possible to separate out the moisture effects and acetylation from these plots. Thybring suggested that instead of examining the conventional moisture content, MC_O should instead be analyzed. ΔMC_O represents the moisture generated by the fungus and, in theory, should be linearly related to the mass loss with a slope of 0.56 g g^{-1} [18].

We plot ΔMC_O as a function of mass loss in the third row of Figure 5. We define ΔMC_O using the following equation:

$$\Delta MC_O = MC_O - MC^{ster} = \frac{m^{inoc}_{post} - m^{inoc}_{od,post}}{m^{inoc}_{od,pre}} - \frac{m^{ster}_{post} - m^{ster}_{od,post}}{m^{ster}_{od,post}} \quad (6)$$

where the superscripts and subscripts have the same meanings as Equations (1)–(5). ΔMC is calculated here the same way as Thybring [18]. The theoretical slope of 0.56 g g^{-1} is shown in each of the subfigures in the third row for reference. The data roughly follow the theoretical line for the 124% SWHC condition, however, the data at 120% and 140% SWHC deviate strongly from the expected values.

It is worth further discussion of why Thybring suggested that ΔMC_O should be proportional to the mass loss and why this was not observed in some of our experiments. The calculations of Thybring assume that biomass is converted to water and carbon dioxide and that this extra water is maintained within the wood. Clearly, this is true instantaneously as the stoichiometry of the reaction dictates the amount of water generated by the biodegradation. However, the composition of the polymers changes during degradation becoming less hydrophilic and as degradation progresses there are more fungal hyphae in the wood which are hydrated by the fungi. Furthermore, the post-test moisture content is measured at the end of an eight-week experiment and both the fungus and the wood are passively and actively transporting moisture throughout the experiment. Peterson and Cowling have already demonstrated that fungi actively lower the wood moisture content in soil bottle experiments on unmodified wood [19]. While the relationship between ΔMC_O and mass loss must be true instantaneously, this analysis does not account for moisture redistribution and therefore cannot reliably predict the moisture content at the end of the test. Therefore, one possible explanation of the data is that some of the moisture generated by the fungus was translocated by the fungus before the end of the test. If that were the case, however, it is unclear why the data for the 124% SWHC experiment closely follow the expected behavior.

Although every effort was made to examine the effect of wood moisture content on mass loss independently of fungal moisture generation, it was not possible to separate out the effect of acetylation from fungal moisture generation and to a lesser extent fungal activity in different soil moisture contents. While higher amounts of acetylation were correlated with both lower moisture contents and lower mass losses, it is impossible to fully separate the effect of reduced hygroscopicity on decay resistance from these experiments.

4. Conclusions

The pre-wetting of the wood was examined as a potential way to better understand the effect of the moisture excluding properties of acetylation on the decay process. However, pre-wetting had only minor effects on the mass loss except for the 18% WPG samples.

As observed in previous studies, the moisture content at the end of the test is linearly related to the amount of mass loss. However, it was not possible to separate out the moisture generation from the fungus from the reduced hygroscopicity of the wood for acetylated samples.

The parameter ΔMC_O, proposed by Thybring represents the moisture generated by the fungus and should in theory be linearly proportional to mass loss [18]. While the data roughly follow the theoretical line for the 124% SWHC condition, the data at 120% and 140% SWHC deviate from the expected values. Similar deviations were also observed by Peterson and Cowling who observed that ΔMC_O overestimated moisture generation due to the fungus by roughly three times [19].

While the reduction in wood moisture content caused by acetylation may play a role in the decay resistance of acetylated wood, the fungus actively controls the moisture in the wood and it is not possible to independently control the wood moisture content during the decay process. Because of active water transport by the fungi, it is not possible to understand the role of moisture reduction in the decay process of acetylated wood by altering the soil bottle moisture content.

Author Contributions: Conceptualization, S.L.Z. and G.T.K..; methodology, A.B.B., S.L.Z., and G.T.K. formal analysis, S.L.Z, A.B.B., and S.V.G.; investigation, A.B.B.; writing—original draft preparation, S.L.Z.; writing—review and editing, S.L.Z., G.T.K., A.B.B., and S.V.G.; visualization, S.L.Z., A.B.B., and S.V.G. All authors have read and agreed to the published version of the manuscript.

Acknowledgments: We thank Forest Products Laboratory chemist Linda Lorenz for acetylating the samples. This work was inspired through discussions with Rebecka Ringman and Annica Pilgård of RISE (Sweden), and further refined through heated debates with Emil Thybring of the University of Copenhagen.

Conflicts of Interest: The authors declare no conflict of interest.

References

1. Hill, C.A. *Wood Modification: Chemical, Thermal and other Processes*; John Wiley & Sons: West Sussex, UK, 2006; Volume 5.
2. Ringman, R.; Pilgård, A.; Brischke, C.; Richter, K. Mode of action of brown rot decay resistance in modified wood: A review. *Holzforschung* **2014**, *68*, 239–246. [CrossRef]
3. Jakes, J.E.; Plaza, N.; Stone, D.S.; Hunt, C.G.; Glass, S.V.; Zelinka, S.L. Mechanism of transport through wood cell wall polymers. *J. For. Prod. Ind.* **2013**, *2*, 10–13.
4. Jakes, J.E.; Hunt, C.G.; Zelinka, S.L.; Ciesielski, P.N.; Plaza, N.Z. Effects of Moisture on Diffusion in Unmodified Wood Cell Walls: A Phenomenological Polymer Science Approach. *Forests* **2019**, *10*, 1084. [CrossRef]
5. Zelinka, S.L.; Ringman, R.; Pilgård, A.; Thybring, E.E.; Jakes, J.E.; Richter, K. The role of chemical transport in the brown-rot decay resistance of modified wood. *Int. Wood Prod. J.* **2016**, *7*, 66–70. [CrossRef]
6. Hunt, C.G.; Zelinka, S.L.; Frihart, C.R.; Lorenz, L.; Yelle, D.; Gleber, S.-C.; Vogt, S.; Jakes, J.E. Acetylation increases relative humidity threshold for ion transport in wood cell walls–A means to understanding decay resistance. *Int. Biodeterior. Biodegrad.* **2018**, *133*, 230–237. [CrossRef]

7. Ringman, R.; Beck, G.; Pilgård, A. The Importance of Moisture for Brown Rot Degradation of Modified Wood: A Critical Discussion. *Forests* **2019**, *10*, 522. [CrossRef]
8. Anon. *CEN EN 113—Wood Preservatives Test Method for Determining the Protective Effectiveness against Wood Destroying Basidiomycetes Determination of the Toxic Values*; CEN—European Committee for Standardization: Brussels, Belgium, 1996; p. 28.
9. Anon. *AWPA E-10: Standard Method for Testing of Wood Preservatives by Laboratory Soil-Block Cultures*; American Wood Protection Association: Selma, AL, USA, 2012.
10. Forster, S. *The Decay Resistance of Chemically Modified Softwood*; University of Wales: Bangor, UK, 1998.
11. Brelid, P.L.; Simonson, R.; Bergman, Ö.; Nilsson, T. Resistance of acetylated wood to biological degradation. *Holz als Roh-und Werkstoff* **2000**, *58*, 331–337. [CrossRef]
12. Hill, C.A.; Curling, S.F.; Kwon, J.H.; Marty, V. Decay resistance of acetylated and hexanoylated hardwood and softwood species exposed to Coniophora puteana. *Holzforschung* **2009**, *63*, 619–625. [CrossRef]
13. Meyer, L.; Brischke, C.; Treu, A.; Larsson-Brelid, P. Critical moisture conditions for fungal decay of modified wood by basidiomycetes as detected by pile tests. *Holzforschung* **2016**, *70*, 331–339. [CrossRef]
14. Rowell, R.M.; Ibach, R.E.; McSweeny, J.; Nilsson, T. Understanding decay resistance, dimensional stability and strength changes in heat-treated and acetylated wood. *Wood Mater. Sci. Eng.* **2009**, *4*, 14–22. [CrossRef]
15. Griffin, D. Water potential and wood-decay fungi. *Annu. Rev. Phytopathol.* **1977**, *15*, 319–329. [CrossRef]
16. Zelinka, S.L.; Glass, S.V.; Boardman, C.R.; Derome, D. Moisture storage and transport properties of preservative treated and untreated southern pine wood. *Wood Mater. Sci. Eng.* **2016**, *11*, 228–238. [CrossRef]
17. Kirker, G.T.; Bishell, A.B.; Zelinka, S.L. Electrical properties of wood colonized by Gloeophyllum trabeum. *Int. Biodeterior. Biodegrad.* **2016**, *114*, 110–115. [CrossRef]
18. Thybring, E.E. Water relations in untreated and modified wood under brown-rot and white-rot decay. *Int. Biodeterior. Biodegrad.* **2017**, *118*, 134–142. [CrossRef]
19. Peterson, C.A.; Cowling, E.B. Influence of Various Initial Moisture Contents on Decay of Sitka Spruce and and Sweetgum Sapwood by Polyporus versicolor in the Soil-Block Test. *Phytopathology* **1973**, *63*, 235–237. [CrossRef]
20. Passarini, L.; Zelinka, S.L.; Glass, S.V.; Hunt, C.G. Effect of weight percent gain and experimental method on fiber saturation point of acetylated wood determined by differential scanning calorimetry. *Wood Sci. Technol.* **2017**, *51*, 1291–1305. [CrossRef]
21. Tarkow, H. *Decay Resistance of Acetylated Balsa*; USDA Forest Service, Forest Products Laboratory: Madison, WI, USA, 1945.

Article

Tensile and Impact Bending Properties of Chemically Modified Scots Pine

Susanne Bollmus *, Cara Beeretz and Holger Militz

Wood Biology and Wood Products, University of Goettingen, Buesgenweg 4; D-37077 Goettingen, Germany; cara.beeretz@gmail.com (C.B.); holger.militz@uni-goettingen.de (H.M.)

* Correspondence: susanne.bollmus@uni.goettingen.de; Tel.: +49-551-392-9514

Received: 28 November 2019; Accepted: 6 January 2020; Published: 9 January 2020

Abstract: This study deals with the influence of chemical modification on elasto-mechanical properties of Scots pine (*Pinus sylvestris* L.). The elasto-mechanical properties examined were impact bending strength, determined by impact bending test; tensile strength; and work to maximum load in traction, determined by tensile tests. The modification agents used were one melamine-formaldehyde resin (MF), one low molecular weight phenol-formaldehyde resin, one higher molecular weight phenol-formaldehyde resin, and a dimethylol dihydroxyethyleneurea (DMDHEU). Special attention was paid to the influence of the solution concentration (0.5%, 5%, and 20%). With an increase in the concentration of each modification agent, the elasto-mechanical properties decreased as compared to the control specimens. Especially impact bending strength decreased greatly by modifications with the 0.5% solutions of each agent (by 37% to 47%). Modification with DMDHEU resulted in the highest overall reduction of the elasto-mechanical properties examined (up to 81% in work to maximum load in traction at 20% solution concentration). The results indicate that embrittlement is not primarily related to the degree of modification depended on used solution concentration. It is therefore assumed that molecular size and the resulting ability to penetrate into the cell wall could be crucial. The results show that, in the application of chemically modified wood, impact and tensile loads should be avoided even after treatment with low concentrations.

Keywords: wood modification; melamine-formaldehyde resin; phenol-formaldehyde resin; DMDHEU; impact bending strength; tensile strength

---

1. Introduction

Among others the main objectives of chemical modifications of wood are to improve dimensional stability and resistance against wood-decaying fungi in order to offer an alternative to tropical hardwoods. The modification may, however, also cause undesirable changes of other wood characteristics, mainly elasto-mechanical properties. The reasons why chemical modification systems influence biological and physical properties of wood have been the subject of numerous studies [1–4]. There is still a lack of knowledge about the mode of action of the resins. However, mechanisms via which impregnation modifications work according to Hill [5] are: (1) permanent swelling by incorporating the resin in the cell wall structure (bulking); (2) moisture content is reduced; (3) cell wall pores with a size of 2–4 nm are blocked and it is assumed that a chemical reactions between modification agents and cell wall polymers (cross-linking) occurs.

Impregnation modifications based on reactive resins have been shown to result in embrittlement of wood [6–8]. The embrittlement is most evident in a decrease of tensile properties and dynamic strength properties like impact bending strength [9,10].

Studies have shown that a treatment with melamine-formaldehyde (MF) resin, besides providing decay resistance and dimensional stability, increases compressive strength and hardness, while decreasing tensile and impact bending strength compared to untreated wood [11–13].

As early as the 1930s, Stamm and Seborg [7] modified wood with phenol-formaldehyde (PF) resins and thereby enhanced the dimensional stability as well as the durability of the wood. Bicke et al. [9] found bending strength and modulus of elasticity in bending of plywood increased after modification with a PF-resin. At the same time, the impact bending strength decreases. Alike modification with MF-resins, tensile strength of wood modified with PF-resins also exhibits a lower tensile strength [14]. Furuno, Imamura, and Kajita [15] as well as [16] examined the influence of the molecular weight of the PF-resin on its ability to penetrate the wooden cell wall using microscopy. They stated that, with an increase of the molecular weight, a higher share of the resins remained in the cell wall lumen.

It has been shown that wood modification with dimethylol dihydroxyethyleneurea (DMDHEU) increases resistance against wood decaying fungi, dimensional stability, compressive strength, and hardness while reducing tensile and impact bending strength [6,17–19]. Bollmus [20] showed that wood modification with DMDHEU causes a large reduction of the impact bending strength already at a very low concentration of 0.5%.

In this study, the influence of modifications with curing resins on the most susceptible elasto-mechanical properties was examined. In addition to tensile strength and impact bending strength, work to maximum load in traction of modified wood was investigated in order to be able to compare tensile and dynamic bending properties. Four different modification systems were applied in three different concentrations to Scots pine sapwood.

The main objectives of this study were to investigate the impact of (1) the concentration of the modification solution and (2) the modification agent itself on impact bending strength, tensile strength and work to maximum load in traction.

2. Materials and Methods

2.1. Specimens and Modification Agents

The specimens were made from sapwood of Scots pine (*Pinus sylvestris* L.). The modification was conducted on specimens of the dimensions 28 × 28 × 470 mm^3 (radial × tangential × axial) for tensile tests and 10 × 10 × 150 mm^3 for impact bending tests. The agents for modification were a methyl-etherified melamine formaldehyde resin (MF), a low molecular weight phenol-formaldehyde resin (Phenol1), a higher molecular weight phenol-formaldehyde resin (Phenol2) as well as dimethylol dihydroxyethyleneurea (DMDHEU, Table 1). Each of these modification agents was applied at concentrations (mass-%) of 0.5%, 5%, and 20%. Untreated specimens functioned as controls.

Table 1. Properties of stock resins according to manufacturer specifications.

Resin	pH-Value	Solid Content (%)	Molar Weight (g/mol)	Solvent	Catalyst	Additive
MF	10–11	73.5	840	Water		Triethanolamin, 1% *
Phenol1	9.2	43.5	159	Water		
Phenol2	8–10	54.9	452	20% Ethanol Water mixture		
DMDHEU	5–6	71.8		Water	Magnesium Nitrate, 2% *	

* Concentration based on resin.

2.2. Oven-Dry Density

Specimens were sorted according to their oven-dry density before modification in order to minimize the influence of the density on the examined elasto-mechanical properties. Specimens for the tensile test showed a maximum deviation from the median oven-dry density of ±15%. Specimens for the impact bending test showed a maximum deviation from the median oven-dry density of ±5%.

2.3. Equilibrium Moisture Content (EMC)

Specimens were conditioned at 20 °C and 65% relative humidity (RH) in order to minimize the influence of the moisture content on the examined elasto-mechanical properties. The EMC of control specimens was calculated as

$$EMC = \frac{(m_1 - m_0)}{m_0} \times 100\ (\%) \quad (1)$$

where EMC = equilibrium moisture content, %; m_1 = mass of untreated, conditioned specimen, g; m_0 = mass of untreated, oven-dry specimen, g.

There are two possible reference values to calculate EMC of modified wood. Material moisture content MC is defined as the proportion of moisture of wood based on the dry mass of the modified wood. Reduced equilibrium moisture content EMC_R is defined as the proportion of moisture in wood based on the oven-dry mass of the untreated wood, thus eliminating the influence of the weight added by the resin.

$$MC = \frac{(m_3 - m_2)}{m_2} \times 100\ (\%) \quad (2)$$

$$EMC_R = \frac{(m_3 - m_2)}{m_0} \times 100\ (\%) \quad (3)$$

where MC = material moisture content, %; m_3 = mass of modified, conditioned specimen, g; m_2 = mass of modified, dry specimen, g; EMC_R = reduced equilibrium moisture content, %; m_0 = mass of untreated, oven-dry specimen, g.

EMC_R of treated and EMC of untreated specimens were reached if the mass has not changed more than 0.1% within an interval of 24 h after storage at 20 °C and 65% RH.

2.4. Modification

The modification was performed identically for all modification agents. The first modification step, the impregnation, was conducted in a vacuum pressure impregnation plant. The three phases of the impregnation were (a) vacuum (−0.096 Mpa, 1 h), (b) pressure (1.2 Mpa, 2 h), and (c) diffusion (no pressure, 1 h). The second modification step, the curing, was conducted by slowly increasing the temperature up to 120 °C in a drying oven using the following drying scheme (Table 2).

Table 2. Drying scheme for modification.

Temperature (°C)	Climate Chamber/Drying Oven	Time (h)
20	climate chamber	168
40	drying oven	4
80	drying oven	4
103	drying oven	4
120	drying oven	36

Specimens for tensile testing were further processed into a bone-like shape only after modification in order to achieve a high manufacturing accuracy.

2.5. Weight Percent Gain (WPG)

The deposition of the modification agent in the wood caused a permanent weight percent gain. This is an indicator of the degree of modification. The WPG was examined in each specimen after the curing and calculated as

$$WPG = \frac{(m_2 - m_0)}{m_0} \times 100\ (\%) \quad (4)$$

where WPG = weight percent gain, %; m_2 = mass of modified, dry specimen, g; m_0 = mass of untreated, dry specimen, g.

2.6. Elasto-Mechanical Testing

Specimens that were warped or showed cracks after modification were excluded, which lead to different numbers of specimens. All specimens were conditioned at 20 °C and 65% RH before testing.

2.6.1. Impact Bending Strength

Impact bending strength ω was tested with the hammer method (load of XXXJ) in tangential direction following DIN 52 189-1 [21] by means of dynamic impact bending tests. The specimen dimensions were deviant from the standard 10 × 10 × 150 mm^3. The tests were performed using the Resil Impactor (CEAST) with a supporting width of 115 mm. Numbers of specimens varied between 23 and 37 (Table 3).

$$\omega = \frac{1000 \times W}{a \times b} \left(\mathrm{kJ/\ m^2}\right) \tag{5}$$

where ω = impact bending strength, kJ/ m^2; W = work necessary to break the specimen, J; a = height of conditioned specimen, mm; b = width of conditioned specimen, mm.

Table 3. Number of specimens depending on the modification agent and concentration.

	Number of Specimens							
	Impact Bending Specimens				**Tensile Test Specimens**			
Treatment	Control	0.5%	5%	20%	Control	0.5%	5%	20%
Control	23				35			
MF		37	37	37		33	33	33
Phenol1		36	37	37		33	31	33
Phenol2		37	37	36		33	33	33
DMDHEU		37	37	36		33	33	32

2.6.2. Tensile Strength Parallel to Grain and Work to Maximum Load in Traction

Tensile strength β_Z was tested parallel to grain following DIN 52 188 [22]. The tests were performed using a Zmart.Pro with a 100 kN load cell (ZWICK ROELL, Ulm, Germany). Specimens were of 470 mm length. The clamping surface was 50 × 50 mm^2 with a height of 15 mm. The tapered part of the specimens showed a height a (tangential) of 6 mm and a width b (radial) of 20 mm in the area of the smallest cross-section (numbers of specimens varied between 31 and 35) (Table 3).

The software TestExpert II (ZWICK ROELL, Ulm, Germany) calculated the tensile strength β_Z according to the formula

$$\beta_Z = \frac{F_{max}}{a \times b} \left(\mathrm{N/mm^2}\right) \tag{6}$$

where β_Z = tensile strength, N/mm^2; F_{max} = maximum force necessary to tear the specimen, N; a = height of conditioned specimen in the area of the smallest cross-section, mm; b = width of conditioned specimen in the area of the smallest cross-section, mm.

Strain to failure (mm) was measured by the VideoExtens camera system (ZWICK ROELL, Ulm, Germany). Work to maximum load in traction (kJ/m^2) was calculated by TestExpert II (ZWICK ROELL, Ulm, Germany) as the integral of the stress–strain curve from the origin to the maximum force F_{max}.

2.7. Statistical Analysis

Each group of modified specimens was tested for significant difference in contrast to the control group using ANOVA and Tukey HSD tests at an error level of $\alpha = 0.05$. Both ANOVA and Tukey HSD are parametric hypothesis tests and as such are based on the assumptions of normal distribution and of homogeneity of variances. Most groups fulfilled the criterion of normal distribution, some did not fulfill the criterion of homogeneity of variances. However, homogeneity of variance is considered less important for balanced samples. As the numbers of specimens were balanced (Table 3) and due to the higher power of parametric tests, it was decided to apply parametric hypothesis test (i.e., ANOVA and Tukey HSD).

3. Results and Discussion

3.1. Density and Moisture Content

Density and moisture content (below fiber saturation) of native wooden specimens have a strong influence on their elasto-mechanical properties [23]. In this study, density and moisture content were examined based on the impact bending specimens.

Table 4 shows the oven-dry density of the specimens. Since the specimens were sorted according to their oven-dry density before modification, the maximum deviation from the mean oven-dry density within each group was only ±15% for tensile specimens and ±5% for impact bending specimens.

Table 4. Native and modified oven-dry density (g/cm^3) of impact bending specimens depending on the modification agent and the concentration (mean).

	Native Oven-Dry Density (kg/m^3)				Modified Oven-Dry Density (k g/m^3)		
Treatment	Control	0.5%	5%	20%	0.5%	5%	20%
Control	471.76						
Melamine		476.23	471.74	474.74	490.43	493.78	598.35
Phenol1		475.11	470.06	470.48	470.38	490.58	595.31
Phenol2		469.33	471.25	478.82	465.50	489.91	601.02
DMDHEU		472.92	475.77	470.89	467.56	488.44	578.68

Material moisture content (MC) decreased with increasing intensity of the modification (Figure 1a). This was to be expected as the additional weight due to the modification changed the basis on which the MC is calculated. Reduced equilibrium moisture content EMC_R, however, avoids this effect by calculating the moisture content on the basis of the dry weight before modification [5,17]). In contrast to MC, EMC_R decreased only after modification with either one of the phenol-resins (Figure 1b). Since EMC_R values varied between 8.9% and 12.8%, an influence of the moisture content on the examined elasto-mechanical properties cannot be eliminated.

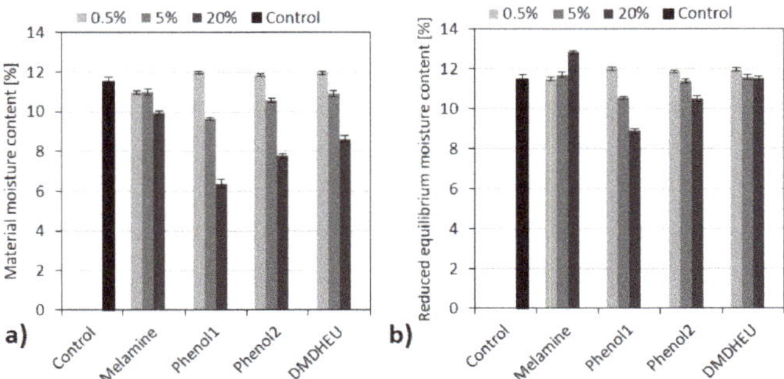

Figure 1. (a) Material moisture content MC (%) and (b) reduced equilibrium moisture content EMC_R (%) at 20 °C and 65% RH of impact bending specimens depending on the modification agent (mean and SD).

Hosseinpourpia et al. [24] stated that modification with a phenol-resin (molecular weight of 191 g/mol) decreased EMC_R stronger than modification with MF. It is therefore assumed that the molecular weight and therefore the cell wall penetration, irrespective of the resin, has a great influence on the penetration and thus on wood properties. Xie [25] reported a slightly increased EMC_R also after modification with DMDHEU.

3.2. Weight Percent Gain (WPG)

Figure 2 shows an increasing WPG with increasing concentration for each modification agent tested. Modifications with 0.5% solutions of both phenol-resins and DMDHEU led to negligible WPG values of less than 0.6%. The highest WPG (ca. 39%) was caused by the lower molecular weight Phenol1 at a concentration of 20%. There was no clear difference in the WPG between smaller impact bending specimens and larger tensile specimens (slats) which demonstrates good penetration into wood species.

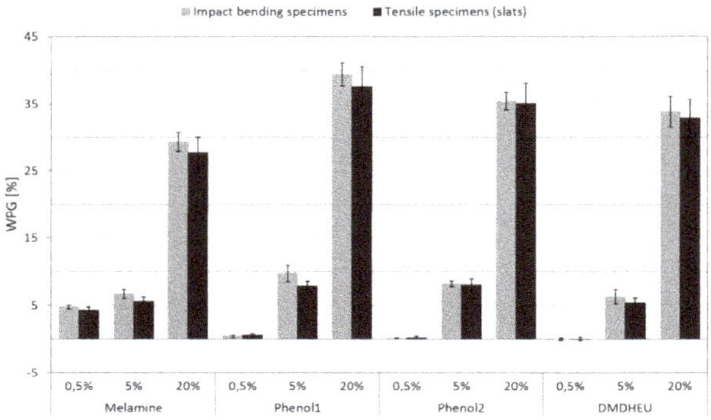

Figure 2. Weight Percent Gain (%) of impact bending and tensile specimens depending on modification agent and concentration (mean and SD).

3.3. Elasto-Mechanical Properties

3.3.1. Impact Bending Testing

The impact bending strength of defect-free control specimens (Table 5) was rather low. Wagenführ [26] states values of 15 ... 40 ... 130 kJ/m^2 for Scots pine. An explanation for the low values might lie in the smaller specimen dimensions and shorter supporting width of the testing device than stated in DIN 25 189-1 [21]. According to Krech [27] impact bending strength increases with increasing specimen dimensions and increasing supporting width.

Table 5. Impact bending strength (kJ/m^2) and change compared to control (%) depending on the modification agent and the concentration (mean ± SD).

	Impact Bending Strength						
Treatment	Control	0.5%		5%		20%	
	(kJ/m^2)	(kJ/m^2)	(%)	(kJ/m^2)	(%)	(kJ/m^2)	(%)
Control	18.8 ± 3.3						
MF		11.9 ± 2.0	−37 [1]	10.8 ± 2.5	−43 [1]	8.8 ± 3.0	−53 [1]
Phenol1		9.9 ± 2.1	−47 [1]	7.1 ± 2.0	−62 [1]	5.8 ± 1.2	−69 [1]
Phenol2		10.4 ± 1.9	−44 [1]	8.6 ± 2.3	−54 [1]	7.7 ± 2.4	−59 [1]
DMDHEU		10.9 ± 3.1	−42 [1]	7.3 ± 2.2	−61 [1]	5.0 ± 1.8	−74 [1]

[1] indicates that the value is significantly different from control at $\alpha = 0.05$.

Modified specimens exhibited significantly lower impact bending strength values than control specimens (Table 4). The values decreased with increasing concentrations of the agent for each modification. Specimens modified with 0.5% solutions already showed a decrease in the impact bending strength of up to 47% (Phenol1). Other authors also found the impact bending strength of chemically modified wood to be reduced [8,19]

Modification caused an increase of density due to deposition of the resin (Table 3). Results from the impact bending test indicate, that this increase in density did not have the same effect on elasto-mechanical properties as higher density in native wood (Figure 3).

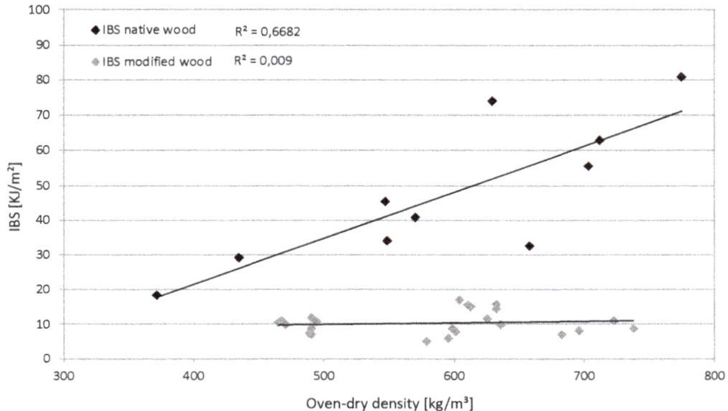

Figure 3. Relationship between impact bending strength (IBS) (kJ/m^2) and oven-dry density (kg/m^3) of native wood species (data from Brischke [28]) and differently modified wood with varying WPG.

Impact bending strength depends on the force applied as well as the deflection the specimen performs before breaking. Modified specimen absorbed similar amounts of force as control specimens (Figure 4). The deflection of modified specimens at F_{max} showed a decreasing tendency with increasing solution concentrations for each modification agent. This shows that the reduction of impact bending strength is mainly caused by the reduction of resiliency as the modification increases the stiffness of the specimens [8].

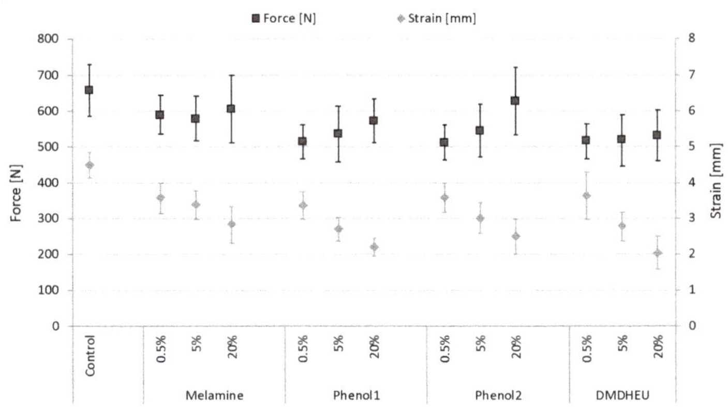

Figure 4. Maximum force (N) and deflection (mm) in impact bending depending on modification agent and concentration (mean and SD).

3.3.2. Tensile Testing Parallel to Grain

Tensile strength parallel to grain of control specimens was within the range [26] stated for Scots pine (35 ... 104 ... 196 N/mm^2). Modified specimens showed lower tensile strength values than control specimens (Table 6). The values decreased with increasing concentrations of the agents. The strongest decrease was observed in the specimens modified with DMDHEU (51%). Other authors also found the tensile strength of chemically modified wood to be reduced [9,13,29].

Table 6. Tensile strength (N/mm^2) and change compared to control (%) depending on the modification agent and the concentration (mean ± SD).

	Tensile Strength						
Treatment	Control	0.5%		5%		20%	
	(N/mm^2)	(N/mm^2)	(%)	(N/mm^2)	(%)	(N/mm^2)	(%)
Control	93.9 ± 25.1						
MF		79.3 ± 22.2	−16 [0]	75.6 ± 23.7	−19 [1]	69.1 ± 18.8	−26 [1]
Phenol1		75.7 ± 22.9	−19 [1]	72.1 ± 17.9	−23 [1]	59.0 ± 13.2	−37 [1]
Phenol2		85.8 ± 22.4	−9 [0]	70.6 ± 18.1	−25 [1]	70.7 ± 13.7	−25 [1]
DMDHEU		84.7 ± 21.3	−10 [0]	60.6 ± 18.6	−35 [1]	45.6 ± 11.4	−51 [1]

[1] indicates that the value is significantly different from control at $\alpha = 0.05$, [0] indicates that the value is not significantly different from control at $\alpha = 0.05$.

Modified specimens exhibited significantly lower values of work to maximum load in traction than control specimens (Table 7). For each modification, values decreased with increasing concentrations of the modification agents. Similar to impact bending strength, specimens modified with 0.5% solutions already showed a significant decrease in the work to maximum load in traction of up to 36% (Phenol1). Again, the strongest decrease over-all was observed in the specimens modified with DMDHEU (81%).

Table 7. Work to maximum load in traction (kJ/m^2) and change compared to control (%) depending on the modification agent and the concentration (mean ± SD).

	Work to Maximum Load in Traction (kJ/m^2)						
Treatment	Control	0.5%		5%		20%	
	(kJ/m^2)	(kJ/m^2)	(%)	(kJ/m^2)	(%)	(kJ/m^2)	(%)
Control	28.2 ± 9.6						
MF		19.3 ± 7.2	−32 [1]	17.0 ± 6.9	−40 [1]	13.5 ± 4.3	−52 [1]
Phenol1		17.9 ± 7.0	−36 [1]	13.7 ± 4.7	−51 [1]	8.8 ± 2.4	−69 [1]
Phenol2		20.4 ± 7.3	−27 [1]	14.4 ± 4.5	−49 [1]	11.7 ± 3.2	−58 [1]
DMDHEU		21.0 ± 8.4	−25 [1]	10.2 ± 4.6	−64 [1]	5.4 ± 2.3	−81 [1]

[1] indicates that the value is significantly different from control at $\alpha = 0.05$.

In contrast to results from impact bending tests, both F_{max} and maximum strain in traction were reduced with increasing concentrations of each agent (Figure 5).

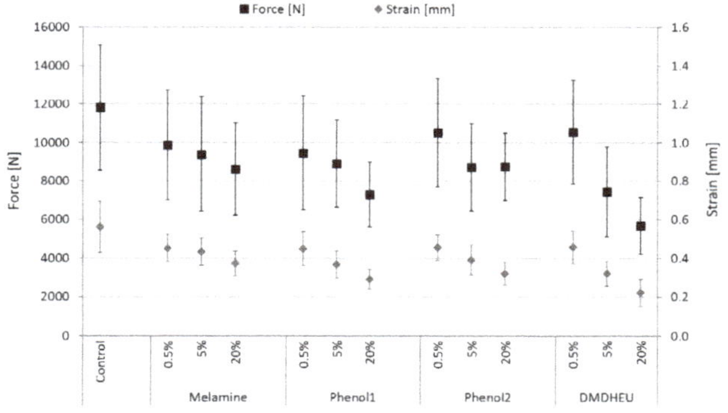

Figure 5. Maximum force (N) and strain (mm) in traction depending on modification agent and concentration (mean and SD).

Tensile properties are mainly determined by the properties of the single cellulose fiber [23,30]. Tests among others by [19] suggest that chemical modification reduces tensile properties of the fibers by embrittlement as a result of deposition of the agent in the fiber cell walls.

3.4. Influence of (Low) Concentration

The influence of the solution concentration was evident in all examined elasto-mechanical properties and for each modification agent. The higher the concentration, the stronger was the reduction. As stated by Bollmus [20] for DMDHEU, this study affirmed the strong influence of chemical modifications, especially on impact bending strength, and work to maximum load in traction already at a concentrations as low as 0.5%.

As the curing temperature was the same regardless of the modification concentration, it thus might be a possible explanation for the strong influence of the low concentration modification on the elasto-mechanical properties. Fengel and Wegener [31] as well as Nicholas and Williams [32] state that under certain circumstances (duration, pressure, moisture content) temperatures above 100 °C might lead to mass losses and reduced elasto-mechanical properties. Thus, it was assumed that the curing temperature of 120 °C, applied for 24 h, could have caused slight decreases in elasto-mechanical properties. However, it is highly unlikely that this curing process caused a decrease of up to 47% of impact bending strength (Table 3). Further investigations should be conducted on the strong influence of very low modification concentrations on elasto-mechanical properties of wood.

3.5. Influence of Modification Agents

Reasons for reduced elasto-mechanical properties of chemically modified wood have been discussed widely [23,33,34]. In which manner, a modification agent affects the elasto-mechanical properties of wood is determined by various factors. The most important factors are: penetration in to the cell wall, cross-linking, pH value, and formaldehyde content. To assess the influence of each factor separately is very difficult, and as the factors might inter- or counteract, the topic is very complex.

One reason for the strong influence of Phenol1 on the elasto-mechanical properties already at a low concentration could be its low molecular weight compared to Phenol2 and MF. This enables the resin to penetrate into the cell wall through nano-pores, whereas only a fraction of those modification systems with a higher molecular weight was able to enter the cell wall [15,16,20,35,36]. However, at a concentration of 0.5%, even if the modification system penetrated the cell wall completely, a reduction as strong as measured in this study was probably not solely due to the modification agent.

Penetration of the cell wall is a requirement for the modification agent to be able to affect EMC_R of wood. Lower EMC_R of specimens modified with phenol-formaldehyde resins, especially Phenol1 at 20%, could have had a positive influence on elasto-mechanical properties [20,23]. However, the results of this study did not show a positive effect of a lower EMC_R on elasto-mechanical properties as specimens modified with Phenol1 exhibited strong reductions of properties just like specimens modified with other modification agents.

During curing, the molecules of the modification agent build a network. If reactions between the modification agent and wood cell wall polymers take place, this is called cross-linking. Cross-linking restricts the flexibility of the wood and thus decreases the ability to relieve mechanic stress through strain [10,32,37]. It has been discussed whether the modification agents applied in this study are able to cross-link with the cell wall [6,38–40]. However, it is rather unlikely that cross-linking occurs at a concentration as low as 0.5%. Even if no cross-linking takes place, the rigid structure of the cured resin itself stiffens the wood to some extent [5]. Since embrittlement of chemically modified wood is one of the biggest drawbacks, further investigations regarding cross-linking are necessary.

The pH-values of the modification agents at initial composition varied (Table 1). The pH-values at the concentrations used for modification were not measured, but it was assumed that they varied between the modifications agents as well. Depending on the pH-value in combination with temperature, different chemical reactions take place [1]. Which milieu occurred during the modification and how it affected the elasto-mechanical properties [10,28] should be subject to further investigations. As DMDHEU was the only modification agent with an acidic pH value, this could be a reason for the fact that it caused the strongest reductions of all elasto-mechanical properties examined. It is known that treatment with high or low pH values has an effect on wood properties.

All of the modification systems applied in this study contained formaldehyde (methanal, CH2O) at different concentrations. Formaldehyde tends to form networks as it is bifunctional [41]. According to Burmester [42] as well as Rowell [43], treatment of wood with formaldehyde alone leads to a considerable embrittlement of the wood. Reasons for this are cross-linking and hydrolysis of cellulose. A higher formaldehyde content of DMDHEU compared to the other modification agents could be a reason for the strong reduction of elasto-mechanical properties of wood modified with DMDHEU. Research in the past focused on formaldehyde containing resins, which have the advantage of faster and more complete curing. However, more recent developments show [44] that even without formaldehyde, some resins do show good potential for wood modification.

4. Conclusions

This study investigated the influence of four chemical modification agents (MF, Phenol1, Phenol2, DMDHEU) in three concentrations (0.5%, 5%, 20%) on impact bending strength, tensile strength, and work to maximum load in traction of Scots pine. The main findings were:

1. Influence of the concentration of the modification solution. A clear concentration-dependent reduction was evident in impact bending strength, tensile strength and work to maximum load in traction of the modified specimens of each modification agent. The significant reduction in impact bending strength and work to maximum load in traction already after modifications with 0.5% solutions of each agent was remarkable.
2. Influence of the modification agent. There was a difference in the intensity of the influence, but inherently each modification caused a reduction of the examined elasto-mechanical properties. In fact, the modification agents showed similar influences regardless of differences in molecular weight or pH-value. Each modification agent caused a striking reduction of elasto-mechanical properties already at a very low concentration of 0.5%. Presumably, the strongest influence of the modifications on the elasto-mechanical properties was through embrittlement, which restricts mechanical stress relief through strain.

The results show that in the application of chemically modified wood, impact and tensile loads should be avoided even after treatment with low concentrations.

Author Contributions: C.B., together with S.B. and H.M., were mainly responsible for the conceptualization, methodology used, and data evaluation; C.B. and S.B. were mainly responsible for the data validation, and formal analysis; Investigations and data curation were conducted by C.B.; The original draft of this article was prepared by S.B. who was also responsible for the review and editing process of this article; C.B., together with S.B. and H.M., oversaw the visualization; S.B. was responsible for funding acquisition and project administration and H.M. supervised the project. All authors have read and agreed to the published version of the manuscript.

Funding: This research was funded by the German Federal Ministry of Food and Agriculture through the Agency for Renewable Resources (FNR) grant number FKZ 22024211.

Acknowledgments: The authors gratefully acknowledge support with Phenol modification from Sacha Bicke and support with MF and DMDHEU modification as well as curing process from Georg Behr.

Conflicts of Interest: The authors declare no conflict of interest.

References

1. Devallencourt, C.; Saiter, J.M.; Capitaine, D. Reactions between melamine formaldehyde resin and cellulose: Influence of pH. *J. Appl. Polym. Sci.* **2000**, *78*, 1884–1896. [CrossRef]
2. Hansmann, C.; Deka, M.; Wimmer, R.; Gindl, W. Artificial weathering of wood surfaces modified by melamine formaldehyde resins. *Holz Als Roh-Und Werkstoff* **2006**, *64*, 198–203. [CrossRef]

3. Lukowsky, D.; Büschelberger, F.; Schmidt, O. In situ testing the influence of melamine resins on the enzymatic activity of basidiomycetes. In Proceedings of the IRG Annual Meeting, Rosenheim, Germany, 6–11 June 1999. IRG/WP 99-30194.
4. Stamm, A.J.; Seborg, R.M. Minimizing wood shrinkage and swelling. *Eng. Chem. Res.* **1936**, *28*, 1164–1169. [CrossRef]
5. Hill, C.A.S. *Wood Modification: Chemical, Thermal and Other Processes*; John Wiley & Sons Ltd.: Bath, UK, 2006.
6. Krause, A. Wood Modification with N-Methylol Crosslinking Agents. Ph.D. Thesis, University of Gottingen, Gottingen, Germany, 2006.
7. Militz, H. Treatment of timber with water soluble dimethylol resins to improve their dimensional stability and durability. *Wood Sci. Technol.* **1993**, *27*, 347–355. [CrossRef]
8. Stamm, A.J.; Seborg, R.M. Resin-treated plywood. *Ind. Eng. Chem. Res. 1939 31*, 897–902. [CrossRef]
9. Bicke, S.; Mai, C.; Militz, H. Modification of beech veneers with low molecular weight phenol formaldehyde for the production of plywood: Durability and mechanical properties. In Proceedings of the European Conference on Wood Modification, Ljubljana, Slovenia, 17–18 September 2012; pp. 363–366.
10. Mai, C.; Xie, Y.; Xiao, Z.; Bollmus, S.; Vetter, G.; Krause, A.; Militz, H. Influence of the modification with different aldehydebased agents on the tensile strength of wood. In Proceedings of the European Conference on Wood Modification, Cardiff, UK, 15–16 October 2007; pp. 49–56.
11. Lukowsky, D. Wood Protection with Melamine Resins. Ph.D. Thesis, University of Hamburg, Berlin/Heidelberg, Germany, 1999.
12. Mahnert, K.-C. Development of a Non-Bearing Flooring for Shipbuilding Based on Selected Wood Modification Processes. Ph.D. Thesis, University of Gottingen, Gottingen, Germany, 2013.
13. Pittman, C.U.J.; Kim, M.G.; Nicholas, D.D.; Wang, L.; Kabir, F.A.; Schultz, T.P.; Ingram, L.L.J. Wood enhancement treatments I. Impregnation of southern yellow pine with melamine-formaldehyde and melamine-ammeline-formaldehyde resins. *J. Wood Chem. Technol.* **1994**, *14*, 577–603. [CrossRef]
14. Evans, P.D.; Kraushaar Gibson, S.; Cullis, I.; Liu, C.; Sèbe, G. Photostabilization of wood using low molecular weight phenol formaldehyde resin and hindered amine light stabilizer. *Polym. Degrad. Stab.* **2013**, *98*, 158–168. [CrossRef]
15. Furuno, T.; Imamura, Y.; Kajita, H. The modification of wood by treatment with low molecular weight phenol-formaldehyde resin: A properties enhancement with neutralized phenolic-resin and resin penetration into wood cell walls. *Wood Sci. Technol.* **2004**, *37*, 349–361. [CrossRef]
16. Biziks, V.; Bicke, S.; Militz, H. Penetration of phenol formaldehyde (PF) resin into beech wood studied by light microscopy. In Proceedings of the IRG Annual Meeting, Vina del Mar, Chile, 10–14 May 2015. IRG/WP 15-20558.
17. Dieste Märkl, A. Wood-Water Relationships in Wood Modified with 1,3-dimethylol-4,5-dihydroxy Ethylene Urea (DMDHEU). Ph.D. Thesis, University of Gottingen, Gottingen, Germany, 2009.
18. Dieste Märkl, A.; Krause, A.; Bollmus, S.; Militz, H. Physical and mechanical properties of plywood produced with 1.3-dimethylol-4.5-dihydroxyethyleneurea (DMDHEU)-modified veneers of *Betula sp.* and *Fagus sylvatica*. *Holz Als Roh-Und Werkstoff* **2008**, *66*, 281–287. [CrossRef]
19. Xie, Y.; Krause, A.; Militz, H.; Turkulin, H.; Richter, K.; Mai, C. Effect of treatments with 1,3-dimethylol-4,5-dihydroxy-ethyleneurea (DMDHEU) on the tensile properties of wood. *Holzforschung* **2007**, *61*, 43–50. [CrossRef]
20. Bollmus, S. Biological and Technological Properties of Beech Wood after Modification with 1,3-dimethylol-4,5-dihydroxy Ethylene Urea (DMDHEU). Ph.D. Thesis, University of Gottingen, Gottingen, Germany, 2011.
21. DIN 52 189. *Testing of Wood; Determination of Impact Bending Strength*; Deutsches Institut für Normung e.V.: Berlin, Germany, 1981.
22. DIN 52 188. *Testing of Wood; Determination of Ultimate Tensile Stress Parallel to Grain*; Deutsches Institut für Normung e.V.: Berlin, Germany, 1979.
23. Kollmann, F. *Technology of Wood and Wood-Based Materials*; Springer: Berlin/Heidelberg, Germany, 1951.
24. Hosseinpourpia, R.; Adamopoulos, S.; Mai, C. Dynamic vapour sorption of wood and holocellulose modified with thermosetting resins. *Wood Sci. Technol.* **2016**, *50*, 165–178. [CrossRef]
25. Xie, Y. Surface Properties of Wood Modified with Cyclic N-Methylol Compounds. Ph.D. Thesis, University of Gottingen, Gottingen, Germany, 2005.

26. Wagenführ, R. *Wood Atlas (6)*; Fachbuchverlag: München, Germany; Leibzig, Germany, 2007.
27. Krech, H. Amount and time characteristics of force and deflection in the impact bending test of wood and their relationship to the impact bending strength. *Holz Als Roh-Und Werkstoff* **1960**, *18*, 95–105. [CrossRef]
28. Brischke, C. Interrelationship between static and dynamic strength properties of wood and its structural integrity [Međusobna ovisnost statičkih i dinamičkih svojstava čvrstoće drva i njegova strukturnog integriteta]. *Drvna Industrija* **2017**, *68*, 53–60. [CrossRef]
29. Hosseinpourpia, R.; Adamopoulos, S.; Mai, C. Tensile strength of handsheets from recovered fibers treated with N-methylol melamine and 1,3-dimethylol-4,5-dihydroxyethyleneurea. *J. Appl. Polym. Sci.* **2015**, 132. [CrossRef]
30. Bodig, J.; Jayne, B.A. *Mechanics of Wood and Wood Composites*; Van Nostrand Reinhold Company Inc.: New York, NY, USA, 1982.
31. Fengel, D.; Wegener, G. *Wood: Chemistry, Ultrastructure, Reactions*; Kessel Verlag: Remagen, Germany, 2003.
32. Nicholas, D.D.; Williams, A.D. Dimensional stabilization of wood with dimethylol compounds. In Proceedings of the IRG Annual Meeting, Honey Harbour, ON, Canada, 17–22 May 1987; p. 8, IRG/WP 1987\IRG 3412.
33. Lukowsky, D. Influence of the formaldehyde content of waterbased melamine formaldehyde resins on physical properties of Scots pine impregnated therewith. *Holz Als Roh-Und Werkstoff* **2002**, *60*, 349–355. [CrossRef]
34. Stamm, A.J. *Wood and Cellulose Science*; The Ronald Press Company: New York, NY, USA, 1964.
35. Gindl, W.; Zargar-Yaghubi, F.; Wimmer, R. Impregnation of softwood cell walls with melamine-formaldehyde resin. *Bioresour. Technol.* **2003**, *87*, 325–330. [CrossRef]
36. Huang, Y.; Fei, B.; Zhao, R. Investigation of low-molecular weight phenol formaldehyde distribution in tracheid cell walls of Chinese fir wood. *BioResources* **2014**, *9*, 4150–4158. [CrossRef]
37. Rowell, R.M. Property Enhanced Natural Fiber Composite Materials Based on Chemical Modification. In *Science and Technology of Polymers and Advanced Materials*; Prasad, P.N., Mark, J.E., Kandil, S.H., Kafafi, Z.H., Eds.; Springer: Boston, MA, USA, 1998. [CrossRef]
38. He, G.B.; Riedl, B. Curing kinetics of phenol formaldehyde resin and wood-resin interactions in the presence of wood substrates. *Wood Sci. Technol.* **2004**, *38*, 69–81. [CrossRef]
39. Niemz, P. *Wood Anatomy Chemistry Physics. Physics of Wood and Wood-Based Materials*; DRW-Verlag Weinbrenner GmbH & Co: Leinfelden-Echterdingen, Germany, 1993.
40. Som, N.C.; Mukherjee, A.K. Chemical bond formation in cross linking reaction between jute fibre and dimethyloldihydroxyethylene urea. *Indian J. Fibre Text.* **1989**, *14*, 45–49.
41. Dunky, M.; Niemz, P. *Wood-Based Materials and Glues: Technology and Influencing Factors*; Springer: Berlin/Heidelberg, Germany, 2013.
42. Burmester, A. Tests for wood treatment with monomeric gas of formaldehyde using gamma rays. *Holzforschung* **1967**, *21*, 13–20. [CrossRef]
43. Rowell, R.M. Chemical modification of wood. *For. Prod. Abstr.* **1983**, *6*, 363–382.
44. Berkenheger, O. Changes in Seleted Static and Dynamic Mechanical Properties of Solid Wood Modified with DMDHEU—A Comparative Study. Bachelor's. Thesis, University of Gottingen, Gottingen, Germany, 2019.

Article

Corrosiveness of Thermally Modified Wood

Samuel L. Zelinka [1,*], Leandro Passarini [2], Frederick J. Matt [1] and Grant T. Kirker [1]

[1] US Forest Service Forest Products Laboratory, Madison, WI 53726, USA
[2] Réseau CCNB-INNOV, Grand-Sault, NB E3Y 3W3, Canada
* Correspondence: samuel.l.zelinka@usda.gov; Tel.: +1-608-231-9277

Received: 2 December 2019; Accepted: 24 December 2019; Published: 31 December 2019

Abstract: Thermally modified wood is becoming commercially available in North America for use in outdoor applications. While there have been many studies on how thermal modification affects the dimensional stability, water vapor sorption, and biodeterioration of wood, little is known about whether thermally modified wood is corrosive to metal fasteners and hangers used to hold these members in place. As thermally modified wood is used in outdoor applications, it has the potential to become wet which may lead to corrosion of embedded fasteners. Here, we examine the corrosiveness of thermally modified ash and oak in an exposure test where stainless steel, hot-dip galvanized steel, and carbon steel nails are driven into wood and exposed to a nearly 100% relative humidity environment at 27 °C for one year. The corrosion rates were compared against control specimens of untreated and preservative-treated southern pine. Stainless steel fasteners did not corrode in any specimens regardless of the treatment. The thermal modification increased the corrosiveness of the ash and oak, however, an oil treatment that is commonly applied by the manufacturer to the wood after the heat treatment reduced the corrosiveness. The carbon steel fasteners exhibited higher corrosion rates in the thermally modified hardwoods than in the preservative-treated pine control. Corrosion rates of galvanized fasteners in the hardwoods were much lower than carbon steel fasteners. These data can be used to design for corrosion when building with thermally modified wood, and highlight differences between corrosion of metals embedded in wood products.

Keywords: modified wood; corrosion; stainless steel; hot-dip galvanized steel; heat treatments

---

1. Introduction

Wood is a sustainable biomaterial that has been used as a building material since the beginning of civilizations. Under proper conditions, wood can last for millennia, as exhibited by artifacts such as the Shigir Idol or the coffin of Tutankhamun [1,2]. However, in outdoor applications, it is susceptible to degradation from moisture cycling, ultraviolet radiation, and decay fungi [3].

Preservative treatments have been used for many years in North America to increase the durability of wood in outdoor applications [4]. Preservative treatments protect wood by impregnating it with chemicals that are either fungistatic or fungitoxic and inhibit fungal growth. Frequently, these chemicals are combined with an insecticide to further protect the wood against termites and other wood-boring insects. Waterborne wood preservative treatments are registered pesticides and their ability to be used in the United States is dependent on their ability to maintain their registration with the US Environmental Protection Agency [5].

Currently, there is an interest in using *modified wood* as an alternative to preservative-treated wood in certain outdoor applications. *Modified wood* is wood whose chemistry and/or structure is altered to achieve desired properties through thermal or chemical treatments [6]. In contrast to preservative-treated wood, the decay resistance in modified wood is a result of non-toxic changes to the wood structure which make the wood harder for the fungi to colonize. The mechanisms through which modified wood achieves its decay resistance are still not fully understood, although it is realized that all wood modifications affect how water is associated with the wood cell wall [7]. One current hypothesis is that wood modifications may inhibit diffusion of fungal decay agents through the cell wall [8–12].

Thermally modified wood is a modification process where the wood properties are changed by heating wood in a non-oxidizing environment. While many different thermal modification profiles have been used, typical treatment temperatures are between 160 °C and the char temperature of wood (300 °C). Thermally modified wood has been studied for over 100 years [13], however, it was not available commercially in North America until recently. In the thermal modification process, hemicelluloses and celluloses are degraded through pyrolysis and other chemical reactions. The amount of the induced chemical changes to the wood are typically measured through the mass loss that occurs during the thermal modification process. The degradation of cellulose and hemicelluloses causes a reduction in the mechanical strength of thermally modified wood [14]. However, the remaining, semi-pyrolyzed material has increased dimensional stability and lower equilibrium moisture content at a given relative humidity [15,16]. Thermally modified wood has also been shown to be more decay resistant than untreated wood [17–21].

Given that thermally modified wood has improved decay resistance and moisture properties, there is interest in using it in place of preservative-treated wood in certain outdoor environments. Prior to widespread commercial adoption, it is necessary to characterize the performance of thermally modified wood in laboratory tests. While much work has already been carried out on the decay resistance of thermally modified wood, there are few published data on the corrosiveness of modified wood to metal fasteners [22–24] and even less data on the corrosiveness of thermally modified wood [25]. While not typically considered a corrosive environment, when wood is above 15% moisture content, fastener corrosion can occur [26–28].

The corrosion of metal fasteners has been widely studied since a 2004 change in the registration of wood preservatives in the United States [5]. At that time, corrosion failures were seen in service as new wood preservatives entered the marketplace [29,30]. As a result of these corrosion concerns, an extensive test program was developed. It was found that wood moisture content has a large effect on the corrosion rate of embedded metals; below 15% corrosion does not occur. As the moisture content is increased, the corrosion rate increases until fiber saturation [26–28]. Corrosion of embedded fasteners was found to proceed at a constant rate with time [31,32]. Most of the previous corrosion testing was performed in an environment at 27 °C and near 100% relative humidity (RH) conditions [22,28,32–38]. It was found that the corrosion rates under these conditions were as high or higher than those measured in the fully saturated wood state [28]. Therefore, corrosion rates measured in service should be less than or equal to those measured in the 100% RH environment.

In this paper, we examine the corrosiveness of thermally modified hardwoods in a year-long exposure test at 27 °C and near 100% RH. The data are compared against controls of untreated and preservative-treated southern pine. In addition to providing new properties of thermally modified wood, the experiments also provide valuable information on the corrosivity of hardwood species, of which little data exists.

2. Materials and Methods

2.1. Thermally Modified Wood

The thermally modified wood was provided from a commercial supplier. The thermal modification followed the Finish Thermowood process [39,40]. Wood was equilibrated at 120 °C to facilitate drying and then heated to 190 °C for 3 h before being quenched with water. Two different species were tested, red oak (*Quercus rubra* L.) and white ash (*Fraxinus americana* L., Sp. Pl.). For each species, three different conditions were tested: "control", without thermal modification, thermally modified, and thermally modified wood with an oil coating. The oil coating is typically applied by the company to their commercial products; however, in this study we tested it to see if it had any effect on fastener corrosion.

In addition to these six specimens, two additional groups were added for a comparison: untreated southern pine (*Pinus* spp.) and southern pine commercially treated with a common copper containing wood preservative, micronized copper azole, (MCA) treated to retention of 1 kg m^{-3} (suitable for above ground use).

2.2. Fasteners

Three types of fasteners were tested: hot-dip galvanized steel, plain carbon steel, and stainless steel 16d nails with a length of 90 mm. Ten replicates were tested for each fastener type and wood treatment. The average diameters of the steel, galvanized steel and stainless steel fasteners were 4.1, 3.6, and 4.2 mm respectively. The composition of the steel and stainless steel fasteners was obtained with optical emission spectroscopy, (Table 1). The composition of the carbon steel fastener was consistent with UNS G10180 carbon steel and the stainless steel fastener was consistent with UNS S30400 austenitic stainless steel. The galvanized coating thickness was measured at six different points along one of the fasteners from a scanning electron micrograph. The mean coating thickness was 91 μm with a standard deviation of 36 μm. The composition of the galvanized coating thickness was measured with an X-ray fluorescence analyzer (Table 1).

Table 1. Composition of the fasteners tested, or for the galvanized fastener, the composition of the galvanized coating. Composition is given as a weight percent.

	Carbon Steel	Stainless Steel	Galvanized Coating
Carbon	0.191	0.040	–
Silicon	0.130	0.419	0.415
Manganese	0.750	1.680	–
Phosphorus	0.007	0.025	–
Sulfur	0.007	0.020	–
Chromium	0.022	18.110	–
Nickel	–	8.830	–
Molybdenum	–	0.253	–
Copper	0.040	0.181	–
Cobalt	0.004	0.106	0.037
Tin	0.001	0.013	0.032
Bismuth	0.006	–	0.186
Zinc	–	0.018	balance
Iron	balance	balance	2.815

Prior to exposure, the surface areas of the fasteners were determined optically with the method of Rammer and Zelinka [41,42]. Fasteners were then cleaned in an ultrasonic cleaner with soap solution for 5 min, rinsed under deionized water, dried, and weighed to the nearest 0.1 mg.

2.3. Exposure

Fasteners were driven into holes predrilled in the wood with a diameter of 4.0 mm (5/32"). Given the extremely high densities of the hardwood species tested, it was impossible to drive the fasteners into the wood unless the entire length of the fastener was predrilled to a diameter near that of the fastener. The carbon steel and stainless steel fasteners could not be driven into holes smaller than 4.0 mm, therefore, this resulted in the galvanized fasteners being driven into slightly oversized holes. Fasteners were driven with a pneumatic palm nailer as opposed to a hammer. The pneumatic nailer was necessary to drive the fasteners into the high density hardwoods without bending the fasteners. For each treatment group, one board was tested; all three types of nails were driven into the same board with a space of at least 25 mm between fasteners. The sample geometry and location of replicates closely followed ASTM standard G198 [43]. Previous work has shown that the area of interaction of the fastener with the wood is localized to a region less than 1 mm from the fastener surface [44,45].

Once the fasteners were driven into the wood, the boards were placed in a sealed container for one year. The boards were placed above a reservoir of water, which created a local environment of close to 100% RH inside of the container; the temperature of the room was 27 °C. These conditions closely match previous corrosion tests on preservative-treated wood [28,33–36,38]. In this experiment however, the containers were inadvertently moved at some point during the exposure. As a result, some of the water from reservoirs made contact with the boards which increased the moisture content of some of the specimens. The moisture content was measured at the end of the experiment by cutting small cross sections of the board throughout its width and gravimetrically determining the moisture content. Moisture contents are listed in Table 2. Although the thermal modifications likely affect the hygroscopicity and thus equilibrium moisture content, the conditions of the corrosion tests involved condensation and, in some cases, splashing, and therefore the final moisture contents are more a result of the environment than the treatment. However, the corrosion rates of fasteners have been found to not vary from the 100% relative humidity condition to full saturation, so these differences are unlikely to affect the reported corrosion rates [28].

Table 2. Final wood moisture contents of the different wood species and treatments tested.

Species	Treatment	Final Moisture Content (Standard Deviation)
Ash	Untreated	36% (8%)
	Thermally Modified	25% (8%)
	Thermally Modified w/Oil Treatment	32% (4%)
Oak	Untreated	18% (3%)
	Thermally Modified	16% (1%)
	Thermally Modified w/Oil Treatment	16% (1%)
Pine	Untreated	27% (1%)
	Preservative-Treated	49% (6%)

The fasteners were originally set to be exposed for one year. However, due to the US Government shut down, the experiment could not be accessed. Instead, fasteners were exposed for slightly longer than 1 year (at most, 11,256 h of total exposure). However, the corrosion of metals in wood has shown to increase linearly with time (constant corrosion rate), so these slight differences in exposure times should not affect the reported corrosion rate [34,46,47].

2.4. Post-Test Cleaning Procedure

Following the exposure, the fasteners were removed from the wood. Fasteners were removed by making cross-cuts in the wood near the fastener. The thin amount of wood on both sides of the fastener could then easily be removed by hand. Larger sections between the fasteners were retained and used to measure the wood moisture content. Fasteners were then cleaned for 60 min in an ultrasonic cleaner with a 50:50 solution (volume ratio) of a proprietary chelating agent (EvapoRust™ Orison Marketing LLC, Abilene, TX, USA) and deionized water. Following the cleaning, the nails were wiped with a paper towel, allowed to air dry and weighed to the nearest 0.1 mg. The mass change caused by the cleaning process (m_c) itself was measured by cleaning uncorroded fasteners using the same process. The change in mass of the corroded fasteners (Δm) was calculated as

$$\Delta m = m_f - m_i + m_c, \tag{1}$$

where m_f and m_i were the initial and final masses, respectively.

2.5. Determination of the pH and Tannin Content

Water extracts of the wood were directly analyzed for pH and tannin concentrations. Since corrosion is an aqueous process, water extracts should mimic the corrosive environment near the fastener and several studies have shown good correlation between corrosion measurements in water extracts of wood and solid wood [48–50]. The extracts were made by the method of Zelinka, Rammer and Stone [50]. Wood was ground into sawdust and then mixed with reverse osmosis water in a 1:10 (wood:water) weight ratio and allowed to sit at room temperature for one week before filtration. The pH was measured in the extract using a pH probe.

Total phenolics and tannins were determined using a lab procedure published by The Food and Agriculture organization of the United Nations (FAO/IAEA) [51,52]. The method was largely based on the published work of Makkar, et al. [53]. This 2-step procedure first uses the Folin test to determine total phenols. Then, polyvinyl polypyrolidone (PVPP) is added to precipitate tannin-sized polyphenols. The Folin test is run again to measure remaining polyphenols, and tannins are calculated as the difference between total and remaining polyphenols (after PVPP precipitation).

In this study, 2 mL aliquots of the water extracts were taken and passed through a 0.45 uM centrifuge filter to remove particulates. A standard curve was prepared using Sigma-Tannic Acid (Sigma-Aldrich, 403040, Saint Louis, MO, USA) at 5 concentrations (including a blank). Then, the Folin test reagents: Distilled Water, Sodium Carbonate (Sigma-Aldrich, 791768, Saint Louis, MO, USA) and Folin & Ciocalteu's phenol reagent (Sigma–Aldrich, F9252, Saint Louis, MO, USA) were mixed together in test tubes for color development for 40 min. Aliquots of the filtered water extracts were treated identically. Sample concentrations were measured (at 725 nm) against the standard curve using a Thermo Scientific Gensys 180 UV-Vis spectrophotometer after subtracting a blank (consisting of dist. H_2O and Folin reagents). Afterwards, 100 mg of PVPP (Sigma-Aldrich, P-2472, Saint Louis, MO, USA) was weighed into fresh test tubes and dissolved in 1 mL of dist. water. A 1 mL aliquot of filtered water extract sample was added and thoroughly mixed. Samples were then placed in a 4 °C refrigerator for 15 min and then spun for 10 min at 10,000 rpm through a 0.45 μM centrifuge filter and analyzed using the Folin test as described above.

3. Results and Discussion

3.1. Stainless Steel Fasteners

The corrosion rate of the stainless steel nails was essentially zero. In all but one case, the corrosion rates were less than 0.3 μm year^{-1} and the standard deviations were bigger than the mean corrosion rate. In one case, the average corrosion rate measured was 1 μm year^{-1}. However, in this case, the standard deviation was 3 μm year^{-1}. Therefore, it can be safely concluded that stainless steel nails do not corrode in thermally modified ash or oak. This is in line with previous results that have shown that stainless steel fasteners exhibit little to no corrosion in preservative-treated and untreated softwoods [34].

3.2. Carbon Steel Fasteners

The corrosion rates of carbon steel fasteners and hot dip galvanized fasteners are presented in Figure 1. In general, the error bars showing the standard deviations are very large. The standard deviations are higher than our previous work on preservative-treated softwoods [35]; these differences are attributed to difficulties in driving the fasteners into the wood without causing splitting near the fastener or otherwise damaging the fastener. Despite the fact that error bars are too large to find statistically significant differences in most cases, the mean corrosion rate appears to exhibit some trends. For instance, for the untreated wood species, it appears that the corrosiveness of ash and pine are similar and that they are both less corrosive than oak.

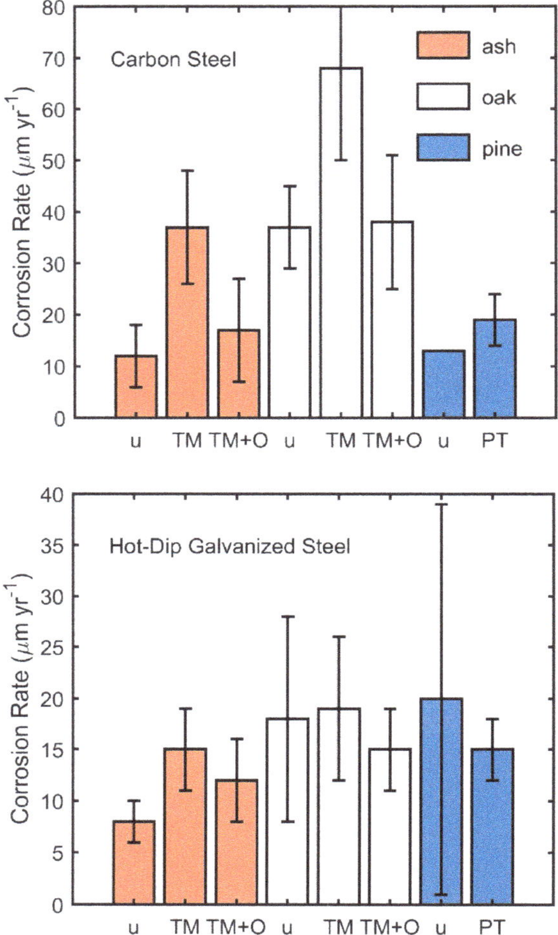

Figure 1. Corrosion rates measured for hot-dip galvanized and plain carbon steel fasteners in ash, oak, and pine. Legend: u = untreated, TM = Thermally Modified, TM + O = Thermally Modified with Oil treatment, PT = Preservative Treatment. The error bars represent the standard deviation. Note the different y-axis scales.

The measured values for corrosion of steel nails in untreated pine is slightly higher than previously measured in previous studies under the same conditions (13 μm year^{-1} as opposed to 5 μm year^{-1}) [22,46]. Likewise, the measured corrosion rate of 20 μm year^{-1} for the preservative-treated wood (a micronized copper azole) is higher than previously measured values of 11–13 μm year^{-1} for copper azole and micronized copper quaternary preservatives [35]. There are slight metallurgical differences in the carbon steel fasteners used in these studies; in the fasteners used in the present study, they contain more carbon than those used in the previous study (UNS G10180 as opposed to a UNS G10140). However, according to Kodama [54] different carbon steel alloys do not exhibit "remarkable (differences) in corrosion behavior". Therefore, it is likely that the corrosion differences are a result of different exposure conditions such as differences in moisture content, preservative treatment formulations, or preservative treatment retentions.

In addition to trends between wood species, the effect of thermal modification on corrosion can be observed by examining a single wood species. For both the ash and the oak, the thermal modification process increased the corrosivity of the wood. The mean corrosion rate of ash nearly tripled from 12 to 37 μm $year^{-1}$ when comparing the untreated and thermally modified wood. Likewise, the corrosion rate of the oak also increased by more than 50% from 37 to 68 μm $year^{-1}$. The post-thermal modification oil treatment also appears to have an effect on the corrosivity of the fasteners, decreasing the measured corrosion rates of both the ash and the oak. The mean corrosion rates for the oil-treated specimens are 17 and 38 μm $year^{-1}$ for the ash and oak specimens, respectively. These corrosion rates are very close to the mean values of the corrosion rates measured for the untreated ash and oak specimens.

3.3. Hot-Dip Galvanized Fasteners

The corrosion rate data for the hot-dip galvanized fasteners are also presented in Figure 1. Similar trends can be observed across the species and treatments as for the carbon steel fasteners; however, their effects are less pronounced. Similar to the carbon steel fasteners, the measured corrosion rate of untreated ash is less than that of oak. Furthermore, the oil treatment results in reducing the corrosiveness of the thermally modified wood. However, unlike the steel fasteners, the galvanized fasteners show less of an increase in corrosion with the thermal modification treatment. The mean corrosion rate of ash increased from 8 to 15 μm $year^{-1}$ between the untreated and thermal modification treatment. In oak, the mean corrosion rates of the galvanized fasteners were 19 μm $year^{-1}$ for both the untreated and thermal modification treatment.

3.4. Comparison of the Corrosion Rates of Hot-Dip Galvanized Fasteners and Carbon Steel Fasteners

In previous work on preservative-treated softwoods, it has been observed that galvanized fasteners exhibit a higher corrosion rate than carbon steel fasteners [34,35,46,50,55]. This is in contrast to atmospheric corrosion where hot-dip galvanized products corrode much more slowly than steel. The local environment makes a large difference in whether or not galvanized products corrode more slowly than carbon steel. For atmospheric corrosion, drying cycles allow a passive film to form which protects the remaining zinc coating from rapid corrosion [56,57]. In previous corrosion testing of galvanized fasteners in softwoods, these corrosion products were not found, and galvanized fasteners corroded more rapidly than steel [35].

It appears in the hardwoods that galvanized steel corrodes more slowly than carbon steel. This can be seen most clearly for oak, where the corrosion rate for carbon steel (37 μm $year^{-1}$) was nearly double that of galvanized steel (19 μm $year^{-1}$). Galvanized steel also corroded more slowly than carbon steel across the thermal modification and oil treatments (Note that in this study, the mean corrosion rate of galvanized steel was slightly lower than that of carbon steel, however, the results are not statistically different and contradict the previous literature. The remainder of the discussion comparing the different treatments and species compares the results to the literature values in treated pine). Therefore, one important finding of this study is that galvanized steel corrodes more slowly than carbon steel in hardwoods whereas it generally corrodes more rapidly in preservative-treated wood. The corrosion mechanism in treated wood involves the reduction of cupric ions from the wood preservative. Since zinc is less thermodynamically stable than steel in the presence of cupric ions, it could be that this larger driving potential is accelerating the kinetics in preservative-treated wood. Limited data has shown that galvanized steel also corroded more rapidly than carbon steel in untreated pine, however, the absolute values of these corrosion rates were small, the error bars high, and the results statistically uncertain. This study has highlighted how little is known about our understanding of the corrosion rates in different metals in untreated wood species. Beyond scientific importance, quantifying the relative corrosion rates of different metals is incredibly important for materials selection, as materials that work well in pressure-treated pine may not work well in hardwoods.

3.5. Comparison of the Wood Species and the Effect of pH and Tannins

Zelinka and Stone [48] developed a model to explain differences in the corrosion of metals embedded in untreated wood of different species. Their model was based on corrosion rates measured electrochemically in water extracts of wood. While this method was shown to give similar corrosion rates to those measured in solid wood for preservative-treated wood, corrosion rates in the extracts are much higher in untreated wood. Despite this, Zelinka and Stone developed a model that could predict the *relative* corrosiveness of the wood species from the total amount of tannins in the wood and the pH of the wood, as measured by the pH of the extract. Tannins were included in the model because they were shown to act as corrosion inhibitors. The model treated the effects of tannins and pH as orthogonal and showed that the pH only increases the corrosion rate below a pH of 5.

Both the pH and tannin concentrations were measured on water extracts of the different wood species and treatments; results are shown in Figure 2. For the unmodified woods, oak was more acidic than the ash but also contained more tannins. The increase in the corrosiveness of the oak compared to the ash can be explained by the lower pH.

The pH and tannin concentrations also correlated nicely with the corrosion data across the thermal modification and oil treatments. The thermal modification results in the wood becoming more acidic and at the same time, removes or destroys some of the soluble tannins available in the untreated wood. Both of these would be expected to create a more corrosive environment towards embedded fasteners, which was observed for both steel and galvanized steel in both wood species (Figure 1). Furthermore, the oil treatment greatly raised the pH of both the oak and the ash so that the pH was higher than the unmodified wood. The oil treatment also appears to raise the tannin level of the wood, either through solubilizing polyphenols in the thermally modified wood or the oil treatment which contains tannins itself. As a result, the pH of both the ash and oak was higher than the modified wood and the observed amount of tannins were much greater than the thermally modified material. As expected, this correlates with a decrease in the observed corrosion rates with the oil treatments (Figure 1).

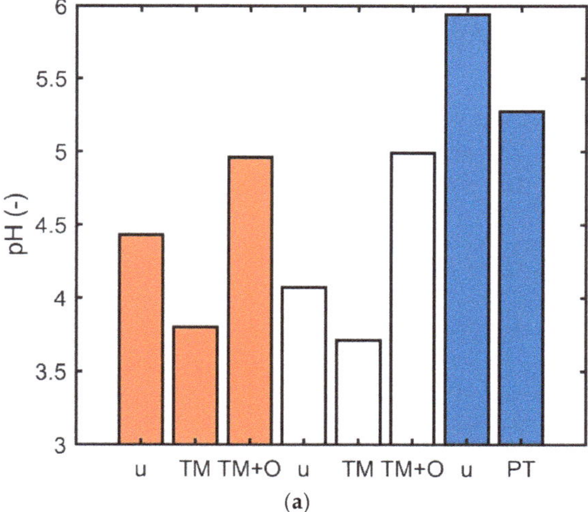

(a)

Figure 2. *Cont.*

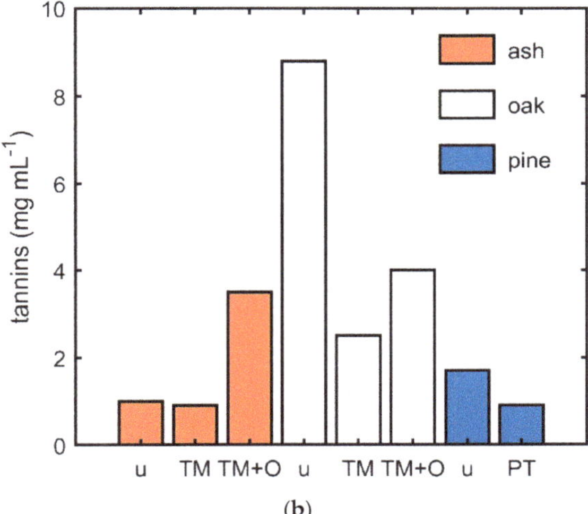

(b)

Figure 2. pH (**a**) and concentration of tannins (**b**) found in the different wood treatments. Legend: u = untreated, TM = Thermally Modified, TM + O = Thermally Modified with Oil treatment, PT = Preservative Treatment.

Beyond the aforementioned study, it is difficult to place the corrosion data collected in this study in the broader literature as very little quantitative corrosion data has been published for hardwoods. Smith [58] published a table ranking wood species from "most corrosive" to "least corrosive but did not list corrosion rates or a methodology of how the table was constructed. Similar qualitative rankings were developed by Farmer [59] and Bartel-Kornacka [60]. Knotkova-Cermakova and Vlckova [61] did compare the corrosiveness of oak and ash; however, in their study, they did not examine the corrosion of embedded metals. Instead, they examined the corrosion of metals in sealed containers with high humidity and wood veneers. They found that in these conditions, the corrosion rate of steel in oak vapors was 120 μm year^{-1} and that of ash was 35 μm year^{-1}. The ratio between the corrosion rates of these two species (3.5) was similar to the ratio of the corrosion rates measured in this study (2.9). Given that the results of [61] were attributed to acid vapors produced by the wood, and similar results were observed between the two species groups in both studies, it suggests that the acidity of the wood may play a large role in the corrosiveness of embedded fasteners.

4. Conclusions

This paper examined the effect of a thermal modification and oil treatment on the corrosion of embedded fasteners for their potential use in outdoor structures. The data show that thermal modification makes the wood more acidic and increases the corrosiveness of both ash and oak. However, the post-thermal modification oil treatment reduces the corrosivity of the wood of the thermally modified wood by raising the pH and increasing the water soluble tannin content. For galvanized steel, the range of corrosion rates in thermally modified wood was similar to those measured in pressure-treated southern pine. However, carbon steel fasteners in thermally modified oak and ash exhibited higher corrosion rates than in pressure-treated wood.

In all wood species and treatments, stainless steel fasteners exhibited negligible corrosion. It is likely that stainless steel fasteners will not corrode in thermally modified wood in service. For the hardwoods, galvanized steel fasteners exhibited lower corrosion rates than steel fasteners, regardless of the thermal modification or oil treatments. This is in contrast to previous studies on preservative-treated wood where galvanized fasteners corroded significantly faster than steel fasteners. From a materials selection standpoint, if stainless steel fasteners cannot be used, hot-dip galvanized fasteners are preferable to carbon steel fasteners.

Author Contributions: Conceptualization, S.L.Z. and G.T.K.; methodology, S.L.Z., G.T.K, L.P., F.J.M.; writing—original draft preparation, S.L.Z.; writing—review and editing, S.L.Z. All authors have read and agreed to the published version of the manuscript.

Funding: This research was funded by the Forest Products Laboratory. Additional research funding was provided by the Tournesol Siteworks, LLC.

Acknowledgments: The authors acknowledge the assistance of the Forest Products Laboratory carpenter shop and Forest Products Laboratory employee, Steve Halverson, for their assistance in sample preparation.

Conflicts of Interest: The funders had no role in the design of the study; in the collection, analyses, or interpretation of data; in the writing of the manuscript, or in the decision to publish the results.

References

1. Savchenko, S.; Lillie, M.C.; Zhilin, M.G.; Budd, C.E. New AMS dating of bone and antler weapons from the shigir collections housed in the sverdlovsk regional museum, Urals, Russia. In *Proceedings of the Prehistoric Society*; Cambridge University Press: Cambridge, UK, 2015; pp. 265–281.
2. Rifai, M.M.; El Hadidi, N.M. Investigation and analysis of three gilded wood samples from the tomb of Tutankhamun. In *Decorated Surfaces on Ancient Egyptian Objects, Technology, Deterioration and Conservation*; Dawson, J., Rozeik, C., Wright, M.M., Eds.; Cairo University: Giza, Egypt, 2010; pp. 16–24.
3. Williams, R.S. Weathering of wood. *Handb. Wood Chem. Wood Compos.* **2005**, *7*, 139–185.
4. Lebow, S. Wood Preservation. In *Wood Handbook- Wood as an Engineering Material*; U.S. Department of Agriculture, Forest Service, Forest Products Laboratory: Madison, WI, USA, 2010; p. 508.
5. Lebow, S. *Alternatives to Chromated Copper Arsenate for Residential Construction*; Res. Pap. FPL-RP-618; U.S. Department of Agriculture, Forest Service, Forest Products Laboratory: Madison, WI, USA, 2004; p. 9.
6. Hill, C.A. *Wood Modification: Chemical, Thermal and Other Processes*; John Wiley & Sons: West Sussex, UK, 2006; Volume 5.
7. Ringman, R.; Pilgård, A.; Brischke, C.; Richter, K. Mode of action of brown rot decay resistance in modified wood: A review. *Holzforschung* **2014**, *68*, 239–246. [CrossRef]
8. Zelinka, S.L. Role of transport in wood damage mechanisms–WSE 2017 keynote lecture. *Int. Wood Prod. J.* **2018**, *9*, 50–57. [CrossRef]
9. Zelinka, S.L.; Ringman, R.; Pilgård, A.; Thybring, E.E.; Jakes, J.E.; Richter, K. The role of chemical transport in the brown-rot decay resistance of modified wood. *Int. Wood Prod. J.* **2016**, *7*, 66–70. [CrossRef]
10. Jakes, J.E.; Plaza, N.; Stone, D.S.; Hunt, C.G.; Glass, S.V.; Zelinka, S.L. Mechanism of transport through wood cell wall polymers. *J. For. Prod. Ind.* **2013**, *2*, 10–13.
11. Hunt, C.G.; Zelinka, S.L.; Frihart, C.R.; Lorenz, L.; Yelle, D.; Gleber, S.-C.; Vogt, S.; Jakes, J.E. Acetylation increases relative humidity threshold for ion transport in wood cell walls—A means to understanding decay resistance. *Int. Biodeterior. Biodegrad.* **2018**, *133*, 230–237. [CrossRef]
12. Jakes, J.E.; Hunt, C.G.; Zelinka, S.L.; Ciesielski, P.N.; Plaza, N.Z. Effects of Moisture on Diffusion in Unmodified Wood Cell Walls: A Phenomenological Polymer Science Approach. *Forests* **2019**, *10*, 1084. [CrossRef]
13. Stamm, A.J.; Hansen, L. Minimizing wood shrinkage and swelling effect of heating in various gases. *Ind. Eng. Chem.* **1937**, *29*, 831–833. [CrossRef]
14. Rapp, A.O.; Brischke, C.; Welzbacher, C.R. Interrelationship between the severity of heat treatments and sieve fractions after impact ball milling: A mechanical test for quality control of thermally modified wood. *Holzforschung* **2006**, *60*, 64. [CrossRef]
15. Hill, C.A.S.; Ramsay, J.; Laine, K.; Rautkari, L.; Hughes, M. Water vapour sorption behaviour of thermally modified wood. *Int. Wood Prod. J.* **2013**, *4*, 191–196. [CrossRef]
16. Jalaludin, Z.; Hill, C.A.S.; Xie, Y.; Samsi, H.W.; Husain, H.; Awang, K.; Curling, S.F. Analysis of the water vapour sorption isotherms of thermally modified acacia and sesendok. *Wood Mater. Sci. Eng.* **2010**, *5*, 194–203. [CrossRef]
17. Calonego, F.W.; Severo, E.T.D.; Furtado, E.L. Decay resistance of thermally-modified Eucalyptus grandis wood at 140 C, 160 C, 180 C, 200 C and 220 C. *Bioresour. Technol.* **2010**, *101*, 9391–9394. [CrossRef] [PubMed]

18. Shi, J.L.; Kocaefe, D.; Amburgey, T.; Zhang, J. A comparative study on brown-rot fungus decay and subterranean termite resistance of thermally-modified and ACQ-C-treated wood. *Holz Als Roh Und Werkst.* **2007**, *65*, 353–358. [CrossRef]
19. Candelier, K.; Thevenon, M.-F.; Petrissans, A.; Dumarcay, S.; Gerardin, P.; Petrissans, M. Control of wood thermal treatment and its effects on decay resistance: A review. *Ann. For. Sci.* **2016**, *73*, 571–583. [CrossRef]
20. Rapp, A.O.; Brischke, C.; Welzbacher, C.R.; Jazayeri, L. Increased resistance of thermally modified Norway spruce timber (TMT) against brown rot decay by Oligoporus placenta–a study on the mode of protective action. *Wood Res.* **2008**, *53*, 13–25.
21. Chaouch, M.; Pétrissans, M.; Pétrissans, A.; Gérardin, P. Use of wood elemental composition to predict heat treatment intensity and decay resistance of different softwood and hardwood species. *Polym. Degrad. Stab.* **2010**, *95*, 2255–2259. [CrossRef]
22. Zelinka, S.L.; Passarini, L. Corrosion of metal fasteners embedded in acetylated and untreated wood at different moisture contents. *Wood Mater. Sci. Eng.* **2018**, 1–18. [CrossRef]
23. Humar, M.; Brischke, C.; Meyer, L.; Lesar, S.; Thaler, N.; Jones, D.; Bardage, S.; Belloncle, C.; Bulcke, J.; Abascal, M. Introduction of the COST FP 1303 cooperative performance test. IRG/WP 15-20567. In Proceedings of the 46th Annual Meeting of the International Research Group on Wood Protection, Vina del Mar, Chile, 10–14 May 2015.
24. Jermer, J.; Andersson, B.L.; Schalnat, J. Corrosion of fasteners in furfurylated wood- final report after 9 years exposure outdoors. IRG/WP 17-40810. In Proceedings of the 48th Annual Meeting of the International Research Group on Wood Protection, Ghent, Belgium, 4–8 June 2017.
25. Jermer, J.; Andersson, B.L. *Corrosion of Fasteners in Heat-Treated Wood—Progress Report after Two Years' Exposure Outdoors*; IRG/WP 05-40296; IRG Secretariat: Bangalore, India, 2005.
26. Dennis, J.K.; Zou, C.; Short, N.R. Corrosion behaviour of zinc and zinc alloy coated steel in preservative treated timber. *Trans. Inst. Met. Finish.* **1995**, *75*, 96–101. [CrossRef]
27. Short, N.R.; Dennis, J.K. Corrosion resistance of zinc-alloy coated steel in construction industry environments. *Trans. Inst. Met. Finish.* **1997**, *75*, 47–52. [CrossRef]
28. Zelinka, S.L.; Derome, D.; Glass, S.V. The effect of moisture content on the corrosion of fasteners embedded in wood subjected to alkaline copper quaternary treatment. *Corros. Sci.* **2014**, *83*, 67–74. [CrossRef]
29. Burkholder, M. CCA, NFBA, and the post-frame building industry. *Fram. Build. News* **2004**, *16*, 6–12.
30. Rush, F.A. *Deck Collapse- 4403 Ocean. Drive*; Town of Emerald Isle: Emerald Isle, NC, USA, 2015; p. 19.
31. Zelinka, S.L.; Derome, D.; Glass, S.V. Combining hygrothermal and corrosion models to predict corrosion of metal fasteners embedded in wood. *Build. Environ.* **2011**, *46*, 2060–2068. [CrossRef]
32. Baker, A.J. Corrosion of nails in CCA- and ACA-treated wood in two environments. *For. Prod. J.* **1992**, *42*, 39–41.
33. Zelinka, S.L. Uncertainties in corrosion rate measurements of fasteners exposed to treated wood at 100% relative humidity. *ASTM J. Test. Eval.* **2007**, *35*, 106–109.
34. Zelinka, S.L.; Rammer, D.R. Corrosion rates of fasteners in treated wood exposed to 100% relative humidity. *ASCE J. Mater. Civ. Eng.* **2009**, *21*, 758–763. [CrossRef]
35. Zelinka, S.L.; Sichel, R.J.; Stone, D.S. Exposure testing of fasteners in preservative treated wood: Gravimetric corrosion rates and corrosion product analyses. *Corros. Sci.* **2010**, *52*, 3943–3948. [CrossRef]
36. Kear, G.; Wu, H.-Z.; Jones, M.S. Weight loss studies of fastener materials corrosion in contact with timbers treated with copper azole and alkaline copper quaternary compounds. *Corros. Sci.* **2009**, *51*, 252–262. [CrossRef]
37. Kear, G.; Wú, H.Z.; Jones, M.S. *The Corrosion of Metallic Fastener Materials in Untreated*; CCA-, CuAz0 and ACQ-Based Timbers; BRANZ: Porirua, New Zealand, 2006; p. 281.
38. Kear, G.; Wú, H.Z.; Jones, M.S. Non-accelerated weight loss studies of fastener material corrosion in contact with CCA, CuAz, and ACQ-treated timbers. In Proceedings of the Corrosion and Protection 06 Conference, Hobart, Australia, 19–22 November 2006.
39. Anon. *ThermoWood handbook*; International ThermoWood Association: Helsinki, Finland, 2003; p. 66.
40. Siteworks, T. *Boulevard™ Decking Technical Data Sheet*; Tournesol Siteworks: Union City, CA, USA, 2019; Available online: https://cdn.tournesol.com/media/Documents/BL119.pdf (accessed on 23 December 2019).
41. Rammer, D.R.; Zelinka, S.L. Analytical determination of the surface area of a threaded fastener. *ASTM J. Test. Eval.* **2008**, *36*, 80–88.

42. Rammer, D.R.; Zelinka, S.L. Optical method for measuring the surface area of a threaded fastener. *Exp. Tech.* **2010**, *34*, 36–39. [CrossRef]
43. Anon. ASTM G198-17: Standard test. In *Method for Determining the Relative Corrosion Performance of Driven Fasteners in Contact with Treated Wood*; ASTM International: West Conshohocken, PA, USA, 2017.
44. Zelinka, S.L. Comparing the methodologies in ASTM G198 using combined hygrothermal-corrosion modeling. *Corrosion* **2013**, *70*, 206–213. [CrossRef]
45. Zelinka, S.L.; Jakes, J.E.; Kirker, G.T.; Passarini, L.; Hunt, C.G.; Lai, B.; Antipova, O.; Li, L.; Vogt, S. Copper distribution and oxidation states near corroded fasteners in treated wood. *SN Appl. Sci.* **2019**, *1*, 240. [CrossRef]
46. Zelinka, S.L. Corrosion of metals in wood products. In *Developments in Corrosion Protection*; Aliofkhazraei, M., Ed.; InTech: Rijeka, Croatia, 2014; Chapter 23; pp. 568–592.
47. Zelinka, S.L.; Stone, D.S. Corrosion of metals in wood: Comparing the results of a rapid test method with long-term exposure tests across six wood treatments. *Corros. Sci.* **2011**, *53*, 1708–1714. [CrossRef]
48. Zelinka, S.L.; Stone, D.S. The effect of tannins and pH on the corrosion of steel in wood extracts. *Mater. Corros.* **2011**, *62*, 739–744. [CrossRef]
49. Zelinka, S.L.; Rammer, D.R.; Stone, D.S. Corrosion of metals in contact with treated wood: Developing test methods. In Proceedings of the Corrosion Conference 2008, New Orleans, LA, USA, 16–20 March 2008.
50. Zelinka, S.L.; Rammer, D.R.; Stone, D.S. Electrochemical corrosion testing of fasteners in wood extract. *Corros. Sci.* **2008**, *50*, 1251–1257. [CrossRef]
51. Makkar, H. A laboratory manual for the FAO/IAEA co-ordinated research project on use of nuclear and related techniques to develop simple tannin assays for predicting and improving the safety and efficiency of feeding ruminants on tanniniferous tree foliage. In *Animal Production and Health Sub-Progrmme, FAO/IAEA Working Document*; IAEA: Vienna, Austria, 2000.
52. Makkar, H.P. Use of nuclear and related techniques to develop simple tannin assays for predicting and improving the safety and efficiency of feeding ruminants on tanniniferous tree foliage: Achievements, result implications, and future research. *Anim. Feed Sci. Technol.* **2005**, *122*, 3–12. [CrossRef]
53. Makkar, H.P.; Blümmel, M.; Borowy, N.K.; Becker, K. Gravimetric determination of tannins and their correlations with chemical and protein precipitation methods. *J. Sci. Food Agric.* **1993**, *61*, 161–165. [CrossRef]
54. Kodama, T. Corrosion of wrought carbon steels. In *ASM Metals Handbook, Volume 13B Corrosion: Matierals*; Cramer, S.D., Covino, B.S., Eds.; ASM International: Materials Park, OH, USA, 2003; pp. 5–10.
55. Zelinka, S.L.; Glass, S.V.; Boardman, C.R.; Derome, D. Comparison of the corrosion of fasteners embedded in wood measured in outdoor exposure with the predictions from a combined hygrothermal-corrosion model. *Corros. Sci.* **2016**, *102*, 178–185. [CrossRef]
56. Zhang, X.G. Corrosion of Zinc and Zinc Alloys. In *ASM Handbok Volume 13B, Corrosion: Materials*; Cramer, S.D., Covino, B.S., Eds.; ASM International: Materials Park, OH, USA, 2003; pp. 402–417.
57. Zhang, X.G.; Hwang, J.; Wu, W.K. Corrosion testing of steel and zinc. In Proceedings of the 4th International Conference on Zinc and Zinc Alloy Coated Steel Sheet (GALVATECH'98), Makuhari, Japan, 20–23 September 1998.
58. Smith, C.A. Corrosion of metals by wood. *Anti Corros. Methods Mater.* **1982**, *29*, 16–17.
59. Farmer, R.H. Corrosion of metals in association with wood. Part 2. Corrosion of metals in contact with wood. *Wood* **1962**, *27*, 443–446.
60. Bartel-Kornacka, E.T. Corrosion of iron by Ghana timbers. *Wood* **1967**, *32*, 39–42.
61. Knotkova-Cermakova, D.; Vlckova, J. Corrosive effect of plastics, rubber and wood on metals in confined spaces. *Br. Corros. J.* **1971**, *6*, 17–22. [CrossRef]

 *forests*

Article

Effect of Concrete on the pH and Susceptibility of Treated Pine to Decay by Brown-Rot Fungi

Darrel Nicholas *, Amy Rowlen and David Milsted

Department of Sustainable Bioproducts, Mississippi State University, Box 9820, Mississippi State, MS 39762, USA; amy.rowlen@msstate.edu (A.R.); drm522@msstate.edu (D.M.)
* Correspondence: d.nicholas@msstate.edu; Tel.: +1-662-325-8838

Received: 1 December 2019; Accepted: 21 December 2019; Published: 27 December 2019

Abstract: Treated wood timbers employed in ground contact are often installed with a cement collar to firmly fix the structural wood post in place. Few prior studies have determined the effect of concrete on decay efficacy on treated wood, however. Treated wood nominal 4 × 4 posts were installed at four locations, with the upper ground-contact portion of each post encased in concrete, and the samples removed at various times for pH measurements. The wood alkalinity quickly increased at all four sites for the portion of the treated wood in concrete contact compared to the wood in ground contact without concrete. In laboratory decay tests employing three decay fungi, untreated wood which was first exposed or unexposed to concrete had no consistent difference in decay susceptibility. For wood treated with three different commercial copper/organic systems, cement exposure had no effect on wood treated with an amine copper azole system, while treatment with amine copper quat showed a statistically significant fungal efficacy enhancement for cement-exposed samples with both copper-tolerant fungi. Conversely, with a micronized copper azole preservative, cement exposure resulted in reduced fungal efficacy compared to treated samples which were not cement-exposed for all three decay fungi.

Keywords: cement; wood decay; soil block test; wood preservatives

---

1. Introduction

Treated wood products are often used as posts in ground contact to support decks and various other structures. To minimize lateral movement, cement is often applied around the posts. Presently, it is unknown what effect the alkaline cement, which contains high levels of alkaline calcium compounds that would increase the pH of the wood, has on the efficacy of the treated wood against brown-rot decay fungi. If the cement does increase the susceptibility of treated wood to decay, it could reduce the service life of the posts and pose serious issues due to early collapse of structures and resulting injuries.

Early studies indicated that for chromated copper arsenate (CCA) treated wood the inclusion of concrete collars may enhance soft-rot decay in some applications [1–3]. In a more recent lab study using agar block microcosms [4], it was shown that both $CaCl_2$ and $CaSO_4$ inhibited decay of untreated sapwood by *Serpula lacrymans* and *Serpula himantioides*. Another study evaluated the effect of $CaCl_2$ on decay of untreated wood and copper-citrate-treated wood when exposed to *S. lacrymans* in a soil block test [5]. They concluded that $CaCl_2$ inhibited decay by this fungus for the copper-citrate-treated wood, but not for untreated wood. Another study reported that sapwood pine samples in soil block decay tests with 2% $CaCl_2$ added to the soil significantly reduced the extent of decay by both *Gloeophyllum trabeum* and *Rhodonia placenta* [6].

On the basis of the above limited studies it appears that calcium, which is a major component of cement, can in some cases influence the rate of wood decay by common wood-degrading fungi. However, it is not known what effect cement, which contains high levels of CaO and has a pH of 13, has on the performance of wood treated with commercial copper-based wood preservatives currently being used in soil contact applications. To provide some insight into the influence of concrete exposure on treated wood performance, our first goal of this study was to determine to what extent the Ca compounds in cement diffuse into treated wood when a cement collar is applied during installation of posts in soil. The second goal was to determine what effect cement has on the decay resistance of untreated wood or wood treated with commercial copper-based wood preservatives used in residential applications in North America.

2. Materials and Methods

2.1. Cement Diffusion Study

Four 12-foot nominal 4 × 4 southern pine (*Pinus spp.*) posts commercially treated with micronized copper azole (MCA) with a labeled retention of 2.4 kg/m^3 were obtained from a local lumber dealer in Starkville, Mississippi (MS). (Confusingly, in the US the actual size of lumber is smaller than the nominal size. This is due to lumber being sold over 100 years ago in the green and rough state; in later years when lumber was kiln dried then planed before being sold, the original green and rough size was still employed but denotated as nominal size. Further, the US Federal Code specifies that while the dimensions are to be in inches, the inch unit is not given. Thus, the actual size of a nominal 4 × 4 is 3.5 × 3.5 inches, or 89 × 89 mm.) Each of the four posts were then crosscut into four 864 mm long by 89 × 89 mm pieces, with each tagged. Four pieces, one from each of the four posts, were then installed in soil to a depth of 500 mm at each of four outdoor test sites which had soil pH values ranging from acid to alkaline. Three of the selected test sites were in the Starkville MS area and the fourth was at our Saucier MS test site in southern MS. Before backfilling with soil, Sakrete™ cement was applied to an area from the ground line to approximately 60 mm below the ground line. Following this, to determine whether cement components would diffuse into the posts, one post section was removed from each site at three exposure periods ranging from 42 to 340 days (Table 1). These post sections were then crosscut to provide 5-mm-thick samples (longitudinal direction) from both above ground areas and also two ground-contact areas, with and without cement contact. Each of these samples was then cut into sections at three depths, representing the outer 4.5 mm, second 4.5 mm and third 4.5 mm zones. Representative samples from these sections were then ground in a Wiley mill using a 20-mesh screen. One wood meal sample from each section was then evaluated by adding 10 mL of deionized (DI) water to 1 g of wood meal and the pH measured after one hour with a calibrated pH meter.

Table 1. Effect of cement on the pH of micronized or particulate copper azole system (MCA) treated wood after exposure of nominal 4 × 4 posts to soil contact at four different locations in Mississippi.

Post Exposure Site	Exposure Time (Days)	Wood pH at Var. Sample Depths and Vert. Location [1] in Posts											
		0–4.5 (mm)				4.5–9 (mm)				9–13.5 (mm)			
		A	B	C	D	A	B	C	D	A	B	C	D
Dorman	42	5.4	6.3	8.4	5.8	5.3	5.5	6.4	5.4	5.2	5.4	5.5	5.2
Dorman	208	5.0	5.1	8.1	5.8	5.0	5.0	6.8	5.3	4.8	4.9	6.0	5.1
Dorman	320	5.3	6.1	8.5	5.9	5.2	5.4	6.9	5.4	5.0	5.1	5.9	5.1
Saucier	39	5.3	6.4	9.3	5.7	5.3	5.6	8.4	5.5	5.3	5.3	6.6	5.4
Saucier	173	4.8	4.8	8.9	5.7	4.7	7.8	7.8	5.5	4.7	4,7	6.8	5.2
Saucier	340	5.0	5.6	9.1	5.8	5.0	5.1	7.9	5.4	5.0	5.1	7.0	5.5
Hillbrook	42	5.3	5.4	8.0	5.9	5.2	5.2	6.0	5.4	5.1	5.1	5.5	5.3
Hillbrook	177	5.2	5.0	7.3	5.9	5.0	5.0	6.1	5.6	4.9	4.9	5.7	5.4
Hillbrook	320	5.1	5.2	7.8	5.9	5.1	5.2	6.0	5.4	5.0	5.0	5.8	5.5
Longs Lk	42	5.4	5.6	8.3	6.0	5.2	5.5	6.3	5.6	5.1	5.3	5.8	5.4
Longs Lk	173	5.1	5.2	7.4	5.6	5.3	5.0	6.6	6.0	4.9	5.0	6.0	5.2
Longs Lk	320	5.3	5.4	8.0	5.4	5.1	5.3	6.5	5.8	5.0	5.2	6.1	5.4

[1] **A** denotes aboveground sample, **B** denotes aboveground samples adjacent to the ground line, **C** denotes sample in contact with concrete belowground, **D** denotes sample belowground not in contact with concrete.

Since cement contains high levels of calcium oxide, which has an approximate pH of 13, the pH of wood in contact with the cement should increase if appreciable amounts of the alkaline calcium salt diffuse into wood. Consequently, measuring the pH of wood in contact with cement was considered to be a good measure of calcium migration into wood.

2.2. Laboratory Decay Test

To determine the effect of cement on wood decay, a soil block test was performed with three brown-rot fungi. The decay test was carried out in accordance with American Wood Protection Association (AWPA) Standard E22-15 using three different brown-rot fungi, *Gloeophyllum trabeum* (ATCC 11539; American Type Culture Collection, Old Town Manassas, Virgina, VA, USA), *Rhodonia placenta* (ATCC 11538) and *Fibroporia radiculosa* (TFFH 294, USDA FPL). The fungal cultures were maintained on 2% malt extract agar (Difco Laboratories, Detroit, Michigan, MI, USA). A total of eight replicate wafers were used in each test, using four and six weeks of exposure time for each decay test.

The wood test samples for the decay test were produced from three flatsawn pine (*Pinus glabra* Walt.) boards. Three defect-free sapwood sections measuring 19 × 70 × 1120 mm (radial × tangential × longitudinal) were cut from each of the boards, providing one sample for each of the three test fungi. These samples were then rip sawn into three 19 × 19 × 1120 mm sticks which were randomly assigned to each of the three preservative treatments. Each stick was then crosscut into two 560-mm-long pieces to provide end-matched untreated material for controls. The sticks designated for treatment were then pressure treated to a target of the specified ground-contact residential retention with three commercial copper-based systems (with the actual retention obtained shown in parenthesis): the amine copper azole system (CA-C, 2.3 kg/m^3), the amine copper quaternary system (ACQ-D, 6.3 kg/m^3) and micronized or particulate copper azole system (MCA, 2.4 kg/m^3). The full cell pressure treatment used an initial vacuum of 95 kPa for 30 min followed by 1034 kPa pressure for 30 min. The samples were then wiped clean and weighed to determine the actual treating solution retention before air drying the samples. The actual retentions for all three preservatives were within 0.1 kg/m^3 of the specified AWPA UC4A retention.

Both the untreated and treated sticks were then crosscut into two 280-mm-long pieces to provide samples for cement-exposure and non-cement-exposure controls. The sticks designated for cement exposure were end-sealed with a wax emulsion (seal type ISK Biocides, Memphis, Tennessee, TN, USA) and then coated with a thin layer of wet Sakrete™ cement purchased at Lowes, Starkville, Mississippi, MS, USA, approximately 5 mm thick, on all the lateral surfaces and allowed to air dry. Following this, the cement-coated sticks were wrapped in nylon stocking material and placed vertically into plastic buckets containing wet soil obtained from the Dorman MS wood preservation test plot. After 45 days of exposure the sticks were removed and allowed to air dry. The cement was then removed from the sticks followed by a thorough cleaning to remove all visible cement from the surfaces.

Each of the twelve 280-mm-long sticks was then crosscut to provide three groups of 5-mm-thick wafers for the soil block test. Within each stick, one group of wafers was designated as unexposed controls and the other two groups were assigned to the two fungal exposure of four and six weeks. After fungal decay exposure the compression strength was measured as per AWPA E22 Standard, with the extent of deterioration reported as the strength loss of the fungal-exposed samples relative to the unexposed matched controls. On sets where the strength difference between the treated samples that were cement-exposed compared to non-cement-exposed appeared appreciably different at four and six weeks of exposure, a *t*-test was run using the individual eight replicate samples using Minitab at the 95% confidence level.

3. Results and Discussion

The data in Table 1 clearly show that when cement is applied to nominal 4 × 4 posts in soil contact, alkaline components diffuse into the wood. As expected, the alkalinity is greatest in the outer 0–4.5 mm zone for belowground wood adjacent to the cement (samples C) compared to the lower portion of the wood with no cement (samples D), with good diffusion into the wood outer shell occurring at all four test sites. The middle zone of 4.5–9 mm also shows greater alkalinity for cement-exposed C samples as compared to D samples, and the inner 9–13.5 mm zone also shows greater alkalinity with the alkalinity increasing somewhat at the longer exposure times. The data also indicate that diffusion occurs rapidly in the outermost layer, with no appreciable increase with the longer exposure times. With regard to the effect of exposure sites, greater diffusion occurred at the Saucier test site.

Results of the soil block decay test with three brown-rot fungi for samples with and without cement exposure are presented in Tables 2–4. The results are presented as the average percent compression strength loss after exposure to the fungus for four and six weeks of exposure, with greater values indicating higher deterioration. Table 2 represents the copper-intolerant fungus *G. trabeum*; Tables 3 and 4 show the highly aggressive copper-tolerant fungi *R. placenta* and *F. radiculosa*, respectively.

Table 2. Comparative decay resistance of untreated and preservative-treated pine wafers with and without cement infusion and exposed to *Gloeophyllum trabeum* in a soil block test. Each value is the average of eight replicates, with high values representing extensive decay.

Preservative Treatment	Cement Treatment	Percent Compression Strength Lost after Exposure to the Fungus for:	
		4 Weeks	6 Weeks
None	No	97	98
None	Yes	98	98
CA-C	No	6	10
CA-C	Yes	0	0
None	No	98	99
None	Yes	99	100
ACQ-D	No	11	8
ACQ-D	Yes	6	12
None	No	96	97
None	Yes	98	97
MCA	No	22	14
MCA	Yes	27	40

Table 3. Comparative decay resistance of untreated and preservative-treated pine wafers with and without cement infusion and exposed to *Rhodonia placenta* in a soil block test. Each value is the average of eight replicates, with high values representing extensive decay.

Preservative Treatment	Cement Treatment	Percent Compression Strength Lost after Exposure to the Fungus for:	
		4 Weeks	6 Weeks
None	No	88	91
None	Yes	93	94
CA-C	No	91	96
CA-C	Yes	82	96
None	No	96	96
None	Yes	95	97
ACQ-D	No	63	94
ACQ-D	Yes	16	84
None	No	90	96
None	Yes	96	98
MCA	No	36	64
MCA	Yes	46	78

Table 4. Comparative decay resistance of untreated and preservative-treated pine wafers with and without cement infusion and exposed to *Fibroporia radiculosa* in a soil block decay test. Each value is the average of eight replicates, with high values representing extensive decay.

Preservative Treatment	Cement Treatment	Percent Compression Strength Lost after Exposure to the Fungus for:	
		4 Weeks	6 Weeks
None	No	92	95
None	Yes	95	98
CA-C	No	90	95
CA-C	Yes	85	97
None	No	91	94
None	Yes	94	97
ACQ-D	No	82	94
ACQ-D	Yes	26	69
None	No	89	92
None	Yes	84	96
MCA	No	89	91
MCA	Yes	93	96

For the untreated wood samples, extensive deterioration occurred with all three fungi for samples both exposed and unexposed to concrete, even at the shorter exposure time of four weeks. No consistent deterioration effect was observed between the concrete- and non-concrete-exposed untreated samples. Thus, we conclude that exposure to concrete gives no consistent difference in deterioration effect in short duration laboratory tests for pine samples which are untreated and exposed to three common decay fungi.

With the CA-C-treated samples, no or minor deterioration was obtained with the copper-intolerant *G. trabeum* fungus (Table 2). Conversely, with the copper-tolerant fungi *R. placenta* (Table 3) or *F. radiculosa* (Table 4), extensive deterioration occurred. However, with all three fungi no consistent practical differences were observed in the deterioration between the concrete-exposed versus concrete-unexposed CA-C-treated samples.

For the ACQ-D-treated samples, minor deterioration was obtained with the copper-intolerant fungus *G. trabeum* (Table 2) for both cement-exposed and unexposed samples. As expected, greater deterioration was observed with the copper-tolerant fungi *R. placenta* (Table 3) and *F. radiculosa* (Table 4). Interestingly, with both of these two copper-tolerant fungi, samples that were first exposed to concrete exhibited greater decay resistance compared to ACQ-D-treated samples which were not exposed to concrete. A *t*-test comparing cement-exposed versus non-cement-exposed samples showed that the greater efficacy was significant at the 95% or greater level for all four sets at both incubation times with the two copper-tolerant fungi. We hypothesize that greater alkalinity upon concrete exposure results in a stronger complex between the quat cation and the acidic anion groups in wood and, thus, enhanced fungicidal efficacy.

For MCA, or micronized copper azole, the copper is mainly present as submicron-sized particles, with even the smallest particulate copper particles composed of many millions of insoluble and thus nonfungicidal copper atoms; these copper atoms are slowly solubilized over time to form individual and therefore fungicidal copper ions with the solubilization rate increasing as the wood acidity increases. With MCA-treated samples, moderate deterioration was obtained with the copper-intolerant *G. trabeum* fungus (Table 2), with the concrete-exposed samples showing slightly more but not significant deterioration at four weeks and much greater and statistically significant deterioration at six weeks compared to the non-concrete-exposed samples based on a *t*-test. As expected with the aggressive copper-tolerant fungi *R. placenta* (Table 3) or *F. radiculosa* (Table 4), extensive deterioration occurred with the MCA-treated samples. While greater deterioration occurred for the concrete-exposed samples relative to the non-concrete-exposed samples at both four and six weeks for exposure with *R. placenta*, neither difference was statistically significant. With *F. radiculosa*, slightly greater decay was again obtained with the concrete-exposed samples, and while the difference at four weeks was not significant, the decay at six weeks was statistically greater for the concrete-exposed samples. We propose that the greater alkalinity for concrete-exposed samples, as determined in the first part of this study, results in reduced solubilization of the particulate copper and, consequently, lower amounts of soluble and, thus, fungicidal copper ions in concrete-exposed MCA-treated samples as compared to treated wood which is not exposed to concrete.

4. Conclusions

Commercial MCA-treated ground-contact posts surrounded by a concrete collar installed at four outdoor sites quickly increased to a greater pH than the lower portion of the post below the concrete at all four test sites locations. In laboratory soil block decay tests, untreated pine sapwood which was exposed or not exposed to concrete prior to fungal testing showed no deterioration effect upon exposure for 4 and 6 weeks to three different fungi. With wood samples treated with three commercial copper organic preservatives and exposed to three decay fungi, different decay efficacies were obtained for samples which were exposed to concrete depending on the particular preservative employed; for all three systems greater deterioration was always observed with the two copper-tolerant fungi than the copper-intolerant fungus. Specifically, the amine copper azole preservative CA-C showed no difference in deterioration with all three fungi. The amine copper quat system ACQ-D showed significantly greater decay resistance for the cement-exposed samples with both exposure times and the two copper-tolerant decay fungi. Conversely, for micronized copper azole MCA-treated wood, the cement-exposed samples had greater deterioration with all three fungi compared to the non-cement-exposed samples, with some of the greater decay statistically significant. We conclude that for wood treated with copper-based preservatives and then exposed to concrete when installed in ground contact, decay susceptibility may be unaffected, reduced, or enhanced, depending on the particular copper/organic wood preservative employed.

Author Contributions: Conceptualization, D.N.; Methodology, D.N.; Formal Analysis, A.R.; Investigation, D.N. and A.R.; Resources, D.N.; Data Curation, A.R.; Writing—Original Draft Preparation, D.N.; Writing—Review and Editing, D.M.; Project Administration, D.N.; Funding Acquisition, D.N. All authors have read and agreed to the published version of the manuscript.

Funding: This work is/was supported by the United States Department of Agriculture National Institute of Food and Agriculture, McIntire Stennis project #1005755.

Acknowledgments: This publication is a contribution of the Forest and Wildlife Research Center, Mississippi State University.

Conflicts of Interest: The authors declare no conflict of interest.

References

1. Murphy, R.J. *The Influence of Cement and Calcium Compounds on Performance of CCA Preservatives*; IRG/WP 13–3221; International Research Group on Wood Protection: Stockholm, Sweden, 1983. Available online: https://www.irg-wp.com (accessed on 25 December 2019).
2. Murphy, R.J. *Wood in Concrete. Summary of Discussion at IRG 14, Surfers Paradise, Australia*; IRG/WP 14–3264; International Research Group on Wood Protection: Stockholm, Sweden, 1984. Available online: https://www.irg-wp.com (accessed on 25 December 2019).
3. Leightley, L.E.; Willoughby, G.A. *The Effect of Concrete on CCA Treated Hardwood and Softwood Timbers*; IRG/WP 15–3340; International Research Group on Wood Protection: Stockholm, Sweden, 1985. Available online: https://www.irg-wp.com (accessed on 25 December 2019).
4. Schilling, J.S. Effects of calcium-based materials and iron impurities on wood degradation by the brown rot fungus *Serpula lacrymans*. *Holzforschung* **2010**, *64*, 93–99. [CrossRef]

5. Hastrup, A.C.S.; Jensen, B.; Clausen, C.; Green III, F. The effect of $CaCl_2$ on growth rate, wood decay and oxalic acid accumulation in *Serpula lacrymans* and related brown-rot fungi. *Holzforschung* **2006**, *60*, 339–345. [CrossRef]
6. Little, N.S.; Schultz, T.P.; Nicholas, D.D. Effect of different soils and pH amendments on brown-rot decay activity in a soil block test. *Holzforschung* **2010**, *64*, 667–871. [CrossRef]

Article

Study of Gliding Arc Plasma Treatment for Bamboo-Culm Surface Modification

Bin Li †, Jinxing Li †, Xiaojian Zhou *, Jun Zhang, Taohong Li and Guanben Du *

Yunnan Provincial key Laboratory of wood Adhesive and Glued Products, Southwest Forestry University, Kunming 650224, China; touchbinbin@swfu.edu.cn (B.L.); jinxingli@swfu.edu.cn (J.L.); zj8101274@163.com (J.Z.); lith.cool@163.com (T.L.)

* Correspondence: xiaojianzhou@hotmail.com (X.Z.); gongben9@hotmail.com (G.D.)

† These authors contributed equally to this work.

Received: 28 October 2019; Accepted: 26 November 2019; Published: 1 December 2019

Abstract: Plasma treatment was conducted to modify the outer- and inner-layer surfaces of bamboo in a multi-factor experiment, where the surface contact angles and surface energy were measured, followed by investigation on the surface microstructure and functional groups using a scanning electron microscope (SEM) and X-ray photoelectron spectroscopy (XPS), respectively. The result showed that when the power of the gliding arc plasma treatment was 1000 W while the bamboo surface was 3 cm away from the nozzle of the plasma thrower in the plasma flame, the contact angles of the outer- and inner-layer surfaces decreased, whereas the surface energy increased as a function of the treatment time. The 40 s treatment on the outer-layer surface caused the contact angle to reach 40°, and the surface energy accomplished a value of 45 J. Likewise, when the inner-layer surface was exposed for 30 s treatment, its contact angle attained a value of 15°, while the surface energy elevated to 60 J. Surface assessment with scanning electron microscopy (SEM) demonstrated etched microstructures of outer- and inner-layer surfaces of the bamboo culm after the treatment with gliding arc plasma. Moreover, the soaking test performed on the surfaces signified that 2D resin could have adhered more easily to outer- and inner-layer surfaces, which was considered a result of the greater uniformity and smoothness acquired after the treatment. X-ray photoelectron spectroscopic (XPS) analysis revealed that hydrophilic groups (O-CO-N, $-NO^{2-}$,$-NO^{3-}$, C-O-C, C-O-H and O-CO-OH, C-O-C = O) emerged on outer- and inner-layer surfaces of bamboo culms after being treated by gliding arc plasma, which enhanced the interaction of bamboo culms with applied protective coating resins.

Keywords: moso bamboo (*Phyllostachysheterocycle cv. Pubescens*); gliding arc plasma; surface treatment; activated surface; wettability

---

1. Introduction

Moso bamboo (*Phyllostachysheterocycle cv. Pubescens*) is widely used in architecture and furniture industries due to its fast-growing and good structural properties. However, the nutrient-rich parenchyma cells of bamboo, which contain starch and carbohydrates, make it highly susceptible to damages caused by molds and bamboo beetles, resulting in mildew, decay, and cracking. All these outcomes will reduce the service life of bamboo and limit its applications [1–3]. In addition, the existing green bamboo and yellow bamboo have a resulting poor gluability and paintability. There are some reports about the modification of bamboo using different methods [4–7], and a wide range of studies focusing on bamboo protection and modification have been undertaken from perspectives of their biodegradation mechanisms, preservatives development [8,9], and anti-decay methods [10–12].

The last few years have witnessed a considerable expansion of products made of bamboo culms. The absence of horizontal tissues, together with the presence of a wax layer on the hard surface, made it very difficult for regular preservatives and mildew preventive treatments to penetrate smoothly into the bamboo. Further, other problems like the poor adherence of preservatives to bamboo culms and the susceptibility to cracking have not been addressed properly [13], rendering bamboo materials highly vulnerable to mildew and worm problems. Many antiseptic methods and equipment have been developed to address this issue. Among them, one is based on soaking preservatives by exerting pressure on one end of the bamboo [14]. This helped to effectively improve the decay resistance ability. However, these methods often involve utilizing complicated equipment, thus cannot be employed on a large-scale.

Plasma can be used to activate surfaces of materials by introducing new functional groups to the material surface, thereby enhancing materials' properties [15–19]. Exposing bamboo surface to plasma treatment results in enhanced surface wettability and improved the accessibility of coating materials [20,21]. However, most of undergoing research on the application of plasma for surface treatment of bamboo focuses on the mechanism of action and temporal effect of plasma. At present, plasma equipment, involving either radio frequency discharge (RF) [22] or dielectric barrier discharge (DBD) [23], are widely engaged in wood and bamboo treatments. Regrettably, the use of this equipment is restricted in terms of the sample sizes to be treated as they can only process small-sized materials [24,25].

Although gliding arc plasma treatment methods have been used in different fields [26–28], there is no research on bamboo treatment with gliding arc treatment. This shows the importance and challenge of using such existing technology to process large-sized materials like bamboo culms.

The 2D resin (Dimethylol dihydroxy ethylene urea, DMDHEU) is a typical resin of the nitrogen hydroxymethyl compound with low molecular weight, which has excellent properties in low formaldehyde emission, anti-wrinkle and anti-shrink mechanical properties, and stability, and is used to replace formaldehyde resin in the cotton fiber textile industry and wood modification [29–33]. Some researchers have carried out wood modification studies using 2D resin; the self-polymerization and cross-linking reaction of the 2D resin and wood compounds occurred within the cell wall, resulting in a permanent bulking of the cell wall and leading to a reduction in the swelling and shrinkage properties, thus with the dimensional stability considerably increased. In addition, good preservative properties and high resistance against white, brown, and soft rot fungi are obtained. The treatments also enhance the wood's acoustic, weathering, and aging properties, furnishing and gluing performances, as well as mechanical properties [30–33]. However, there are no reports on bamboo culms modification with the 2D resin.

Improving the retention of the 2D resin in bamboo was attempted in the current work after enhancement of the surface reactivity by applying its surfaces to low-temperature gliding arc plasma treatment, which would provide a theoretical basis and technical platform to solve the bamboo decay, mildew, instability and cracking, etc., for expanding the industrial use of bamboo.

2. Materials and Methods

2.1. Experimental Material

A 4-year-old fresh moso bamboo (*Phyllostachys heterocycle cv. Pubescens*) culm, with a height above 8 m, average diameter around 12 cm, minimum diameter no less than 8 cm, and wall thickness close to 10 mm, was selected from Longnan, Ganzhou City, Jiangxi Province, China. To prepare the samples for plasma treatment, bamboo culms with uniformly straight and smooth surfaces were initially cut 10 cm above ground level, then cut sequentially into 2 sections of 1.5 m bamboo tubes with a moisture content of 65%–70% and density of 700–720 kg/m^3. A gliding arc plasma equipment (model: CTD-2000 F; size: 250 × 200 × 360 mm; weight: 12 kg; handheld; Suman plasma, Nanjing, China) with an atmospheric low-temperature plasma torch was employed for surface treatment. It was fully controlled using software with one button through a microcontroller unit (MCU). The rated input power supply for the plasma torch was AC 220 V. Providing 1000 W output power, the plasma torch was capable of treating specimens with a width between 60–80 mm. A lab-made 55%–65% active component dimethylol dihydroxy ethylene urea resin (DMDHEU, also known as 2D resin), with colorless to pale-yellow color and a slightly offensive odor, was used, which can be dissolved in water in any proportion. Analytical-grade diiodomethane and glycerol were utilized for contact angle measurements.

2.2. Plasma Treatment of Outer- and Inner-Layer of Bamboo Culms

In Figure 1, Bamboo culm specimens with the bamboo outer surface up were placed at different distances from the nozzle of the gliding arc plasma emitter (3, 5, and 7 cm), which was operated at a power of 1000 W. The moving speed of the thrower was set at 1 cm/s and the samples were treated repeatedly and evenly for 5, 10, 15, 20, 30, 40, and 60 s, respectively. The treatment of the inner layer or the inner wall surface was performed on bamboo culms cut into strips of 5 mm width and involved exposure to plasma radiation exactly as described for the outer layer.

Figure 1. The plasma treatment of bamboo culms using a plasma generator.

2.3. Resin Soaking after Plasma Treatment

After treatment with cold plasma, the bamboo specimens, including culms and strips, were soaked with the 2D resin for 24 h. Subsequently, they were placed in a drying oven initially set at 60 °C, and the temperature was increased to 100 °C at a rate of 10 °C/min. Finally, the samples were taken out after the temperature subsided gradually to 40 °C.

2.4. Characterizations

2.4.1. Contact Angle Measurements and Surface Energy Calculation

A static solvent contact angle meter (JC2000A, Shanghai Zhongcheng Co. Ltd., Shanghai, China) was utilized for measuring the contact angles of the outer- and inner-layer surfaces using diiodomethane and glycerol as test liquids in the laboratory condition of 25 °C and humidity of 40% RH. According to the sample preparation procedure, photos were taken 2 s after the liquids were dropped on treated bamboo surface, and the angle-measuring method was employed to measure the contact angles of the liquid on specimen surfaces. The measurements were undertaken on 6 points on the outer- as well as inner-layer surfaces of bamboo culms for each solvent, and average values were recorded. After that, the Owens two-liquid method was applied to calculate the surface energy values of the specimens [34].

2.4.2. Surface Microstructure Observation

Both the outer- and inner-layer surfaces of treated bamboo specimens were fixed on the sample holder using conductive adhesive with the outer- or inner-layer to be tested facing upwards. Then, the specimens were placed in the compartment of sputter coating equipment, which was degassed using a vacuum pump for 5 min. The surfaces were sputter-coated with gold to avoid charging when the vacuum level reached 3 Pa, and the surface morphology was examined using SEM (Quanta 200, ESEM, FEI, Hillsboro, OR, USA).

2.4.3. X-Ray Photoelectron Spectroscopy (XPS) Investigation

The treated outer- and inner-layer surfaces treated with gliding arc plasma were scanned with K-Alpha + X-ray Photoelectron Spectroscopy at a vacuum of almost 2×10^{-7} mbar using monochromatic Al Kα X-ray source with energy 1486.6 eV, 6 mA × 12 kV, and a spot size of 400 μm as the light beam. The measurement parameters were incorporated CAE (constant analyzer energy) as a scanning mode and a full-survey spectrum with a pass energy of 100 eV and step size of 1 eV or narrow-survey spectrum with a pass energy of 30 eV and step size of 0.1 eV. The binding energy measurements were calibrated according to C1s surface contamination standard (284.8 eV).

3. Results and Analysis

3.1. Effect of Plasma Treatment on the Surface Characteristics of Bamboo Culm

3.1.1. Effect of Gliding Arc Plasma Treatment of Bamboo Culm on Contact Angles and Surface Energy of Outer-Layer Surface

Figure 2a represents the trend of variation in the surface energy of the outer-layer surface of bamboo culm after exposure to the gliding arc plasma treatment for different time intervals, whereas the samples were mounted at various distances from the nozzle of the plasma emitter, 3, 5, and 7 cm. Initially, it is obvious that the 7 cm distance seems too far for the surface to be influenced by the treatment, and the change was statistically insignificant. The surface energy response in the case of the 5 cm distance was much higher, and maximum energy was achieved within 25 s and then started to decrease again with prolonging the treatment time to reach a minimal level. This may reveal either damage or carbonization of the surface. The 3 cm distance between the plasma emitter and the sample looks much more appropriate for the treatment, as the surface energy continued to increase with the exposure time up to 40 s. Even if the treatment was extended 20 s more, the extent of carbonization was very limited. Careful examination of the associated change in contact angle between bamboo culm and the resin indicates a strong correlation with the developed surface energies (Figure 2b). That is to say, if more active surface groups are born, they would contribute appreciably to enhance the wetting of the surface with the resin. Thus, the optimal contact angles between the bamboo culm and 2D resin can be obtained by verifying the samples exposed at 3 cm away from the nozzle of the plasma emitter while the treatment lasts for 40 s. Under such conditions, the contact angle between the outer-layer and the 2D resin declines remarkably from 100° to about 40°.

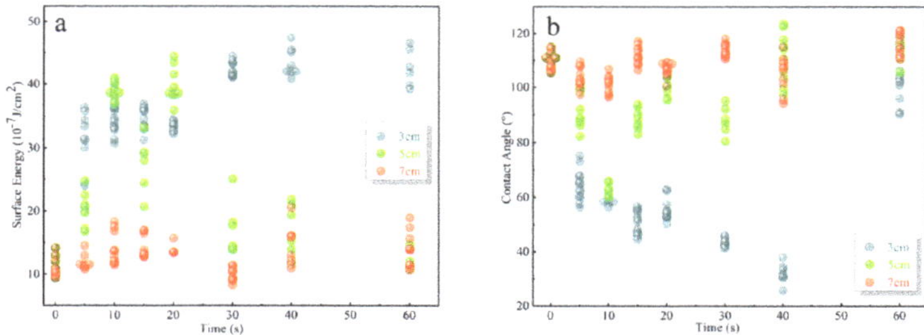

Figure 2. Variations in the surface energy of the outer-layer surface of bamboo culm exposed to treatment with gliding arc plasma (**a**) and the contact angle of the treated outer-layer with 2D resin (**b**) as a function of treatment time.

3.1.2. The Effect of Gliding Arc Plasma Modification on Inner-Layer Surface Properties

Figure 3 provides the trend of variations in surface energy and contact angle of the inner-layer after the gliding arc plasma treatment. It is clear that the surface energy of the inner-layer increased significantly regardless of the distance between the nozzle of the plasma emitter and the inner layer surface. However, the enhancement in surface energy is more sounding as far as the distance is shorter (Figure 3a). In parallel, the contact angle between the inner-layer surface and the 2D resin changes sharply in a reverse direction (Figure 3b). After 30 s, there might be a certain level of carbonization developed over time, which is the reason the contact angle started to re-increase.

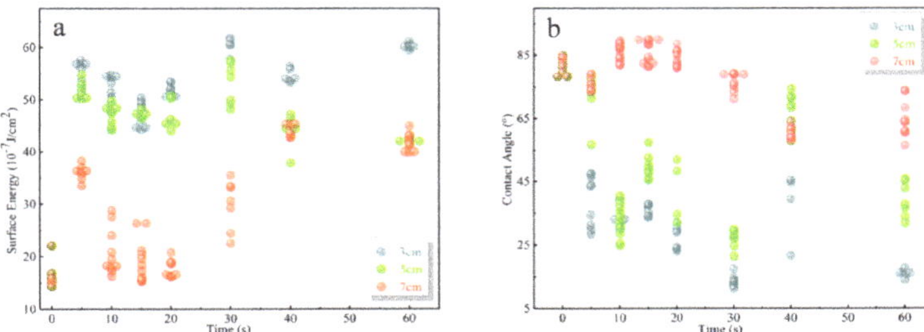

Figure 3. Variations in the surface energy of the inner-layer surface of bamboo culm exposed to gliding arc plasma treatment (**a**) and contact angle of the treated inner-layer with 2D resin (**b**) as a function of treatment time.

3.2. *Effect of Plasma Treatment on Surface Microstructure of Bamboo Culm*

3.2.1. Effect of Gliding Arc Plasma Treatment on the Microstructure of Outer-Layer Surface

As can be seen from the outer-layer surface microstructures shown in Figure 4, a significant difference can be recognized after the plasma treatment of 40 s, and the bamboo culm surface is around 3 cm away from the nozzle of the plasma emitter. Comparing Figure 4a with Figure 4d reveals etching developed on the outer-layer surface after gliding arc plasma treatment, which imposes the resin more liable for the soaking step. While comparing Figure 4b,e against Figure 4c,f dictates a uniform smoother surface, lacking any signs of cracking for the 2D resin soaking surface after the plasma treatment, which was not the case for the surface soaked with 2D resin without qualifying the surface to this step by prior plasma treatment. It is obvious for the latter that the 2D resin adhered more strongly to the outer-layer surface, which allows for an effective protective role for the resin on the bamboo culms.

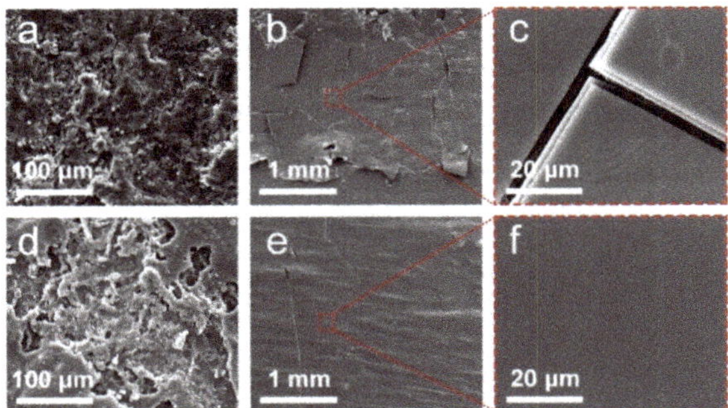

Figure 4. Morphological features of the outer-layer surface: (**a**) before resin soaking, (**b**,**c**) different magnification images after soaking followed by drying, (**d**) after exposing to gliding arc plasma treatment, and (**e**,**f**) different magnification images of soaking and drying following the treatment.

3.2.2. Effect of Gliding Arc Plasma Treatment on the Microstructure of Inner-Layer Surface

Figure 5 shows the morphological features of the inner-layer surface at different magnifications with the plasma treatment of 30 s, and the culm inner surface is around 3 cm away from the nozzle of the plasma emitter. Figure 5a,b exhibit the morphological aspects of the untreated inner-layer surfaces, which is relatively smooth, and no grains could be detected. On the other hand, the treatment with gliding arc plasma rendered the surface a little bit rougher (Figure 5d,e). The comparison between Figures 5c and 5f indicates that the spreading and covering of the 2D resin over the surface exposed in prior to plasma arc treatment proceeded much more effectively, which sheds light on the responsibility of the treatment in generating active sites that play the role of the receptor to the 2D resin and enhances its adhesion to the surface.

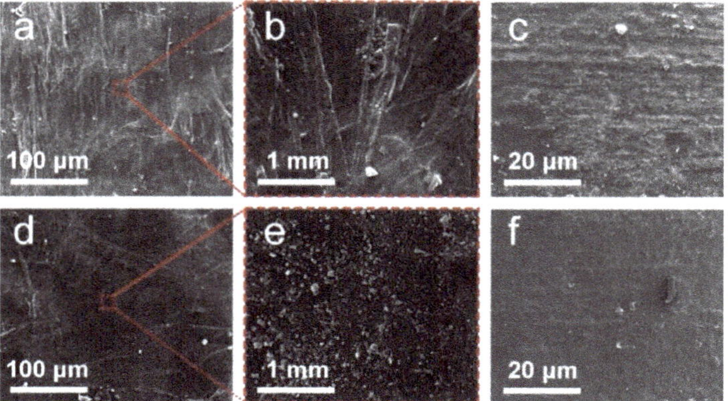

Figure 5. Morphological features of the inner-layer surface at different magnifications before (**a**,**b**) and after soaking followed by drying (**c**), different magnification images of the surface after exposing to gliding arc plasma treatment (**d**,**e**), and subsequent soaking and drying (**f**).

3.3. *XPS Study of the Surface Functional Groups of Bamboo Culm Following Gliding Arc Plasma Treatment*

3.3.1. The Outer-Layer Surface

To verify changes on the outer-layer surface and evolution of polar groups after plasma treatment, XPS spectroscopic analysis was undertaken. Figure 6a–c shows the scans of some elements on the outer-layer surface before treatment with gliding arc plasma, which translates into a few active groups anchored on the surface. After exposure to plasma treatment, new active functional groups emerged on the outer-layer surface, as indicated mostly by changes in C1s environments (Figure 6d). This reveals the transformation of N-C = O into O-CO-N after the gliding arc plasma treatment. Similarly, some associated changes took place in the environments of the other elements, particularly O1s (Figure 6e,f), signifying the appearance of newborn hydrophilic groups. The emergence of these hydrophilic groups justifies the ability of 2D resin to be soaked and adhered more efficiently to the outer-layer surface of the bamboo culm.

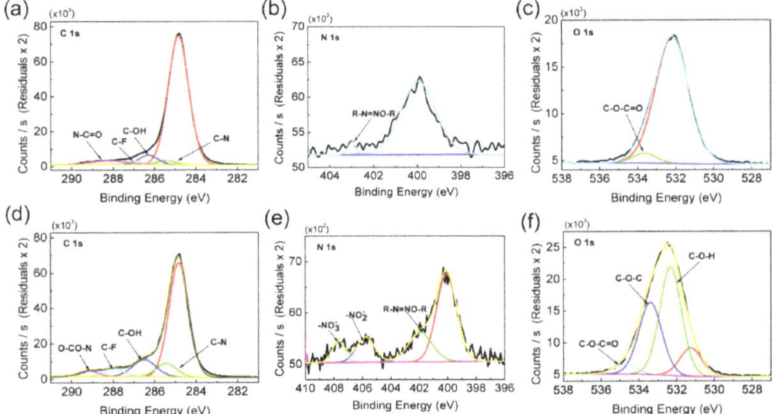

Figure 6. Narrow scan XPS spectra of the outer-layer surface for C1s (**a**), N1s (**b**), and O1s (**c**) elements before and after treatment (**d**–**f**), respectively.

3.3.2. The Inner-Layer Surface

Furthermore, Figure 7 exhibits the comparable XPS spectra recorded for the same groups on the inner-layer surface. It is clear that the extent of variation in the surface environment, especially the C1s, is much higher as compared to the outer surface. It additionally demonstrates that an additional hydrophilic O-CO-OH group appeared on the inner-layer surface after treatment. However, no significant difference was recognized in the case of N1s, whereas the corresponding scans of O1s show the emergence of an additional hydrophilic group C-O-C = O, which ensures enhanced accountability for resin soaking and penetration. The newly emerged groups on both inner as well as outer-layer surfaces of the bamboo culm are summarized in Table 1.

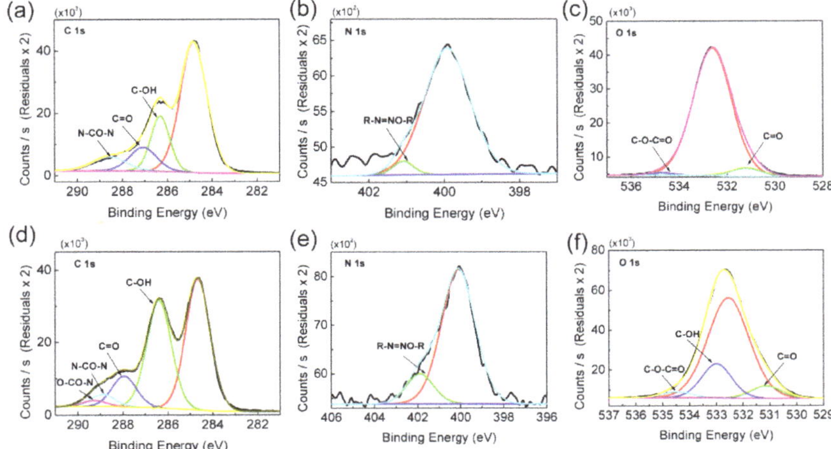

Figure 7. Narrow scan XPS spectra of the inner-layer surface for C1s (**a**), N1s (**b**), and O1s (**c**) elements before and after treatment (**d**–**f**), respectively.

Table 1. A list of the new-born hydrophilic groups on bamboo culm surfaces following gliding arc plasma treatment.

Layers of Bamboo Culm	Evoluted Groups
Outer-layer	O-CO-N, $-NO^{2-}$,$-NO^{3-}$, C-O-C, C-O-H
Inner-layer	O-CO-OH, C-O-C = O

Similar research results have been demonstrated in which new functional groups like CF_3, CHF, CF, C-O-H, C = O, COOH, CO_2, O-C-O, O = C-O, C = O, and C = N were found on the matrix surface after the plasma treatment [35–38]. This could be attributed to the formation of oxidized groups rich in hydroxyl, carbonyl, carboxyl groups, and phenoxy radicals. In this study, the evolution of functional groups can be explained by the fact that gliding arc plasma is a high energy ion, which can decompose the waxy material on the surface of bamboo so that the ester compounds in the waxy layer can be modified into new functional groups.

4. Conclusions

The outer- and inner-layer surfaces of bamboo-culm were activated by exposure to gliding arc plasma under optimized conditions. Maximized enhancement of the surface activity was reached when the bamboo culm surface was no more than 3 cm away from the nozzle of the plasma emitter for about 40 s. These conditions were quite sufficient, as revealed by the measurements of the contact angle and surface energy, 40° and 45 J, respectively. However, the inner-layer surface required shorter time (30 s) to attain the optimized state and became more energetic and wettable as demonstrated by the contact angle and surface energy, 15° and 60 J, respectively. This was achieved via etching and a morphological change of the surface, which enabled the surface to respond more effectively to soaking by 2D resin. The stronger adhesion achieved on the surface between the bamboo culm and 2D resin was a direct result of the emergence of new hydrophilic groups on the surface following the treatment, which caused more wetting and better resin spreading, leading to more effective protective covering of the surface.

Author Contributions: B.L. and J.L. contributed the experiment processing and characterization of the samples. J.Z. and T.L. contributed the XPS testing and analysis. X.Z. and G.D. contributed the design of the experiment and the analysis of the results as well as organize the whole manuscript.

Funding: This work was supported by the National Key Research and Development Program of China (Research of protection technology for bamboo culms, Project No. 2017YFD0600803), the National Natural Science Foundation of China (NSFC 31760187) and Yunnan Provincial Natural Science Foundation (2017FB060) as well as the Yunnan provincial youth and middle-age reserve talents of academic and technical leaders (2019HB026).

Conflicts of Interest: The authors declare no conflict of interest.

References

1. Zhang, Q.S.; Jiang, S.X.; Tang, Y.Y. Industrial utilization on bamboo. In *International Network for Bamboo and Rattan*; People's Republc: Beijing, China, 2002.
2. Liese, W. Bamboo and Rattan in the World. *J. Bamboo Rattan* **2003**, *2*, 189. [CrossRef]
3. Liese, W.; Tang, T.K.H. *Bamboo: The Plant and Its Uses*; Springer International Publishing: Cham, Switzerland, 2015.
4. Liu, Q.S.; Zheng, T.; Li, N.; Wang, P.; Abulikemu, G. Modification of bamboo-based activated carbon using microwave radiation and its effects on the adsorption of methylene blue. *Appl. Surf. Sci.* **2010**, *256*, 3309–3315. [CrossRef]
5. Lu, T.; Jiang, M.; Jiang, Z.; Hui, D.; Wang, Z.; Zhou, Z. Effect of surface modification of bamboo cellulose fibers on mechanical properties of cellulose/epoxy composites. *Compos. Part B Eng.* **2013**, *51*, 28–34. [CrossRef]
6. Biswas, D.; Bose, S.K.; Hossain, M.M. Physical and mechanical properties of urea formaldehyde-bonded particleboard made from bamboo waste. *Int. J. Adhes. Adhes.* **2011**, *31*, 84–87. [CrossRef]

7. Zhou, H.; Wei, X.; Smith, L.M.; Wang, G.; Chen, F. Evaluation of Uniformity of Bamboo Bundle Veneer and Bamboo Bundle Laminated Veneer Lumber (BLVL). *Forests* **2019**, *10*, 921. [CrossRef]
8. Liese, W.; Kumar, S. *Bamboo Preservation Compendium*; Centre for Indian Bamboo Resource and Technology: Vansda, India, 2003.
9. Liese, W.; Shanmughavel, P.; Peddappaiah, R.S.; Liese, W. Research on bamboo. *Wood Sci. Technol.* **1987**, *21*, 189–209.
10. Kaur, P.J.; Satya, S.; Pant, K.K.; Naik, S.N.; Kardam, V. Chemical characterization and decay resistance analysis of smoke treated bamboo species. *Eur. J. Wood Wood Prod.* **2016**, *74*, 1–4. [CrossRef]
11. Umphauk, S. CCA wood preservative absorption of bamboo culms by soaking method. In Proceedings of the Kasetsart University Conference, Bangkok, Thailand, 30 January 2 February 2007.
12. Zhou, Y.; Sun, F.; Bao, B. Field tests for mold resistance with BHT and BTA added to bamboo preservatives. *J. Zhejiang A F Univ.* **2013**, *30*, 385–391.
13. Rao, K.S. *Bamboo Preservation by Sap Displacement*; International Network for Bamboo and Rattan: New Delhi, India, 2001.
14. Wu, K.T. The effect of high-temperature drying on the antisplitting properties of makino bamboo culm (Phyllostachys makinoi Hay.). *Wood Sci. Technol.* **1992**, *26*, 271–277. [CrossRef]
15. Fang, Q.; Cui, H.W.; Du, G.B. Surface wettability, surface free energy, and surface adhesion of microwave plasma-treated Pinus yunnanensis wood. *Wood Sci. Technol.* **2016**, *50*, 285–296. [CrossRef]
16. Nishikawa, K.; Wakatani, M. *Plasma Physics: Basic Theory with Fusion Applications*; Springer Science and Business Media: New York, NY, USA, 2013.
17. Zhou, X.Y.; Tang, L.J.; Zheng, F.; Xue, G.; Du, G.B.; Zhang, W.D.; Lv, C.L.; Yong, Q.; Zhang, R.; Tang, B.J. Oxygen plasma-treated enzymatic hydrolysis lignin as a natural binder for manufacturing biocomposites. *Holzforschung* **2011**, *65*, 829–833. [CrossRef]
18. Köhler, R.; Sauerbier, P.; Ohms, G.; Viöl, W.; Militz, H. Wood Protection through Plasma Powder Deposition-An Alternative Coating Process. *Forests* **2019**, *10*, 898. [CrossRef]
19. Avramidis, G.; Militz, H.; Avar, I.; Viöl, W.; Wolkenhauer, A. Improved absorption characteristics of thermally modified beech veneer produced by plasma treatment. *Eur. J. Wood Wood Prod.* **2012**, *70*, 545–549. [CrossRef]
20. Jamali, A.; Evans, P. Etching of wood surfaces by glow discharge plasma. *Wood Sci. Technol.* **2011**, *45*, 169–182. [CrossRef]
21. Xu, X.; Wang, Y.; Zhang, X.; Jing, G.; Yu, D.; Wang, S. Effects on surface properties of natural bamboo fibers treated with atmospheric pressure argon plasma. *Surf. Interface Anal. Int. J. Devot. Dev. Appl. Tech. Anal. Surf. Interfaces Thin Films* **2006**, *38*, 1211–1217. [CrossRef]
22. Tang, L.; Zhang, R.; Wang, X.; Yang, X.; Zhou, X. Surface modification of poplar veneer by means of radio frequency oxygen plasma (RF-OP) to improve interfacial adhesion with urea-formaldehyde resin. *Holzforschung* **2015**, *69*, 193–198. [CrossRef]
23. Chen, M.; Zhang, R.; Tang, L.; Zhou, X.; Li, Y.; Yang, X. Development of an industrial applicable dielectric barrier discharge (DBD) plasma treatment for improving bondability of poplar veneer. *Holzforschung* **2016**, *70*, 683–690. [CrossRef]
24. Chen, M.; Zhang, R.; Tang, L.; Zhou, X.; Li, Y.; Yang, X. Effect of plasma processing rate on poplar veneer surface and its application in plywood. *BioResources* **2015**, *11*, 1571–1584. [CrossRef]
25. Li, Y.; Yang, X.; Chen, M.; Tang, L.; Chen, Y.; Jiang, S.; Zhou, X. Influence of atmospheric pressure dielectric barrier discharge plasma treatment on the surface properties of wheat straw. *BioResources* **2014**, *10*, 1024–1036. [CrossRef]
26. Burlica, R.; Shih, K.-Y.; Hnatiuc, B.; Locke, B.R. Hydrogen generation by pulsed gliding arc discharge plasma with sprays of alcohol solutions. *Ind. Eng. Chem. Res.* **2011**, *50*, 9466–9470. [CrossRef]
27. Indarto, A.; Yang, D.R.; Choi, J.-W.; Lee, H.; Song, H.K. Gliding arc plasma processing of CO_2 conversion. *J. Hazard. Mater.* **2007**, *146*, 309–315. [CrossRef] [PubMed]
28. Czernichowski, A. Gliding arc: Applications to engineering and environment control. *Pure Appl. Chem.* **1994**, *66*, 1301–1310. [CrossRef]
29. Safdar, F.; Hussain, T.; Nazir, A.; Iqbal, K. Improving dimensional stability of cotton knits through resin finishing. *J. Eng. Fibers Fabrics* **2014**, *9*, 28–35. [CrossRef]
30. Militz, H.; Schaffert, S.; Peters, B.C.; Fitzgerald, C.J. Termite resistance of DMDHEU-treated wood. *Wood Sci. Technol.* **2011**, *45*, 547–557. [CrossRef]

31. Krause, A.; Wepner, F.; Xie, Y.; Militz, H. Wood protection with dimethyloldihydroxy-ethyleneurea and its derivatives. Development of commercial wood preservatives. Efficacy, environmental, and health issues. *Am. Chem. Soc.* **2008**, *2008*, 356–371.
32. Jiang, T.; Gao, H.; Sun, J.; Xie, Y.; Li, X. Impact of DMDHEU resin treatment on the mechanical properties of poplar. *Polym. Polym. Compos.* **2014**, *22*, 669–674. [CrossRef]
33. Yasuda, R.; Minato, K.; Norimoto, M. Chemical modification of wood by non-formaldehyde cross-linking reagents. 2. Moisture adsorption and creep properties. *Wood Sci. Technol.* **1994**, *28*, 209–218. [CrossRef]
34. Van Oss, C.J.; Chaudhury, M.K.; Good, R.J. Interfacial Lifshitz-van der Waals and polar interactions in macroscopic systems. *Chem. Rev.* **1988**, *88*, 927–941. [CrossRef]
35. Vesel, A.; Mozetic, M.; Zalar, A. XPS characterization of PTFE after treatment with RF oxygen and nitrogen plasma. *Surf. Interface Anal. Int. J. Devot. Dev. Appl. Tech. Anal. Surf. Interfaces Thin Films* **2008**, *40*, 661–663. [CrossRef]
36. Kim, Y.; Lee, Y.; Han, S.; Kim, K.J. Improvement of hydrophobic properties of polymer surfaces by plasma source ion implantation. *Surf. Coat. Technol.* **2006**, *200*, 4763–4769. [CrossRef]
37. Aydin, I.; Demirkir, C. Activation of spruce wood surfaces by plasma treatment after long terms of natural surface inactivation. *Plasma Chem. Plasma Process.* **2010**, *30*, 697–706. [CrossRef]
38. Altgen, D.; Avramidis, G.; Viöl, W.; Mai, C. The effect of air plasma treatment at atmospheric pressure on thermally modified wood surfaces. *Wood Sci. Technol.* **2016**, *50*, 1227–1241. [CrossRef]

Article

Functionalized Surface Layer on Poplar Wood Fabricated by Fire Retardant and Thermal Densification. Part 1: Compression Recovery and Flammability

Demiao Chu [1,2], Jun Mu [2,*], Stavros Avramidis [3], Sohrab Rahimi [3], Shengquan Liu [1] and Zongyuan Lai [2]

1 School of Forestry and Landscape Architecture, Anhui Agricultural University, Hefei 230036, China; 13chudemiao@sina.cn (D.C.); liusq@ahau.edu.cn (S.L.)
2 Key Laboratory of Wood Material Science and Utilization (Beijing Forestry University), Ministry of Education, Beijing 100083, China; lzy2015@bjfu.edu.cn
3 Department of Wood Science, University of British Columbia, Vancouver, BC V6T 1Z4, Canada; stavros.avramidis@ubc.ca (S.A.); sohrab.rahimi@alumni.ubc.ca (S.R.)
* Correspondence: mujun@bjfu.edu.cn; Tel.: +86-010-62336053

Received: 9 October 2019; Accepted: 23 October 2019; Published: 26 October 2019

Abstract: To enhance compression stability and fire retardancy of densified wood, a new modification method i.e., combined nitrogen–phosphorus (NP) fire retardant pre-impregnation with surface thermo-mechanical densification is used to fabricate a certain thickness of functionalized surface layer on poplar. This combined treated wood is investigated via vertical density profile (VDP), and the compression stability is revealed by both soaking test and cone analysis. Results demonstrate that the combined treatment hardened the surface of wood and reformed the interface combination of the NP with the wood cell wall, thus making the surface tissue more close-grained. Fire retardancy was also enhanced; the total heat release and CO generation values decreased by 21.9% and 68.4%, respectively, when compared with that of solely NP-treated wood. Moreover, surface hardness increased by 15.8%, and the recovery of surface hardness and thickness were 56.8% and 77.2% lower than that of simply densified wood. It appears that this NP-involved thermal densification could be considered as an alternative approach to enhance both the compression stability and fire resistance of wood.

Keywords: wood; fire retardancy; heat treatment; compression recovery; surface layer

1. Introduction

Wood has been considered as one natural composite material for a long time, and it is one of the most popular materials in construction, interior decoration, and the furniture industry mainly because of its strength-to-weight ratio, renewability, and versatility [1]. Furthermore, small-diameter timbers or some remains can be transformed into particles or fibers, which are commonly used to manufacture different composites with other materials like polymers, cement, etc. [2,3].

Since the conservation mandates for natural forests were improved in China, the demand for plantation trees such as poplar, eucalypt, and pine has increased, making them a reliable alternative for industrial timbers [4,5]. Nevertheless, the drawback of using plantation timbers is mostly related to their low density, mechanical strength, and dimensional stability [6]. Densification has been reported for over a century as an efficient method for enhancing wood mechanical properties by reducing porosity and increasing density [7]. Densification is proven to improve mechanical properties such as modulus of elasticity (MOE), surface hardness, and mechanical strength of wood, especially for low-density species [8]. Song et al. (2018) successfully created a high-performance structural compressed wood via partial removal of lignin and hemicellulose and hot-pressing into highly aligned cellulose nanofibers, resulting in specific strength higher than that of most structural metals and alloys [9].

The conventional wood densification is usually carried out by thermo-mechanical (TM) or thermo-hydro-mechanical (THM) means [3,10]. Büyüksarı found that the thermally compressed veneer could be used to laminate with MDF panels for structural purposes due to the improvement of hardness [8]. Nonetheless, the wood is incompletely densified and the state of compression of TM/THM wood is sensitive to moisture and temperature that could cause utilization problems [7,11,12]. Researchers have studied different methods such as resin pre-impregnation, steam treatment, microwave treatment, and pre-acetylation in order to improve the compression stability of densified wood [13–15].

Recently, researchers focused on the combined effects of heat treatment (HT) and wood densification on the assumption that the post-thermal treatment could release the inner stresses and increase the stability of the compressed wood [8,16]. HT has been employed at various temperatures (160–260 °C) and various media reactions such as vacuum, nitrogen, steam, or oil [17,18]. Heat-treated wood (W_{HT}) relatively enhanced dimensional stability, biological resistance, and aesthetic characteristics [19–22]. The degradation of hemicellulose, conformational arrangement changes of wood biopolymers, and the plasticization of lignin during HT could result in lower number of hydrophilic groups and, consequently, improved dimensional stability [20,23,24]. Fang et al. (2011, 2012) combined oil-heat treatment with wood densification and claimed that the HT could efficiently improve dimensional stability, mechanical properties, and compression set recovery of THM densified veneers [25,26]. Avila et al. (2012) studied selected properties of THM wood using nano-indentations analysis and found an improvement of mechanical properties attributed to an increase in wood density. Furthermore, they reported an increase in the mechanical properties of the cell wall [27]. Zhan and Avramidis (2016) investigated the equilibrium moisture content and radial swelling strain ratios of the untreated and surface densification and thermal post-treatment combined treated wood and discovered that hygroscopicity and radial swelling deformation are significantly decreased [12].

The HT process requires rigorous conditions; that is, relatively high temperature and long holding time under the oxygen-deficient environment (oxygen content below 3%). Since HT is a chemical-free process, the main modification mechanism is the autocatalytic reactions of the cell-wall constituents at high temperatures. Acetyl groups of the hemicellulose are partly cleaved, which leads to the formation of acidic intermediate products that further promote the changes of other chemical composition [17,23]. Lastly, the HT process does not make wood fireproof [28,29].

Nitrogen–phosphorus (NP) is an environmentally friendly, non-toxic, relatively inexpensive fire retardant [30]. Furthermore, the catalytic dehydration and catalytic charring of the NP wood fire resistance occurs between heating of 115 °C and 246 °C [31–33]. Hygroscopicity and low leachability resistance are possible limitations for a practical application of the NP, and studies for enhancing these properties can be found in published papers [29,31–33]. In former studies, NP has been used as an acidic-accelerating medium to intensify the effect of HT on wood. It brought about significant property improvement for the NP pre-impregnated and heated wood at a low treatment temperature for a short period of time [29,33]. More relevant studies also proved that acid pre-treatments can accelerate the thermal degradation of wood components and led to a treated wood moisture content lower than conventional HT, under the same treatment weight loss level [34,35].

There is a research gap regarding the effects of the NP fire retardant on THM densification of wood. In the present study, the NP fire retardant is used to reinforce poplar wood together with the thermal densification, aimed at fabricating a functionalized surface layer to enhance fire retardancy and compression stability of combined treated wood. This study provides a novel combined modification, which can functionalize the surface layer and endow multiple functions to the treated wood. It will be useful for promoting the usage of plantation timber.

2. Materials and Methods

2.1. Treatment

Poplar (*Populus beijingensis* W. Y. Hsu) at about 10% moisture content was purchased from a timber supplier in Beijing, China. Defect-free sapwood samples with length, width, and thickness of 300 mm × 150 mm × 19 mm, respectively, were prepared. All samples were polished with 240-grit sandpaper to ensure uniform surface roughness prior to modification. Then, they were divided into six groups (Table 1), each comprised of 15 samples.

Table 1. Treatment classes and the treatment condition of treated samples.

Groups	Pre-Treatment	Compression Ratios (%)	Compression Temperature (°C)
$W_{NP\text{-}TM}$	10% NP	21	180
W_{TM}	No	21	180
$W_{HT\text{-}TM}$	HT220	21	180
W_{HT}	HT220	No	No
W_{NP}	10% NP	No	No
W_C	No	No	No

Note: 10% NP means 10% nitrogen phosphorus fire-retardant aqueous solution impregnation; HT220 means heat treatment under 220 °C for 120 min.

The NP impregnation and HT conduction protocols were kept the same as in former studies [29,33]. Samples in groups $W_{NP\text{-}TM}$, W_{TM}, and $W_{HT\text{-}TM}$ were pre-heated using a thermo-compressor (MLG68-S, Zhengzhou, China) for 5 min, then thermo-mechanically pressed to 14 mm final thickness using a stopper for 20 min at constant temperature.

2.2. Properties of the Functionalized Surface Layer

The surface colorimetric parameters were evaluated based on the CIELAB color coordinates using a photometry instrument (Dataflash 110 Datacolor, Lawrenceville, NJ, USA) with illuminant D65 and 10° standard observer. The mean values of five positions on each sample were recorded and the total color difference (ΔE^*) was calculated.

Five samples (50 mm × 50 mm × thickness) were cut from the middle of each of the untreated and treated groups. A vertical density profile tester (CreCon's X-ray densitometer, Martinsried, Germany) was used to measure the vertical density profile (VDP) of each sample and the average density profile was calculated. The average density profile accurately reflects density change throughout the sample thickness.

2.3. Compression Recovery via Soaking Test

Compression recovery was measured by soaking samples in a deionized water bath (20 ± 2 °C) for 120 min and then oven-drying. Their axial surfaces (20 mm × 20 mm × thickness) were covered by resin before the test. The thickness and surface hardness (*HD*) of each sample were measured before and after the soaking test. Thickness was measured three times from each sample at the center by digital caliper and *HD* was tested by a Shore D Hardness Tester (TH210, Beijing). All measurements were conducted after conditioning the samples to 10% equilibrium moisture content. Surface hardness recovery (HD_r) and thickness recovery (T_r) were calculated as follows:

$$HD = 100 - L/0.025 \quad (1)$$

$$HD_r = (HD_{fd} - HD_{fs})/(HD_{fd} - HD_{bd}) \times 100\% \quad (2)$$

$$Tr = (T_{fd} - T_{fs})/(T_{fd} - T_{bd}) \times 100\% \quad (3)$$

where L is the displacement of the needle when the pressure foot surface is in full contact with the sample surface; HD_{bd} and T_{bd} are the surface hardness and thickness before densification; HD_{fd} and T_{fd} are the surface hardness and thickness after densification; and HD_{fs} and T_{fs} are the surface hardness and thickness after soaking test.

2.4. Cone Analysis

Cone calorimeter tests were conducted according to ASTM 1354 (2004) with a cone calorimeter (FTT0242, West Sussex, UK). Three samples in the groups $W_{NP\text{-}TM}$, W_{TM}, $W_{HT\text{-}TM}$, and W_C were selected and wrapped with aluminum foil leaving a 100 mm square surface. During the test, the sample was exposed to a constant external heating flux, which was located at 250 mm over the surface, at an irradiance level of 50 kW m^{-2}.

2.5. Clustering and Morphological Analysis

Hierarchical cluster analysis (HCA) was performed using IBM SPSS Statistics 22.0 by between group linkage method [36]. The heat release and smoke production values from the cone test and compression recovery values from soaking test were taken into consideration, respectively. Microscopy and microanalysis of the groups Wc, $W_{NP\text{-}TM}$, W_{TM}, $W_{HT\text{-}TM}$, and W_{NP} were conducted with a SEM-EDS energy-dispersive X-ray spectrometer (JSM-6510, Tokyo, Japan). Samples were covered by galvanic gold deposition with an MC1000 ion sputter (FEI, Tokyo, Japan) working at a current of 5 mA for 45 s. All analyses were performed with 20 kV acceleration voltages.

3. Results and Discussion

3.1. Properties of the Functionalized Surface Layer

Photos and vertical density profile (VDP) curves are shown in Figure 1; the CIEL*a*b* and VDP values on different regions of the untreated and treated samples are shown in Table 2. It is evident that the $W_{NP\text{-}TM}$ exhibits a uniform light-brown surface whereas the W_{TM} one presents a light surface. It can be deduced that the combined process could play an important role in the color generation of the NP-TM-treated wood because the merely NP impregnation process cannot change the surface color significantly [37].

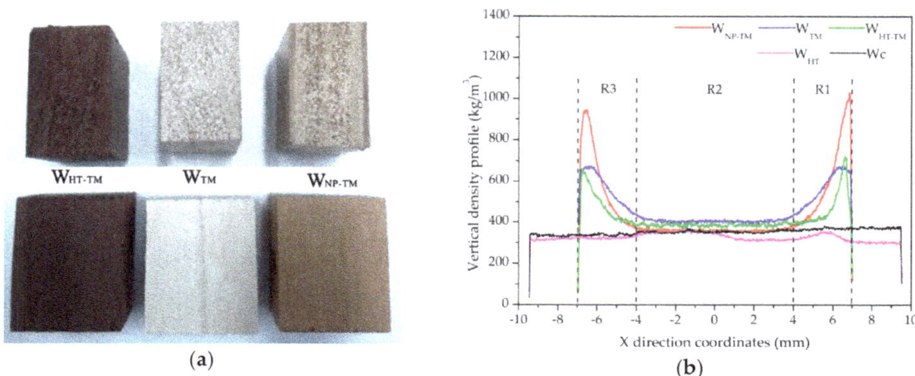

Figure 1. Photo (**a**) and vertical density profile (VDP) curves (**b**) of the selected samples.

Table 2. Color in CIE L*a*b* system and density of the selected samples.

Groups	Color in CIE L*a*b* System				Vertical Density Profile (kg m^{-3})				
	L^*	a^*	b^*	ΔE^*	R1	R2	R3	Maximum (× coordinate)	Average Value
$W_{NP\text{-}TM}$	54.99 ± 3.41	13.26 ± 1.68	29.19 ± 0.51	33.37 ± 1.38	675.68	365.48	569.97	1030.22 (6.83)	453.67
W_{TM}	80.33 ± 3.76	3.49 ± 0.60	21.40 ± 0.36	6.09 ± 2.03	578.65	410.62	567.90	673.43 (−6.43)	473.61
$W_{HT\text{-}TM}$	44.56 ± 1.07	9.36 ± 0.51	19.40 ± 0.43	40.21 ± 1.56	391.69	386.55	484.60	720.29 (6.66)	420.21
W_{HT}	47.18 ± 4.23	9.87 ± 0.42	20.80 ± 2.02	37.85 ± 2.72	297.25	330.58	313.76	365.29 (−1.74)	324.41
Wc	84.16 ± 5.79	2.95 ± 1.48	16.69 ± 1.83	-	372.18	354.10	331.02	382.76 (5.59)	352.38

According to Table 2, the a*, b*, and L* values of W_{TM} and $W_{HT\text{-}TM}$ changed slightly compared with that of Wc and W_{HT}. That is to say, the only-TM process under 180 °C could not change the surface color of poplar wood. Interestingly, the b* and a* values of the $W_{NP\text{-}TM}$ increased by 74.9% and 349.5%, respectively. The ΔE^* values of the $W_{NP\text{-}TM}$ are higher than 33.37, further indicating that NP-TM treatment with relatively a short period (20 min) is quite sufficient for changing the color of poplar. After the NP-TM, poplar possessed a warmer and more attractive surface color, which is considered to be essential for furniture and interior decoration materials.

Figure 1b indicates that the VDP of untreated poplar was relatively uniform throughout the thickness with an average density of 352.38 kg/m^3. The average density of W_{HT} decreased by 7.94% because of the thermal degradation of wood, which is consistent with some previous studies [17,22,38]. All the treated samples were divided into three regions according to the density changes; namely two surface layers (R1 and R3) and one core layer (R2). After the TM, the VDP values of the surface layers on $W_{NP\text{-}TM}$, W_{TM}, and $W_{HT\text{-}TM}$ increased significantly. The NP-TM process brought the highest VDP values to the $W_{NP\text{-}TM}$ on the surface layers, wherein the VDP value of R1 was 16.77% higher than that of W_{TM}. Meanwhile, the location of maximum VDP value on the $W_{NP\text{-}TM}$ was closer to the surface and density of the R2 was almost kept the same level as Wc. However, the maximum VDP values of the W_{TM} and $W_{HT\text{-}TM}$ were much lower than $W_{NP\text{-}TM}$, and compression was also affected in the core section. This indicates that the compression was mostly in the surface layer (about 3 mm). It has been reported that the NP has a catalytic dehydration effect on wood polysaccharides during HT [32], and this could also make the wood surface layer more likely to be compressed because of the softening effect at higher moisture and temperature conditions [25]. From NP fire retardants can be produced softener-NH_3, which reduces glass transition temperature of lignin, i.e., decrease Tg from approximately 170 °C [39]. Furthermore, the impregnated NP partly increased the density as most of it remained in the surface layer.

3.2. Compression Recovery of the Functionalized Surface Layer during Soaking Test

Normally, the compression state of TM wood is sensitive to moisture, which limits the usage of it to certain condition [12]. The thickness and surface hardness (*HD*) change of treated wood before and after the TM process, as well as compression recovery (HD_r and T_r), are displayed in Figure 2. The NP-TM treatment resulted in a better surface strength and the compression-fixing ability for poplar. $W_{NP\text{-}TM}$ has higher *HD* value than that of W_{TM} before and after the water soaking test.

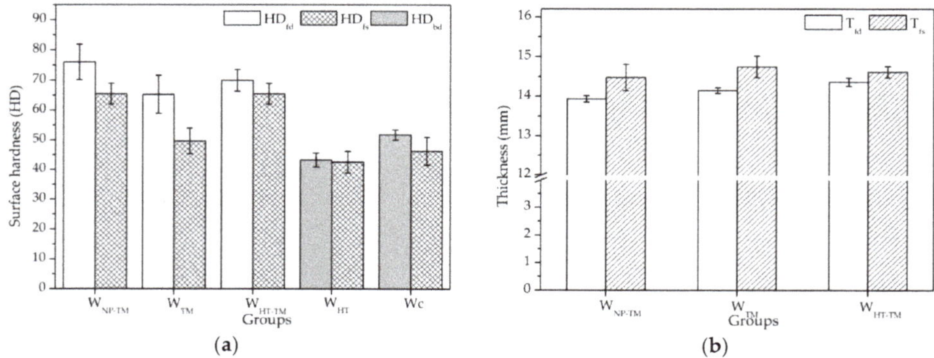

Figure 2. Surface hardness (**a**) and thickness changes (**b**) of selected samples.

Because of the TM, the HD_{fd} value of W_{TM} was 65.23, which is 20.59% higher than untreated poplar; the HD_{fd} value of $W_{HT\text{-}TM}$ increased to 69.92, namely 61.66% higher than W_{HT}. After the NP-TM, the HD_{fd} values of $W_{NP\text{-}TM}$ became 75.98, that is, 16.48% higher than W_{TM}. In other words, the NP-involving compression could cause larger surface hardness improvement on wood.

After water soaking test, all densified samples swelled to some extent, resulting in recovery of thickness and *HD*. Even for Wc, the surface hardness after soaking (HD_{fs}) decreased by 10.66% (Figure 2a). The W_{TM} had a poor compression ability; HD_r and T_r were the highest of all. The surface layer absorbed water and recovered to the status before the TM. While the $W_{HT\text{-}TM}$ had the best compression ability, HD_r and T_r values of $W_{HT\text{-}TM}$ were the lowest. The reason is that the HT increased the hydrophobicity and the TM process further enhanced the surface density of $W_{HT\text{-}TM}$. The compression stability of $W_{HT\text{-}TM}$ samples remarkably enhanced due to the NP pre-treatment and the acceleration effect of NP on wood thermal degradation during the thermo-densification process [29,33]. When compared with that of W_{TM}, the HD_{fs} value was 31.60% higher and the T_r value decreased from 26.65% to 6.07%.

3.3. Effect of the Functionalized Surface Layer on Heat and Flue Gas Release via Cone Analyses

As shown in Figure 3 and Table 3, the $W_{NP\text{-}TM}$ samples had the most moderate heat and flue gas release rate. W_{TM} and $W_{HT\text{-}TM}$ samples were also enhanced by the TM process to some extent, especially in the preliminary stage of burning during the Cone test.

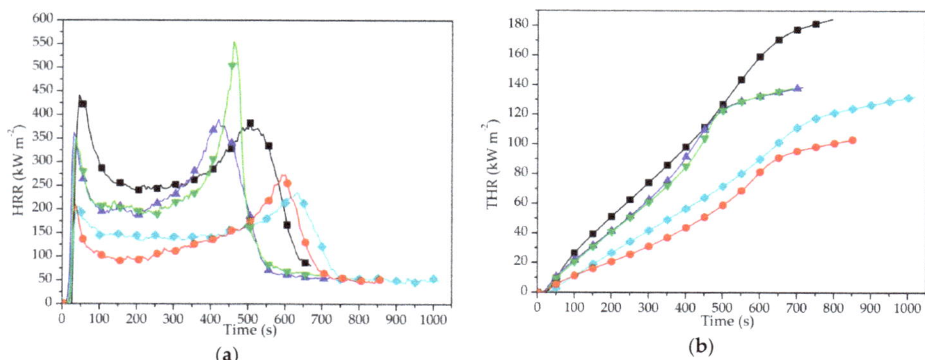

Figure 3. *Cont.*

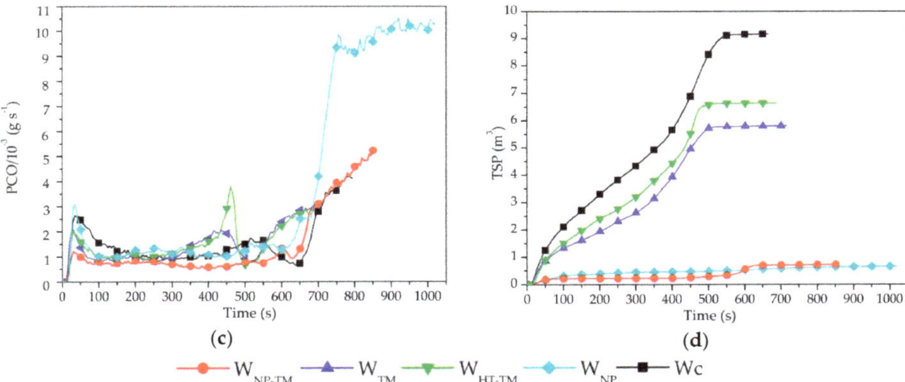

Figure 3. Heat release rate (**a**) and total heat release (**b**), generation rate of carbon monoxide (**c**), and total smoke production (**d**) of the selected samples.

Table 3. Heat release and flue gas release of selected samples during cone analysis.

Item	Groups				
	W_{NP-TM}	W_{TM}	W_{HT-TM}	W_{NP}	Wc
HRR (kW m^{-2})	122.49	196.54	204.57	131.69	236.08
THR (MJ m^{-2})	103.50	138.60	138.14	132.44	184.24
HRR (60 s) (kW m^{-2})	133.74	259.51	254.70	122.00	311.26
HRR (360 s) (kW m^{-2})	110.27	225.66	217.24	146.72	255.06
Peak *HRR1* (kW m^{-2})	208.84	362.82	353.31	172.83	437.16
Peak *HRR2* (kW m^{-2})	273.71	390.40	555.30	234.94	345.39
Time to Peak *HRR1* (s)	30	25	30	45	45
Time to Peak *HRR2* (s)	585	415	455	645	520
mean EHC (MJ kg^{-1})	14.64	21.83	21.03	18.78	27.38
TSP (m^2)	0.72	5.81	6.64	0.65	7.86
TSR (m^2 m^{-2})	81.33	657.39	751.48	73.02	888.95
COY/10^{-3} (kg kg^{-1})	20.20	19.10	17.80	63.90	20.70
COY (60 s) /10^{-3} (kg kg^{-1})	10.70	12.40	12.60	23.00	18.30
COY (360 s) /10^{-3} (kg kg^{-1})	10.60	12.90	11.60	16.50	14.60
CO_2Y (kg kg^{-1})	1.29	1.96	1.88	1.59	2.37

Note: *HRR*, heat release rate; *THR*, total heat release; *HRR* (60 s), heat release rate within 60 s; *HRR* (360 s) heat release rate within 360 s; Peak *HRR1*, peak of the heat release rate at lower temperature; Peak *HRR2*, peak of the heat release rate at higher temperature; *EHC*, effective heat of combustion; *TSP*, total smoke production; *TSR*, total smoke release; *COY*, CO yield; *COY* (60 s), CO yield within 60 s; *COY* (360 s), CO yield within 360 s; CO_2Y, CO_2 yield.

The *HRR* curve in Figure 3a reveals that the Wc had two sharp exothermic peaks; the burning of combustible gas emitted by the decomposition of the surface formed the peak *HRR1* at 45 s, then the oxidation reaction of internal wood material formed peak *HRR2* at 520 s because of the consuming and cracking of the surface charcoal layer.

According to Table 3, the *THR* and *TSP* values of W_{TM} and W_{HT-TM} were decreased because of the TM process. However, the *HRR* curves of W_{TM} and W_{HT-TM} caught up with untreated wood and then reached the peak *HRR2*, which was 13.0% and 60.8% higher than that of Wc (Figure 3a,b). Furthermore, the time to peak *HRR2* appeared 105 s and 65 s ahead; meanwhile, the surface cracking increased the CO release, as shown in Figure 3c. This could be explained that the higher density of formed charcoal provided a better protection for inner material at the initial stage, but the recovery of the compressed wood and cracking of surface charcoal under high temperature cause a more violent combustion reaction. Therefore, the TM treatment may also bring potential risk to the treated wood.

Compared with that of Wc, the *HRR* curve of W_{NP-TM} declined remarkably in the whole combustion process. The peak *HRR1* decreased by 52.2% and appeared 15 s ahead, then remained at a continuously lower *HRR* value. It points to the fact that the W_{NP-TM} formed a protective surface charcoal rapidly with less heat release and the resistance of the charcoal layer enhanced. Different from the simply TM wood, the NP-TM process made the W_{NP-TM} more stable during the burning. The extent of surface cracking on W_{NP-TM} was much lower; the peak *HRR2* decreased by 29.9% and 50.7%, and the relevant time delayed for 170 s and 130 s, when compared with that of W_{TM} and W_{HT-TM}. It is worth mentioning that the *THR*, *TSP*, and *COP* values of W_{NP-TM} were also lower than solely NP-treated wood.

The NP-TM made the surface layer of W_{NP-TM} denser, which could be in favor of the oxygen and heat insulation performances of treated wood. Hence, the generation rate of CO and total smoking production were the lowest within the first 600 s. The surface cracking occurred at the faint-flame stage and increased the contact of charcoal with oxygen. The CO release declined sharply at about 600 s and then kept a relatively lower level after 700 s. In other words, the swelling of densified wood and cracking of surface charcoal could be effectively controlled by the NP-TM method. The enhancement could be explained by the synergistic effects of density-increase and carbon-forming by NP fire retardant. Therefore, it can be concluded that the fire retardancy of the W_{NP-TM} was significantly improved by the functionalized surface layer.

3.4. Compression Recovery of the Functionalized Surface Layer during Cone Analyses

In order to uncover the surface compression recovery of the treated wood during burning and its effects on the smoke release, the correlation between the *HRR* and specific mass loss rate (*SMLR*), carbon monoxide yield (CO yield), and smoke production rate (*SPR*) in different stages were also taken into account, as shown in Figure 4.

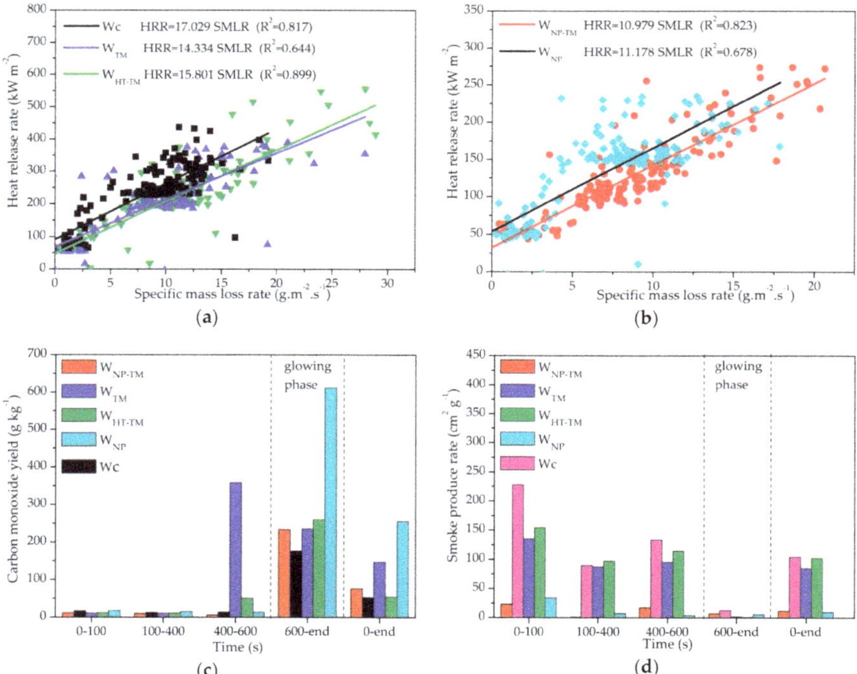

Figure 4. Correlation of heat release rate with specific mass loss rate (**a**, **b**), the CO yield (**c**), and smoke production rate (**d**) of the selected samples.

Equations of the linear statistical dependence of the *HHR* on the specific mass loss rate are listed in Figure 4a,b. Under perfect combustion, the *HRR* is a function of specific mass loss rate only ($R^2 = 1$); the slope of fitting line was the smallest and the R^2 value of $W_{NP\text{-}TM}$ was much higher than that of W_{NP}, showing that the $W_{NP\text{-}TM}$ possesses higher combustion efficiency. Considering the lowest *mean EHC* (Table 3) and lowest heat release (Figure 3a,b), it can be concluded the combustion reaction of $W_{NP\text{-}TM}$ was the most moderate and stable.

The oxygen-deficient combustion produces more CO; the main emission stage is faint flame combustion (glowing phase) of surface charcoal, after nearly 600 s (Figures 3c and 4c). The imperfect combustion causes more smoke release; the main emission stage is the flame combustion stage (Figures 3d and 4d).

Taking an overall consideration of the cone testing results, we can find that surface cracking of W_{TM} and $W_{HT\text{-}TM}$ occurred at high temperature, resulted in shifting forward of the Peak *HRR2* (Figure 3a) and CO release (Figure 3b). Therefore, the surface cracking of W_{TM} and $W_{HT\text{-}TM}$ was before 400 s and 450 s. The peak of CO release appeared with the surface cracking, caused by the anoxic burning of the inner material. Similarly, the surface cracking of $W_{HT\text{-}TM}$ was around 600 s, closing to the glowing phase. As shown in Figure 4c, the CO yield of W_{TM} and $W_{HT\text{-}TM}$ started to increase from the interval of 400–600 s, and it started after 600 s in case of the $W_{NP\text{-}TM}$.

It is worth noting that the CO release of $W_{HT\text{-}TM}$ declined sharply after the surface cracking and kept the same level as Wc, much lower than that of W_{NP}. This could be explained by the fact that most of the inner materials have transformed into charcoal under protection of the functionalized surface layer, then the surface cracking of $W_{NP\text{-}TM}$ took place in the glowing stage and increased the oxygen-contact area of char surface. This indicates that the NP-TM process enhanced the compression stability as well as CO inhibition of the treated wood in the high-temperature condition.

The surface layer of $W_{NP\text{-}TM}$ was filled with NP fire retardant and compressed to a high density, which has an accelerating effect on the formation of charcoal and strengthening effect on the protective char layer at the early stage of burning. Then, the compression-stable charcoal protected the inner material from burning because of the heat or oxygen-barrier properties. Therefore, the heat and CO release of $W_{NP\text{-}TM}$ declined and the surface cracking delayed when compared with that of W_{TM}. At the glowing stage, the surface cracking increased the oxygen contact and reduced CO release from oxidization of charcoal (Figures 3c and 4c).

3.5. Clustering and SEM-EDS Analysis

Cluster analysis of the NP-TM and untreated samples was done based on values of color in CIE $L^*a^*b^*$ system (Table 2), heat release and flue gas release (Table 3), and compression recovery (Figure 2); the dendrogram of those clusters is shown in Figure 5.

The dendrogram of HCA for treated samples in regard to surface color (Figure 5a) shows two main clusters; the $W_{NP\text{-}TM}$ was classified into the first group, which also contains W_{HT} and $W_{HT\text{-}TM}$, while the Wc and W_{TM} were classified into the second group, further proving that the NP-TM process can lead to color generation.

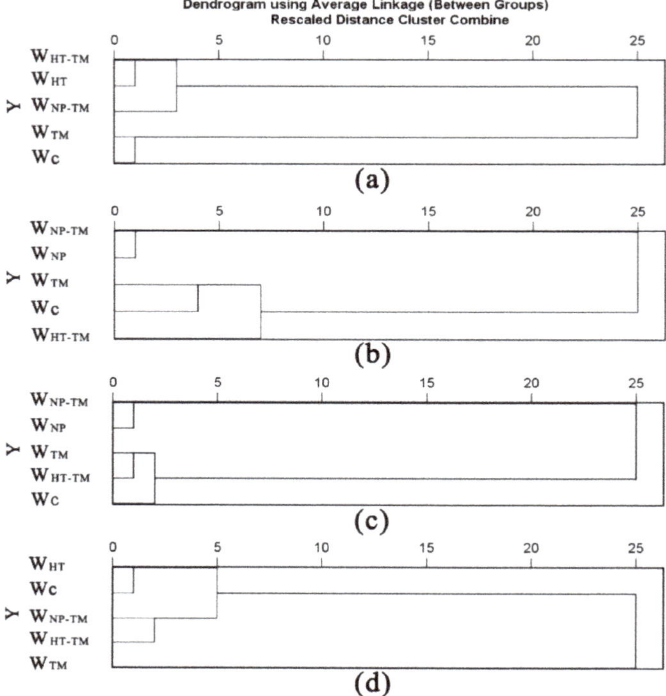

Figure 5. Hierarchical cluster analysis (HCA) of surface color (**a**), heat release (**b**), flue gas release (**c**), and compression recovery (**d**) based on different variables. Note: The horizontal axis shows the distances of clusters and the vertical axis shows combine treatment and the control group.

In terms of dendrogram for heat release (Figure 5b) and flue gas release (Figure 5c), it is observed that the $W_{NP\text{-}TM}$ was classified into the same group with the W_{NP}. It verified that the functionalized surface layer, created by NP-TM process, can endow the treated wood with flame retardant and smoke suppression properties. However, the only-TM treatment could not render raw poplar or heat-treated poplar fireproof. This result is in accordance with that of the cone analyses discussed in Sections 3.3 and 3.4. For the compression recovery, dendrogram of HCA are grouped into two blocks: W_{TM} and other groups. The first cluster was separated into two subgroups (Figure 5d). The compression recovery improvement of the $W_{NP\text{-}TM}$ could elucidate that the NP fire retardant acts as an acid accelerator, which could intensify the effectivity of HT [29,33], as the $W_{HT\text{-}TM}$ and $W_{NP\text{-}TM}$ are classified into the same subgroups.

The SEM micrographs and EDXA spectra of untreated and NP-TM-treated samples are displayed in Figure 6.

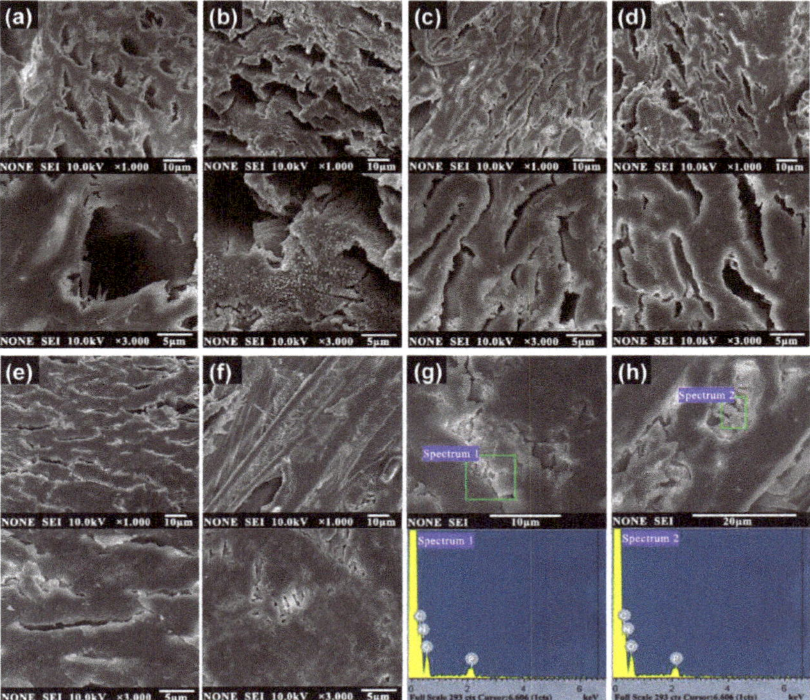

Figure 6. Scanning electron microscopy (SEM) micrographs of treated samples: (**a**) Wc; (**b**) W_{NP}; (**c**) W_{TM}; (**d**) $W_{HT\text{-}TM}$; (**e**) $W_{NP\text{-}TM}$ on transverse section; (**f**) $W_{NP\text{-}TM}$ on tangential section; (**g**,**h**) energy-dispersive X-ray analysis (EDXA) spectra of $W_{NP\text{-}TM}$.

The wood cell wall structure of Wc was intact and the inner faces are flat and smooth. After the TM, cell lumens (particularly distinct in the vessels) of W_{TM} and $W_{HT\text{-}TM}$ samples were deformed, and the volume of void spaces declined sharply. However, the lumens of W_{TM} were partly open and had many fractures on the cell walls. This could increase the water absorption and cell wall swelling, finally causing compression recovery. The $W_{HT\text{-}TM}$ also had some fractures on the cell walls, while the less hydrophilic caused by HT process could prevent it from swelling and recovery [40].

For the W_{NP}, the NP fire retardant distributed over the inner surface separately, as represented in Figure 6b. Because of the NP-TM process, it is quite clear that a large number of vessels are almost closed up and the cell walls are deformed on the $W_{NP\text{-}TM}$. Moreover, there are a few fractures that occurred on the cell walls (Figure 6e). According to Figure 6g,h, some crystalline particles fill between small gaps on the cell wall, and the crystallized-substance integrated the compressed and flattened cell lumens of the $W_{NP\text{-}TM}$. EDXA spectra further confirm that the filler gathered in pore spaces both on transverse (Figure 6g) and tangential (Figure 6h) sections is composed of NP fire retardant. It is speculated that the low molecular ammonium polyphosphate (APP) in NP fire retardant decomposed during the TM process and the intermediate products like H_3PO_4 further promote the catalytic dehydration of wood polysaccharides and enhanced the plasticity of the cell walls. Furthermore, the NP entered micro-voids, forming a connective crystal between the cell walls. Compared to that of W_{NP}, the connection of the NP and cell walls transform from particle to cementing, which will probably be another nonnegligible factor for the enhance of fire resistance and compression stability.

4. Conclusions

This work investigated the fire resistance and compression recovery properties of the treated poplar with a functionalized surface layer, formed by a novel combined treatment of nitrogen–phosphorus fire retardant (NP) pre-impregnation and thermo-mechanical densification (TM). Based on the findings of this study, the combined process could improve the surface aesthetics, fire retardancy, and compression stability of poplar wood. The NP accelerated the thermal degradation of wood and transformed to connective crystal between wood cell walls and cell lumens. This functionalized surface layer is composited by the degradation products of NP and wood. It possesses a superiority of high density and compression stability. The heat release rate of the combined treated wood is 48.1% lower than that of untreated wood, and the CO yield is 68.4% lower than that of only-NP-treated wood. Moreover, the compression recovery in the thickness is 77.2% lower than that of solely TM-treated wood. This study provides a new approach to enhance both fire resistance and compression stability of wood by fabricating a functional surface layer. Further studies in terms of controlling the formation and performance of the surface layer will be of interest in the future.

Author Contributions: Conceptualization, D.C. and J.M.; Methodology, D.C. and J.M.; Software, D.C. and S.R.; Validation, J.M., S.A., and S.L.; Formal Analysis, D.C., J.M., and Z.L.; Investigation, D.C. and J.M.; Resources, D.C. and J.M.; Data Curation, D.C., S.R., and J.M.; Writing-Original Draft Preparation, D.C.; Writing-Review & Editing, J.M., S.A., S.L., S.R., and Z.L.; Visualization, D.C.; Supervision, J.M. and S.A.; Project Administration, J.M.; Funding Acquisition, J.M.

Funding: This research was funded by the Forestry Public Welfare Project Foundation of China, grant number 201404502, and the Fundamental Research Funds for the Central Universities of China, grant number 2015ZCQ-CL-01.

Conflicts of Interest: The authors declare no conflict of interest.

References

1. Hill, C.A.S. *Wood Modification, Chemical, Thermal and Other Processes*; John Wiley & Sons: Chichester, UK, 2006.
2. Quiroga, A.; Marzocchi, V.; Rintoul, I. Influence of wood treatments on mechanical properties of wood–cement composites and of Populus Euroamericana wood fibers. *Compos. B Eng.* **2016**, *84*, 25–32. [CrossRef]
3. Yuan, Q.P.; Su, C.W.; Xiong, J.B. Research Progress of Plantation Wood Machining Properties. *J. Anhui Agric. Sci.* **2011**, *3*, 42–45.
4. Lin, H.; Wang, Y.; Chen, Z.D.; Yang, B.X. Progress & Application Prospects of Modified Wood. *For. Mach. Woodwork. Equip.* **2006**, *34*, 11–13.
5. Kiaei, M.; Naji, H.R.; Abdul-Hamid, H.; Farsi, M. Radial variation of fiber dimensions; annual ring width; and wood density from natural and plantation trees of alder (*alnus glutinosa*) wood. *Drev. Vysk. Wood Res.* **2016**, *61*, 55–64.
6. Bekhta, P.; Proszyk, S.A.; Krystofiak, T. Colour in short-term thermo-mechanically densified veneer of various wood species. *Eur. J. Wood Prod.* **2014**, *72*, 785–797. [CrossRef]
7. Kwon, J.H.; Shin, R.; Ayrilmis, N.; Han, T.H. Properties of solid wood and laminated wood lumber manufactured by cold pressing and heat treatment. *Mater. Des.* **2014**, *62*, 375–381. [CrossRef]
8. Büyüksarı, U. Surface characteristics and hardness of MDF panels laminated with thermally compressed veneer. *Compos. B Eng.* **2013**, *44*, 675–678. [CrossRef]
9. Song, J.; Chen, C.; Zhu, S.; Zhu, M.; Hu, L. Processing bulk natural wood into a high-performance structural material. *Nature* **2018**, *554*, 224–228. [CrossRef]
10. Diouf, P.N.; Stevanovic, T.; Cloutier, A.; Fang, C.; Blanchet, P.; Koubaa, A.; Mariotti, N. Effects of thermo-hygro-mechanical densification on the surface characteristics of trembling aspen and hybrid poplar wood veneers. *Appl. Surf. Sci.* **2011**, *257*, 3558–3564. [CrossRef]
11. Laine, K.; Segerholm, K.; Walinder, M.; Rautkari, L.; Hughes, M. Wood densification and thermal modification, hardness; set-recovery and micromorphology. *Wood Sci. Technol.* **2016**, *50*, 883–894. [CrossRef]
12. Zhan, J.; Avramidis, S. Needle fir wood modified by surface densification and thermal post-treatment, hygroscopicity and swelling behavior. *Eur. J. Wood Prod.* **2016**, *74*, 49–56. [CrossRef]

13. Darwis, A.; Wahyudi, I.; Dwianto, W.; Cahyono, T.D. Densified wood anatomical structure and the effect of heat treatment on the recovery of set. *J. Indian Acad. Wood Sci.* **2017**, 1–8. [CrossRef]
14. Chen, A.T.; Luo, P.P.; Xu, Z.Y.; Hu, Z.J. Effects of furfuryl alcohol impregnation on physical and mechanical properties of densified poplar wood. *J. For. Eng.* **2016**, *1*, 21–25. [CrossRef]
15. Dömény, J.; Čermák, P.; Koiš, V.; Tippner, J.; Rousek, R. Density profile and microstructural analysis of densified beech wood (*Fagus sylvatica L.*) plasticized by microwave treatment. *Eur. J. Wood Prod.* **2017**, 1–7. [CrossRef]
16. Laine, K.; Segerholm, K.; Wålinder, M.; Rautkari, L.; Hughes, M.; Lankveld, C. Surface densification of acetylated wood. *Eur. J. Wood Prod.* **2016**, *74*, 829–835. [CrossRef]
17. Gu, L.B.; Ding, T.; Jiang, N. Development of wood heat treatment research and industrialization. *J. For. Eng.* **2019**, *4*, 1–11. [CrossRef]
18. Salca, E.; Hiziroglu, S. Evaluation of hardness and surface quality of different wood species as function of heat treatment. *Mater. Des.* **2014**, *62*, 416–423. [CrossRef]
19. Ding, T.; Peng, W.W.; Li, T. Mechanism of color change of heat-treated white ash wood by means of FT-IR and XPS analyses. *J. For. Eng.* **2017**, *4*, 25–30. [CrossRef]
20. Altgen, M.; Willems, W.; Hosseinpourpia, R.; Rautkari, L. Hydroxyl accessibility and dimensional changes of Scots pine sapwood affected by alterations in the cell wall ultrastructure during heat-treatment. *Polym. Degrad. Stab.* **2018**, *152*, 244–252. [CrossRef]
21. Zhang, N.N.; Xu, M. Effects of silicon dioxide combined heat treatment on properties of rubber wood. *J. For. Eng.* **2019**, *4*, 38–42. [CrossRef]
22. Wei, Y.X.; Zhang, P.; Liu, Y.; Chen, Y.; Gao, J.M.; Fan, Y.M. Kinetic analysis of the color of larch sapwood and heartwood during heat treatment. *Forests* **2018**, *9*, 289. [CrossRef]
23. Sundqvist, B.; Karlsson, O.; Westermark, U. Determination of formic-acid and acetic acid concentrations formed during hydrothermal treatment of birch wood and its relation to colour; strength and hardness. *Wood Sci. Technol.* **2006**, *40*, 549–561. [CrossRef]
24. Gaff, M.; Babiak, M.; Kačík, F.; Sandberg, D.; Turčani, M.; Hanzlík, P.; Vondrová, V. Plasticity properties of thermally modified timber in bending-The effect of chemical changes during modification of European oak and Norway spruce. *Compos B Eng.* **2019**, *165*, 613–625. [CrossRef]
25. Fang, C.H.; Cloutier, A.; Blanchet, P.; Koubaa, A. Densification of wood veneers combined with oil-heat treatment. Part I, Dimensional stability. *BioRES* **2011**, *6*, 373–385. [CrossRef]
26. Fang, C.H.; Cloutier, A.; Blanchet, P.; Koubaa, A. Densification of wood veneers combined with oil-heat treatment. Part II, hygroscopicity and mechanical properties. *BioRES* **2012**, *7*, 925–935. [CrossRef]
27. Avila, C.B.; Escobar, W.G.; Cloutier, A.; Fang, C.H.; Carrascoa, P.V. Densification of wood veneers combined with oil-heat treatment. Part III, cell wall mechanical properties determined by nanoindentation. *BioRES* **2012**, *7*, 1525–1532. [CrossRef]
28. Şahin, K.H.; Yusuf, S. The thermal conductivity of fir and beech wood heat treated at 170; 180; 190; 200; and 212°C. *J. Appl. Polym. Sci.* **2011**, *121*, 2473–2480. [CrossRef]
29. Chu, D.M.; Mu, J.; Zhang, L.; Li, Y.S. Promotion effect of NP fire retardant pre-treatment on heat-treated poplar wood. Part 1, color generation; dimensional stability; and fire retardancy. *Holzforschung* **2017**, *71*, 207–215. [CrossRef]
30. Yu, L.P.; Li, G.P. Study on processing conditions of spruce with nitrogen-phosphorus-boron environmentally friendly fire retardant. *China Wood Based Panels* **2006**, *13*, 14–16.
31. Gu, B. Study on the Effect of Properties of UF Plywood Treated with BL-Environmental Friendly Flame Retardant. Chapters, Composition Analysis of BL Environmental Protection Flame Retardant. Ph.D. Thesis, Beijing Forestry University, Beijing, China, 2007; pp. 50–56.
32. Zhang, X.T.; Mu, J.; Chu, D.M.; Zhao, Y. Synthesis of fire retardants based on N and P and poly(sodium silicatealuminum dihydrogen phosphate) (PSADP) and testing the flame-retardant properties of PSADP impregnated poplar wood. *Holzforschung* **2015**, *70*, 341–350. [CrossRef]
33. Chu, D.M.; Mu, J.; Zhang, L.; Li, Y.S. Promotion effect of NP fire retardant pre-treatment on heat-treated poplar wood. Part 2, hygroscopicity; leaching resistance; and thermal stability. *Holzforschung* **2017**, *71*, 217–223. [CrossRef]
34. Hosseinpourpia, R.; Adamopoulos, S.; Mai, C. Effects of Acid Pre-Treatments on the Swelling and Vapor Sorption of Thermally Modified Scots Pine (*Pinus sylvestris* L.) Wood. *BioRES* **2018**, *13*, 331–345. [CrossRef]

35. Himmel, S.; Mai, C. Effects of acetylation and formalization on the dynamic water vapor sorption behavior of wood. *Holzforschung* **2015**, *69*, 633–643. [CrossRef]
36. Guo, F.; Huang, R.F.; Lu, J.X.; Chen, Z.J.; Cao, Y.J. Evaluating the effect of heat treating temperature and duration on selected wood properties using comprehensive cluster analysis. *J. Wood Sci.* **2014**, *60*, 255–262. [CrossRef]
37. Li, S.J.; Qian, X.R.; Yu, J.; Wang, Q.W. Influence of FRW Fire Retardant Treatment on the Color and Painting Properties of Wood. *J. Northeast. For. Univ.* **1999**, *6*, 38–40.
38. Boruvka, V.; Zeidler, A.; Holecek, T.; Dudík, R. Elastic and Strength Properties of Heat-Treated Beech and Birch Wood. *Forests* **2018**, *9*, 197. [CrossRef]
39. Ladislav, R. *Wood Deterioration, Protection and Maintenance*; John Wiley & Sons: Chichester, UK, 2016.
40. Wang, W.; Zhu, Y.; Cao, J.Z.; Sun, W.J. Correlation between dynamic wetting behavior and chemical components of thermally modified wood. *Appl. Surf. Sci.* **2015**, *324*, 332–338. [CrossRef]

Article

Functionalized Surface Layer on Poplar Wood Fabricated by Fire Retardant and Thermal Densification. Part 2: Dynamic Wettability and Bonding Strength

Demiao Chu [1,2], Jun Mu [1,*], Stavros Avramidis [3], Sohrab Rahimi [3], Shengquan Liu [2] and Zongyuan Lai [1]

1 Key Laboratory of Wood Material Science and Utilization (Beijing Forestry University), Ministry of Education, Beijing 100083, China; 13chudemiao@sina.cn (D.C.); lzy2015@bjfu.edu.cn (Z.L.)

2 School of Forestry and Landscape Architecture, Anhui Agricultural University, Hefei 230036, China; liusq@ahau.edu.cn

3 Department of Wood Science, University of British Columbia, Vancouver, BC V6T 1Z4, Canada; stavros.avramidis@ubc.ca (S.A.); sohrab.rahimi@alumni.ubc.ca (S.R.)

* Correspondence: mujun@bjfu.edu.cn; Tel.: +86-010-62336053

Received: 9 October 2019; Accepted: 30 October 2019; Published: 5 November 2019

Abstract: In continuation of our former study on a novel combined treatment of nitrogen–phosphorus fire retardant and thermomechanical densification on wood, this study focuses on the dynamic wettability and the bonding strength. The contact angle was measured using the sessile drop method and the surface energy was calculated according to the van Oss method. Water surface penetrating and spreading is analyzed by both the Shi and Gardner model and the droplet volume changing model. The results reveal that the combined treatment increased the surface energy, especially the acid–base component. The contact angle declined and the water droplet spread more easily on the surface. Meanwhile, the rate of relative droplet volume decreased by 32.6% because the surface layer was densified and stabilized by the combined process. Additionally, the surface possesses the lowest roughness and highest abrasion resistance on the tangential section. Thus, the bonding strength of the combined treated poplar decreased by 29.7% compared to that of untreated poplar; however, it is still 53.3% higher than that of 220 °C heat-treated wood.

Keywords: combined treatment; wettability; surface free energy; bonding strength; poplar

1. Introduction

Wood is a renewable bioresource with many applications such as in construction, decoration, furniture, and cabinetry [1]. The demand for wood products in China has dramatically increased not only for esthetic reasons, but also because of government bonuses and mandates for using renewable materials mainly due to environmental concerns [2,3]. As a result, the demand for plantation trees has noticeably increased particularly due to forest product reduction. Nonetheless, the downside of using plantation timbers is substantially related to its low density, mechanical strength, and dimensional stability [4,5]. The service life of wood products depends on chemical or physical modifications. Density of wood is routinely considered to be one of the most important material characteristics on account of its strong correlation with strength [1,6]. Several thermomechanical (TM) methods have been developed for that purpose [7,8]. For further details, see Part 1 of our study [9].

Apart from the compression stability, the TM process may exacerbate the surface characteristics namely roughness and wettability, etc., which impacts bonding and coating performance [10–14].

Surface alteration is an indispensable part in the TM modification of wood; however, its wettability and bonding properties have not been well studied.

Nitrogen–phosphorus (NP) is an ecofriendly, benign, and affordable fire retardant [15]. Our research group attempted to enhance the leachability and hygroscopicity of NP, as well as smoke development during burning [16]. Recent studies revealed that NP also has an intensification effect on high-temperature (HT) treatment of wood [17,18]. In Part 1 of this study, it was reported that the combined process of using NP fire retardant and TM methods improved compression recovery and combustion safety. Since NP is a water-soluble chemical agent, it may enhance the hydrophilia of the treated wood. Here, we attempt to reveal the synergetic effects of this combined method on surface properties and bonding strength. The Shi and Gardner model and the droplet volume changing model are used to characterize the dynamic wetting process, and the surface characters and surface free energy are taken into account to explain bonding performance of the NP–HT combined treated wood.

2. Materials and Methods

2.1. Treatment

The NP–TM method was used to fabricate a certain thickness of functionalized surface layer on poplar (*Populus beijingensis* W. Y. Hsu). The treatment is the same as described in Part 1 [9].

2.2. Surface Free Energy

Surface free energy of wood is calculated mainly based on Young's equation. Two methods have been used for the calculation of the surface free energy, namely the van Oss-Chaudhury-Good (vOCG) theory. The Young equation is

$$\gamma_S = \gamma_L \cos\theta. \tag{1}$$

In the vOCG method, the surface free energy is expressed as

$$\gamma_S = \gamma_S^{LW} + \gamma_S^{AB} + 2(\gamma_S^- \gamma_S^+)^{1/2}. \tag{2}$$

Combining Equation (4) with Young's equation gives

$$\gamma_L (1 + \cos\theta) = 2(\gamma_S^{LW} \gamma_L^{LW})^{1/2} + 2(\gamma_S^+ \gamma_L^-)^{1/2} + 2(\gamma_S^- \gamma_S^+)^{1/2} \tag{3}$$

where γ_S and γ_L are the surface free energy of solids and liquid, respectively, γ_{SL} is the surface tension of the solid–liquid interface, and θ is the contact angle between a solid (S) and a liquid (L). The γ_{Sv} is the total surface energy calculated using the vOCG method, γ_S^{AB} is the acid–base based surface free energy for solids and liquid, γ_S^{LW} and γ_L^{LW} are, respectively, the Lifshitz–van der Waals-based surface free energy for solids and liquid, γ_S^+ and γ_S^- are, in turn, the acid-based surface free energy for solids and liquid, γ_S^- and γ_L^- are, respectively, the base-based surface free energy for solids and liquid. Distilled water, formamide, and diiodomethane, with known energy characteristics (Table 1), were used to calculate the surface free energy of treated wood samples [19].

Table 1. Surface tension and components of three different test liquids.

Liquids	Surface Free Energy (mJ m^{-2})				
	γL	γ_L^{LW}	γ_L^{AB}	γ_L^+	γ_L^-
Distilled water	72.8	21.8	51.0	25.5	25.5
Diiodomethane	50.8	50.8	0	0	0
Formamide	58.0	39.0	19.0	2.28	39.6

2.3. Contact Angles and Dynamic Wettability

A surface contact angle instrument coupled with SCA 20 software (OCA 20 Data Physics Instruments GmbH, Filderstadt, Germany), wherein a video measuring system with a high-resolution CCD camera and a high-performance digitizing adapter that enables instantaneous recording of the image and calculation of the contact angle, was used. Every group contained five replicates (20 mm × 20 mm × thickness) and data were collected randomly from three sites on each sample, using an automatic microsyringe to dispense 1.5 μL drops of testing liquids on the surface.

Images of the droplet on the surface were taken and stored at intervals of 1 s during the first 10 s, then intervals of 5 s until the end of the test. The contact angle (θ), height (h) and liquid–wood interface diameter (d) of each droplet images were measured for further calculation. Thus, the wetting model of the water droplet volume changes (here named the D-V model) during the wetting process, could be calculated as

$$AR_t = 1 - (Vt/V_0) = a\,(1 - \exp\,(-K_a t)) \quad (4)$$

where AR_t is the absorption ratio at t (s), Vt and V_0 is the droplet volume at t (s) and 0 s, respectively; a is a material constant and Ka refers to the decrease rate of intrinsic relative droplet volume.

According to the Shi and Gardner (S-D model), the contact angle changes during the wetting process could be calculated as

$$\theta = (\theta i \cdot \theta e)/\{\theta i + (\theta e - \theta i)\exp[K_\theta(\theta e/(\theta e - \theta i))t]\} \quad (5)$$

where θ is the contact angle at a certain time, θi and θe are the initial (instantaneous) and equilibrium contact angle, respectively; K_θ refers to how fast the liquid spreads and penetrates the porous structure of wood, which is a constant referred to the intrinsic decrease rate of relative contact angle. The t represents wetting time, then the θi, θe and K_θ values could be calculated. As the influence of water absorption, the wetting performance of the tested surface changes with time, and the instant decrease rate of contact angle (IK_θ) in the first 10 s was also calculated.

2.4. Surface Characteristics and Bonding Strength

Mass loss of surface abrasion for the processing layer was measured with a JM-IV instrument (Wuhan Gelaimo Testing Equipment Co., Ltd., Hubei, China), according to international standard ISO 7784-1 (1997). The wheel was coupled with 240-grit sanding paper during testing, and the mass loss values were calculated after 100 rotations. Surface roughness measurements were obtained by the stylus method in the perpendicular direction to the fibers on the wood surface and were carried out using a Taylor Hobson Surtronic 3+ instrument (Metrology Instrument Taylor Hobson Ltd., Leicester, England) at a constant speed of 1 mm/s over 15 mm of tracing length and a 2.5 mm cutoff across the sample grain. Bonding strength of treated wood was determined according to JAS234-2003 and STM D2559 using polyvinyl acetate glue (PVAc). After gluing (200 g m^{-2}), the samples were pressed under 4 kg force (kgft) and maintained for 12 h.

3. Results and Discussion

3.1. Contact Angles and Surface Free Energy

Table 2 shows the contact angles of different test liquids. Calculation of the surface energies was conducted using contact angles of diiodomethane, distilled water, and formamide, wherein the diiodomethane was selected as nonpolar test liquid and the formamide was used as the polar test liquid. The contact angles of the formamide and diiodomethane were lower than that of water. For all samples investigated, the total surface free energy ranged from 39 to 57 mJ m^{-2}, which agrees with previous studies [20].

According to the vOCG theory, the total surface energy (γs) of the $W_{NP\text{-}TM}$ was increased by the NP–HT treatment. The γs^{AB} value increased dramatically, while the γs^{LW} decreased by 7.6%.

The *AB* refers to acid–base interactions and is related to the hydrogen bond component and the *LW* is the London–van der Waals component. Both of the electron donors (γs^-) and acceptors (γs^+) were increased, wherein the γs^+ was proved less affected by the HT [20]. This result means that the treated poplar surface became more hydrophilic after the NP–HT, which helps water-based adhesives to spread and penetrate. The reason is that the nitrogen–phosphorus fire retardant used in the study is a water-soluble chemical, which could cause a small contact angle with distilled water [15,18]. The γs value of the W_{HT} samples decreased because of the HT, where the free reactive hydroxyl groups in hemicelluloses are partly removed [21,22]. The TM process further decreased the surface energy of the W_{HT-TM} samples. That result could be explained by the increased density and enhanced surface smoothness causing a lower contact area of liquid and wood fibrillation. Similarly, the surface smoothness and density enhancement in W_{TM} samples brought a considerable increase of the electron acceptor (γs^+) component of the surface energy.

Table 2. Contact angle and surface free energy of untreated and treated poplar.

Groups	Contact Angle (°)			Surface Free Energy (mJ m^{-2})				
	Diiodomethane	H_2O	Formamide	γs	γs^{LW}	γs^{AB}	γs^+	γs^-
W_{NP-TM}	44.42 ± 2.42	64.98 ± 4.3	71.72 ± 2.68	55.13	37.32	17.81	2.15	36.9
W_{TM}	35.28 ± 4.73	115.24 ± 3.7	97.55 ± 5.28	45.80	41.90	3.90	6.76	0.56
W_{HT-TM}	41.21 ± 4.12	116.88 ± 8.7	73.28 ± 0.62	39.95	39.00	0.95	0.05	4.22
W_{HT}	38.56 ± 3.52	116.56 ± 6.8	69.99 ± 4.30	42.38	40.32	2.06	0.20	5.42
Wc	38.40 ± 1.14	71.58 ± 1.6	59.30 ± 0.83	45.65	40.41	2.18	0.08	15.12

3.2. Dynamic Wetting Process

Representative images of water droplets and contact angle curves of untreated and treated samples are illustrated in Figure 1. It can be seen that all contact angles are decreasing as a function of testing duration.

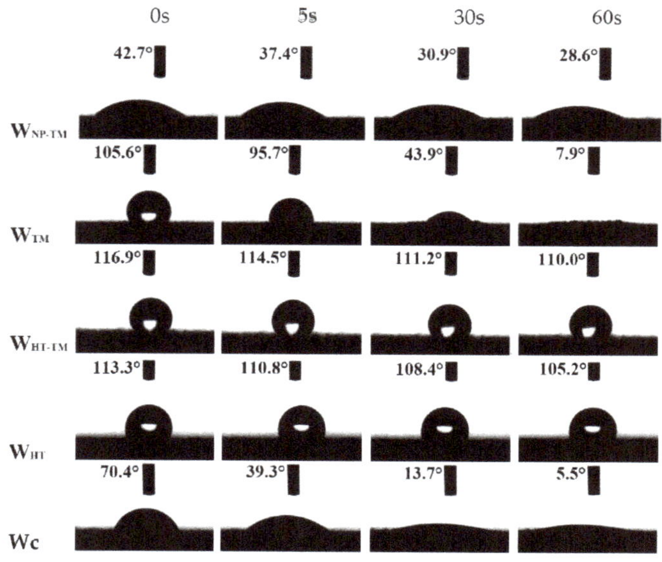

Figure 1. Images of water droplets on surface of untreated and treated poplar.

As seen in Figure 1, the untreated poplar possessed a low contact angle of 70.4° at 0 s, and the volume of the water droplet decreased quickly within 60 s. After 220 °C HT for 2 h, the contact angle of W_{HT} became much higher than that of Wc. The contact angle of W_{HT-TM} was further enhanced because of the TM process. Similarly, the contact angle at 0 s on the W_{TM} was around 50% higher than that of Wc, whereas it decreased sharply as the test time proceeded. It can be clearly observed that the surface swelled from the image at 30 s. In addition, the contact angle was 7.9° and the droplet almost disappeared at 60 s. For the W_{NP-TM}, the contact angle at 0 s was even lower than that of Wc, while it maintained a relatively stable value after 5 s. The contact angle reduced around 33% after 60 s, which was much lower than that of W_{TM} and Wc. Moreover, there was no swelling observed even at the interface of water droplet and wood surface. This could be explained by the high water solubility of the NP fire retardant. Contrary to that of W_{TM}, the contact angle of W_{NP-TM} was much more stable during the whole wetting process, and the enhanced compression stability of the surface relieved the decline of contact angle.

During the wetting process, the decline of the contact angle on the wood surface is caused by both spreading and penetration of the water droplets. The differences existed on both the initial contact angle and the decline rate during the test. Figure 2 illustrates the fitting curves of contact angle, instant decrease rate of contact angle IK_θ value, and the absorption ratio of untreated and treated poplar.

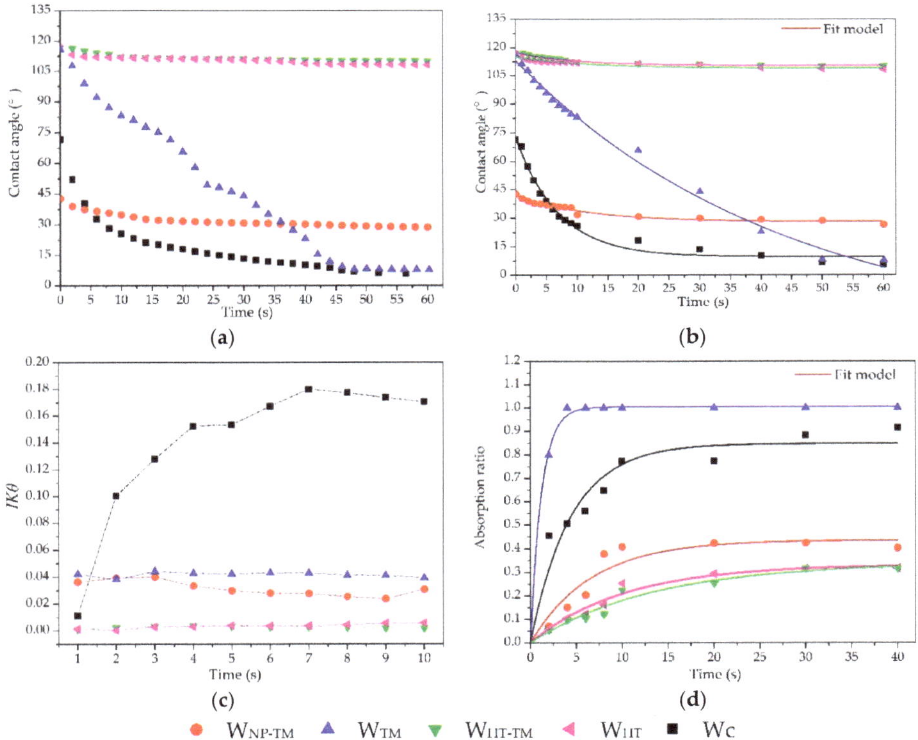

Figure 2. Contact angle curves (**a**), fitting curves of contact angle (**b**), instant decrease rate of contact angle IK_θ (**c**), and absorption ratio of the water droplet (**d**) on the surface of untreated and treated poplar.

The calculation of contact angle and absorption ratio changes was done according to the contact angle changing model (S-D wetting model) and droplet volume changing model (D-V wetting model), respectively, in order to further prove the hypothesis (Table 3). The R^2 values of the wetting models are higher than 0.95 for most samples.

Table 3. Wetting kinetics of water droplets on untreated and combined treated poplar.

Groups	Contact Angle Change				Absorption Ratio Change		
	θ_i (Degrees)	θ_e (Degrees)	K_θ	R^2	K_a	R^2	SE
W_{NP-TM}	41.92	26.24	0.019	0.90	0.153	0.907	0.034
W_{TM}	115.94	0.1 *	0.040	0.99	0.826	0.998	0.036
W_{HT-TM}	116.36	109.60	0.004	0.95	0.076	0.959	0.013
W_{HT}	113.79	105.98	0.002	0.96	0.099	0.966	0.014
Wc	68.73	3.42	0.174	0.92	0.227	0.928	0.039

Note: * means that we presumed the θ_e value of W_{TM} to be 0.1.

In the S-D model, the K_θ value reflected the shape of the wetting curve. The K_θ value of the W_{HT-TM} was 97.70% lower than Wc. The reduction of the K_θ value revealed the decline of the water spread and penetration on the surface. The initial (θ_i) and equilibrium (θ_e) contact angles of Wc was 68.73° and 3.42°, respectively. Due to HT, the θ_i value of W_{HT} increased by 65.56%. Besides, the decrease between θ_i and θ_e was only 6.86%, which indicated that the water droplet changed slightly during the wetting process. For the post-TM-treated samples of W_{HT-TM}, the K_θ values increased. This result could be caused by the enhancement of surface smoothness, thus making the water droplet more easily spread on the surface.

As illustrated in Figure 2a, the curve of contact angle for W_{TM} showed two broad peaks between 10 s to 40 s, the contact angle of W_{TM} declined until the end of the test. The θ_e value was almost zero, here we presumed it to be 0.1. The contact angle of W_{HT} only had a mild decline within the first 10 s and thereafter kept a relatively stable value. There is no obvious difference in the contact angle curve between W_{HT-TM} and W_{HT}, revealing that the post-TM process could not make it more hydrophobic. The increased density could be the reason for the increase of θ_i on the W_{TM}, while the low compression stability caused the decline of the contact angle.

According to Figure 2b, the fitting curves of W_{TM} and Wc decreased sharply at the beginning stage. For the W_{TM}, the contact angle declined until the end of the test. The surface swelling probably alleviated the decrease of the contact angle, causing an inclination to a linear function than the reality. The wettability of the testing surfaces changed as the wetting time increased.

It is worth noting that the θ_i on the W_{NP-TM} decreased significantly, that is 39.15% lower than that of Wc. Nevertheless the contact angle decrease trend and fitting curve assembled with that of W_{HT} and W_{HT-TM} (Figure 2a,b). Besides, the θ_e value was much higher than that of W_{TM} and Wc, revealing that the volume of the water droplet changed less. The instant decrease rate of contact angle value (IK_θ) of Wc increased, while the IK_θ values of W_{NP-TM}, W_{HT-TM}, and W_{HT} were stable, indicating that the water droplets had a weak effect on the surface at the beginning stage (Figure 2c). The alleviation effect on the decrease of the contact angle could also make the IK_θ value of W_{TM} more stable and lower than the reality. The θ_i value of W_{NP-TM} was the lowest because the NP fire retardant used in this study has a high hygroscopicity [15]. The lower decline could be explained by the increased surface density and high compression stability, as stated in Part 1.

In the D-V model, the K_a value reflects the change of absorption ratio; a smaller K_a value reveals a lower decrease rate of the droplet volume. The K_a of the water droplet on all the testing surfaces increased sharply at first, then reached a constant value as the time increased (Figure 2d). The only TM samples of W_{TM} has the highest K_a value, additionally, the absorption ratio curve reached the maximum within 5 s. It could be concluded that the water droplet on W_{TM} was mostly absorbed in the beginning stage. In favor of the result of S-D wetting model, the absorption ratio curve of W_{NP-TM} resembled with that of W_{HT-TM} and W_{HT}. This means that the W_{NP-TM}, W_{HT-TM}, and W_{HT} possessed favorable resistance to water. In other words, the water droplet infiltrated less into the surface layer.

3.3. Surface Characteristics and Bonding Strength

The TM process also exacerbates the surface characteristics and affects the bonding or coating performance [13,14]. Surface roughness, abrasion resistance, and bonding strength of all the treated samples are lower than that of untreated poplar, as illustrated in Figure 3.

As shown in Figure 3a, the TM process significantly enhanced the surface smoothness of W_{HT-TM}, W_{TM}, and W_{HT-TM}; this is because the surface roughness was flattened in these samples. Among those compressed samples, the W_{NP-TM} has the lowest roughness values, including the mean arithmetic deviation of the profile (Ra), root-mean-square roughness (Rq), and mean peak-to-valley height (Rz). The filling effect of the NP probably further enhanced the surface roughness. Besides, HT also enhanced the surface roughness for W_{HT}, which agrees with other studies [23].

Compared to the Wc, the abrasion mass loss of W_{HT} increased sharply (Figure 3b). The abrasion performance related to the shearing strength of the surface; the TM treatment enhanced the surface density mainly by reducing the pore and lumen structures. The degradation of the cell wall chemical components made the surface more brittle; thus, the enhancement of W_{HT-TM} was limited. Additionally, the cracks in cell walls due to the TM may further decrease the abrasion resistance; the mass loss of W_{TM} was higher than that of Wc, although the surface density increased. Since the cell walls were mostly undamaged and the NP recrystallized in the pores, the abrasion resistance of the W_{NP-TM} remained at the same level as the Wc. The slight decrease of abrasion could be explained by the decomposition and catalytic dehydration effects of the NP on the cell wall under hot press conditions [17].

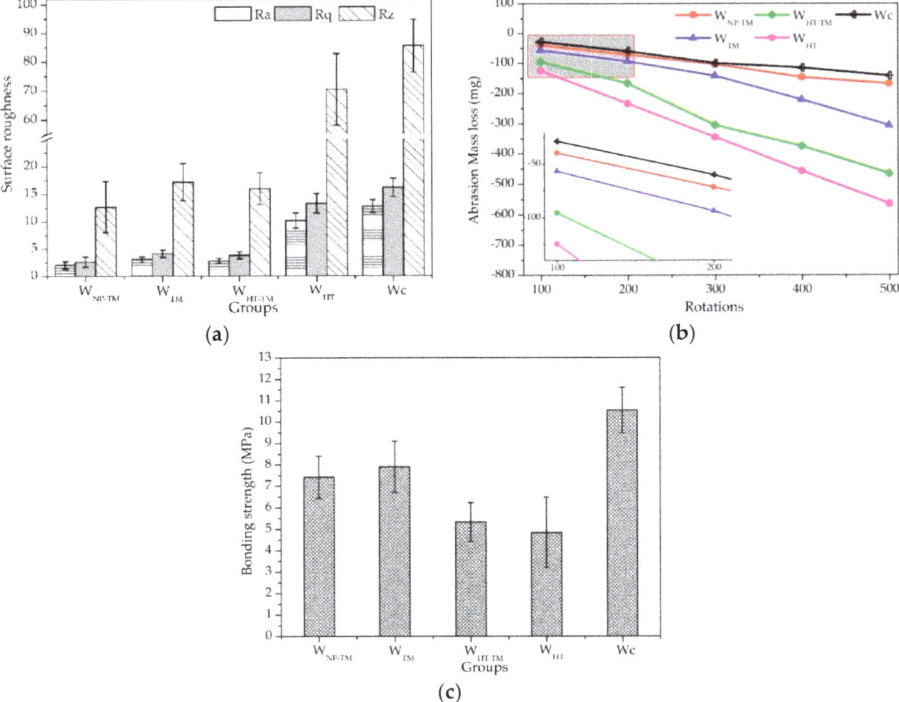

Figure 3. The surface roughness (**a**), abrasion mass loss (**b**), and bonding strength (**c**) of untreated and treated poplar.

As shown in Figure 3c, the bonding strength of all treated poplar decreased, especially the W_{HT} samples. The bonding strength of the W_{HT-TM} increased slightly as the surface hardness and abrasion resistance increased by the TM process, although the wetting properties were not enhanced. For the NP–TM treated poplar, the bonding strength of W_{NP-TM} was 29.67% lower than that of Wc, which could be explained by the low permeability, as discussed according to D-V model. The adhesive hardly goes into the gaps and fewer interactions existed between the wood and adhesive. Conversely, the wetting process was so fast that too much adhesive absorbed into the surface layer and the glue line was weak for W_{TM}.

4. Conclusions

This work further investigated the effect of nitrogen–phosphorus fire retardant (NP) pre-impregnation and the thermomechanical densification (TM) process on poplar wood. This NP–TM combined treatment could increase the surface energy and improve the spreadability of water on wood surfaces, which enhances the contact effect between wood and adhesive. However, the surface layers of the NP–TM combined treated poplar were highly condensed and stabilized, which prevents the permeation of the liquid on the surface. In addition, the NP–TM-treated wood possessed the lowest roughness and highest abrasion resistance on the functionalized surface. Therefore, it conversely influences the interactions of adhesive and pores on the surface. The low compression stability of the TM-only treated poplar causes surface swelling during the wetting process and decreases the bonding strength. Compared with that of wood that was only heat-treated, the surface bonding strength of the NP–TM combined treated poplar increased by 53.3%. In future studies, it is worth investigating possible methods for enhancing wetting performance and bonding strength while keeping the compression stability of this functionalized surface layer, making this new material more useful for the wood industry.

Author Contributions: Conceptualization, D.C. and J.M.; Methodology, D.C. and J.M.; Software, D.C. and S.R.; Validation, J.M., S.A., and S.L.; Formal Analysis, D.C., J.M., and Z.L.; Investigation, D.C. and J.M.; Resources, D.C. and J.M.; Data Curation, D.C., S.R., and J.M.; Writing-Original Draft Preparation, D.C.; Writing-Review & Editing, J.M., S.A., S.L., S.R., and Z.L.; Visualization, D.C.; Supervision, J.M. and S.A.; Project Administration, J.M.; Funding Acquisition, J.M.

Funding: This research was funded by the Forestry Public Welfare Project Foundation of China, grant number 201404502, and the Fundamental Research Funds for the Central Universities of China, grant number 2015ZCQ-CL-01.

Conflicts of Interest: The authors declare no conflict of interest.

References

1. Hill, C.A.S. *Wood Modification: Chemical, Thermal and Other Processes;* John Wiley & Sons: Chichester, UK, 2006.
2. Lin, H.; Wang, Y.; Chen, Z.D.; Yang, B.X. Progress & Application Prospects of Modified Wood. *For. Mach. Woodwork. Equip.* **2006**, *34*, 11–13.
3. Yuan, Q.P.; Su, C.W.; Xiong, J.B. Research Progress of Plantation Wood Machining Properties. *J. Anhui Agric. Sci.* **2011**, *3*, 42–45.
4. Kiaei, M.; Naji, H.R.; Abdul-Hamid, H.; Farsi, M. Radial variation of fiber dimensions, annual ring width, and wood density from natural and plantation trees of alder (*Alnus glutinosa*) wood. *Drev. Vysk. Wood Res.* **2016**, *61*, 55–64.
5. Xu, K.; Lu, J.X.; Li, X.J.; Wu, Y.Q. Effect of heat treatment on dimensional stability of phenolic resin impregnated poplar wood. *J. Beijing For. Univ.* **2015**, *37*, 70–77.
6. Fang, C.H.; Mariotti, N.; Cloutier, A.; Koubaa, A.; Blanchet, P. Densification of wood veneers by compression combined with heat and steam. *Eur. J. Wood Prod.* **2012**, *70*, 155–163. [CrossRef]
7. Laskowska, A. The Influence of Process Parameters on the Density Profile and Hardness of Surface-densified Birch Wood (*Betula pendula Roth*). *BioResources* **2017**, *12*, 6011–6023. [CrossRef]
8. Kariz, M.; Kuzman, M.K.; Sernek, M.; Hughes, M.; Rautkari, L.; Kamke, F.A.; Kutnar, A. Influence of temperature of thermal treatment on surface densification of spruce. *Eur. J. Wood Prod.* **2017**, *75*, 113–123. [CrossRef]

9. Chu, D.M.; Mu, J.; Avramidis, S.; Rahimi, S.; Liu, S.Q.; Lai, Z.Y. Functionalized surface layer on poplar wood fabricated by fire retardant and thermal densification. Part 1: Compression recovery and flammability. *Forests* **2019**, *10*, 955. [CrossRef]
10. Bekhta, P.; Proszyk, S.A.; Krystofiak, T. Colour in short-term thermo-mechanically densified veneer of various wood species. *Eur. J. Wood Prod.* **2014**, *72*, 785–797. [CrossRef]
11. Darwis, A.; Wahyudi, I.; Dwianto, W.; Cahyono, T.D. Densified wood anatomical structure and the effect of heat treatment on the recovery of set. *J. Indian Acad. Wood Sci.* **2017**, 1–8. [CrossRef]
12. Chen, A.T.; Luo, P.P.; Xu, Z.Y.; Hu, Z.J. Effects of silicon dioxide combined heat treatment on properties of rubber wood. *J. For. Eng.* **2016**, *1*, 21–25. (In Chinese) [CrossRef]
13. Liu, H.L.; Shang, J.; Kamke, F.A.; Guo, K.Q. Bonding performance and mechanism of thermal-hydro-mechanical modified veneer. *Wood Sci. Technol.* **2018**, *52*, 343–363. [CrossRef]
14. Pelit, H.; Budakçı, M.; Sönmez, A.; Burdurlu, E. Surface roughness and brightness of scots pine (*Pinus sylvestris*) applied with water-based varnish after densification and heat treatment. *J. Wood Sci.* **2015**, *61*, 586–594. [CrossRef]
15. Gu, B. Study on the Effects of Properties of UF Plywood Treated with BL-Environmental-Friendly Flame Retardant. Composition Analysis of BL Environmental Protection Flame Retardant. Ph.D. Thesis, Beijing Forestry University, Beijing, China, 2007; pp. 50–56.
16. Zhang, X.T.; Mu, J.; Chu, D.M.; Zhao, Y. Synthesis of fire retardants based on N and P and poly (sodium silicate aluminum dihydrogen phosphate) (PSADP) and testing the flame-retardant properties of PSADP impregnated poplar wood. *Holzforschung* **2015**, *70*, 341–350. [CrossRef]
17. Chu, D.M.; Mu, J.; Zhang, L.; Li, Y.S. Promotion effect of NP fire retardant pre-treatment on heat-treated poplar wood. Part 1: Color generation, dimensional stability, and fire retardancy. *Holzforschung* **2017**, *71*, 207–215. [CrossRef]
18. Chu, D.M.; Mu, J.; Zhang, L.; Li, Y.S. Promotion effect of NP fire retardant pre-treatment on heat-treated poplar wood. Part 2: Hygroscopicity, leaching resistance, and thermal stability. *Holzforschung* **2017**, *71*, 217–223. [CrossRef]
19. Van Oss, C.J.; Good, R.J.; Chaudhury, M.K. Additive and nonadditive surface tension components and the interpretation of contact angles. *Langmuir* **1988**, *4*, 884–891. [CrossRef]
20. Gérardin, P.; Petrič, M.; Petrissans, M.; Lambert, J.; Ehrhrardt, J.J. Evolution of wood surface free energy after heat treatment. *Polym. Degrad. Stab.* **2007**, *92*, 653–657. [CrossRef]
21. Wang, W.; Zhu, Y.; Cao, J.Z.; Sun, W.J. Correlation between dynamic wetting behavior and chemical components of thermally modified wood. *Appl. Surf. Sci.* **2015**, *324*, 332–338. [CrossRef]
22. Gao, X.; Zhou, F.; Fu, Z.Y.; Zhou, Y.D. Sorption isotherms characteristics of high temperature heat-treated *Picea abies* and *Pseudotsuga menziesii*. *J. For. Eng.* **2018**, *3*, 25–29. [CrossRef]
23. Salca, E.; Hiziroglu, S. Evaluation of hardness and surface quality of different wood species as function of heat treatment. *Mater. Des.* **2014**, *62*, 416–423. [CrossRef]

Article

Wood Protection through Plasma Powder Deposition—An Alternative Coating Process

Robert Köhler [1,*], Philipp Sauerbier [2], Gisela Ohms [1], Wolfgang Viöl [1] and Holger Militz [2]

1 Laboratory of Laser and Plasma Technologies, University of Applied Sciences and Arts, Von-Ossietzky-Str. 99, 37085 Göttingen, Germany; gisela.ohms@hawk.de (G.O.); wolfgang.vioel@hawk.de (W.V.)

2 Department of Wood Biology and Wood Products, Faculty of Forest Sciences and Forest Ecology, Georg-August-University of Goettingen, Büsgenweg 4, 37077 Göttingen, Germany; philipp.sauerbier@forst.uni-goettingen.de (P.S.); hmilitz@gwdg.de (H.M.)

* Correspondence: robert.koehler@hawk.de; Tel.: +49-551-3705-212

Received: 12 September 2019; Accepted: 9 October 2019; Published: 11 October 2019

Abstract: In contrast to conventional coating processes such as varnishing, plasma powder deposition by means of an atmospheric pressure plasma jet on wood is not yet widely used. A key advantage of this process is that volatile organic compounds and organic solvents are avoided. In the present work, European beech (*Fagus sylvatica* L.) and pine sapwood (*Pinus sylvestris* L.) were coated with polymer (polyester), metal (aluminum coated silver) or metal oxide (bismuth oxide) particles. Furthermore, a layer system consisting of polyester and metal or metal oxide was investigated. The layer thickness and topography were analyzed with a laser scanning microscope and scanning electron microscope, revealing thicknesses of 2–22 µm depending on the coating material. In general, the chemical composition of the layers was determined using X-ray photoelectron spectroscopy and Fourier-transform infrared spectroscopy measurements. The coatings consisting of metal and metal oxide showed a band gap and plasmon resonance in the range of 540 and 450 nm. Through this absorption, the wood may be protected against ultraviolet (UV) radiation. In the water uptake and release tests, the polyester layers exhibited a reduction of water vapor absorption after 24 h in 100% relative humidity (RH) by 53%–66%, whereas the pure metal oxide layers indicated the best desorption performance. The combination of metal oxide and polyester in the one-layer system combines the protection properties of the single coatings against water vapor and UV radiation.

Keywords: plasma powder deposition; moisture; UV protection; coating system; polyester; bismuth oxide; silver; wood protection

1. Introduction

Wood is a hygroscopic and porous material, which is strongly influenced by environmental conditions that change the wood properties. Two main factors are moisture and exposure to ultraviolet (UV) radiation, which influence the mechanical properties [1] and structural changes [2] in wood. In addition, a higher moisture content can promote fungal growth [3]. Therefore, it is necessary to protect the wood from environmental influences.

In order to reduce the moisture absorption of wood, three options are possible: a passive protection by structural measures, a superficial protection, e.g., by lacquers, or a deep protection by wood impregnation procedures.

In many lacquers the binders or pigments are dissolved or dispersed in organic solvents. If the solvents are based on volatile organic compounds, they are harmful to the environment [4]. An alternative could be a powder coating applied with an atmospheric pressure plasma jet (APPJ). In this plasma powder deposition technique, particles on a nm to µm scale are melted in the plasma and applied to the surface. This technique is environmentally friendly by omitting volatile organic compounds and organic solvents. Many organic and inorganic materials can be used as powder for the coating process. Another advantage of this technique is that thermally instable materials such as wood can also be coated.

Wallenhorst et al. [5,6] describe a plasma powder deposition with organic particles (PMMA) on wood. In former research it has been shown that a polyester coating could be deposited on wood and wood composites [7] but the moisture behavior has not yet been investigated. However, it is assumed that a water-repellent layer can be produced by coating with polymer particles.

As already mentioned, another problem is UV radiation. The absorption spectrum of wood, in particular lignin, shows a strong band between 200 and 400 nm in which the light is absorbed [8]. Due to the absorption of UV radiation, there is a radical induced depolymerization in lignin and the cellulose [9], which decomposes the wood surface. In addition to pigmented coatings and organic UV absorbers [10], protective layers based on metals or metal oxides can be used to achieve UV protection for wood.

Many semiconductors exhibit photocatalytic activity; however, this process involves the initial absorption of photons [11]. If a semiconductor with photocatalytic properties is irradiated with light of energy higher than the band gap of the semiconductor, there is a movement of electrons from the valence band to the conduction band in the semiconductor [12].

Due to the absorption, the wood could be protected from decomposition by UV light. Another possibility is the application of metal particles which show a plasmon absorption [13]. Similar to semiconductors, the particles absorb light in a certain spectral range of the light by the plasmon resonance.

For sufficient UV protection, the layers should provide an absorption range of 340–400 nm [14]. Widely used semiconductors with good photocatalytic properties or UV absorption are ZnO [15], TiO_2 [16] or Bi_2O_3 [17]. Materials with plasmon absorption are silver, aluminum and gold [18]. Similar to the water-repellent coatings, metal or metal oxide layers can be applied under atmospheric conditions using the APPJ system. Atmospheric pressure plasma powder depositions were used to apply ZnO [19,20], copper [21,22] or aluminum layers [23] on wood.

To combine the advantages of the metal or metal oxide layers (UV-protection) and the plastic layer (moisture protection), it is evident that a combination of the layer-types is beneficial for wood protection. Such coating systems were produced using an atmospheric pressure plasma chemical vapor deposition (APCVD) process, whereby silver, copper or zinc-SiO_x coatings were produced with hexamethyldisiloxane and a metal salt solution [24,25]. One of the differences between plasma powder deposition and APCVD is the state of the matter of the starting materials to be deposited. In the APCVD, liquid precursors are used. Tshabalala and Sung [26] also describe a hybrid coating of inorganic-organic compounds produced using cold plasma chemical vapor deposition onto wood. The generated films show a lower moisture sorption of the wood samples.

In this study, metal and metal oxide, polyester and a coating system consisting of both were produced using an APPJ on European beech and pine sapwood. Bismuth oxide and aluminum silver particles were used for UV protection due to their band gap and plasmon resonance. The chemical composition of the generated layers was examined with X-ray photoelectron spectroscopy (XPS) and Fourier-transform infrared spectroscopy (FTIR) measurements. The layer thickness and topography were analyzed with laser scanning microscopy (LSM) and scanning electron microscopy (SEM). UV/Vis spectra were recorded to determine the absorbing properties of the layer. Finally, water uptake and release tests were performed to verify the protective effect of the layers.

2. Materials and Methods

2.1. Coating Process

The different coatings were generated using an APPJ system. The system consists of a power supply (high voltage), a self-developed spray nozzle as an electrode and a brush disperser (RBG 2000, Palas GmbH, Karlsruhe, Germany). The plasma was generated between the high voltage electrode and the grounded spraying nozzle as an arc, which is blown out by compressed air to generate a plasma torch. The input power of the APPJ is 2 kW, which is described by the pulsed high voltage of 15 kV (effective voltage 2–3 kV) with a pulse repetition frequency of 50 kHz and a pulse period of 5–10 μs. A simplified structure of the self-developed spray nozzle is shown in Köhler et al. [7].

Three general types of coatings were generated: pure polyester (P) or metal oxide layers (MOL) and a coating system (CS) consisting of polyester and metal oxide.

The powders used for the MOLs and CSs were bismuth oxide (Bi) (grain size < 4 μm) (bismuth (III)-oxide, Asalco GmbH, Lüneburg, Germany) and an aluminum powder coated with silver (eConduct Aluminum 203000, ECKART GmbH, Hartenstein, Germany) (eCon) (D50 = 20–26 μm). The polyester (P) powder (D50 = 45 μm) (Interpon 610 MZ013GF, Akzo Nobel Powder Coatings GmbH, Arnsberg, Germany) was based on iso- and terephthalic acid.

The powders were injected by compressed air via the brush disperser into the plasma jet. The particles were melted by the plasma and deposited on the tangential surface on European beech (*Fagus sylvatica* L.) and pine sapwood (*Pinus sylvestris* L.) (40 (T) × 40 (L) × 10 (R) mm^3), traversing in meander-shaped paths at a distance of 20 mm from the ground electrode. To ensure that the metal oxide was present on the surface of the CSs, the CSs were manufactured in two stages—at first, a polyester layer was sprayed onto the wood samples, afterwards the metal or metal oxide was applied. The parameters used for the coating process refer to the layers described in Köhler et al. [7,27] and are shown in Table 1. Before the coating process, the wood samples were conditioned at 20 °C/65% relative humidity (RH) until constant mass was reached, and the densities of European beech and pine sapwood were measured with a balance (Kern KB 360-3N, KERN & Sohn GmbH, Balingen, Germany)—the density is 690.1 kg/m^3 and 514.8 kg/m^3, respectively.

Table 1. Coating Parameters for polyester, metal oxide and metal powder.

Powder	Plasma Power (%)	Process Gas Flow Rate (L/min)	Substrate Scan Speed (mm/s)	Powder Feed Rate (m^3/h)	Powder Feed Speed (mm/h)
Polyester	100	60	40	4.3	150
Metal oxide/metal	100	60	40	2.2	100

After the coating process, the substrates were annealed for 8 min using an infrared radiation (IR) - heater (ceramic heating element (150 W), RS Components GmbH, Mörfelden-Walldorf, Germany) at a distance of 10 mm. Afterwards, the samples were conditioned at 20 °C/65% RH until constant mass was reached. For the water absorption test, all surfaces except the coating surface of the sample were sealed three times with 2K-PUR boat lacquer (bacuplast Faserverbundtechnik GmbH, Remscheid, Germany). All samples were sterilized with ionizing radiation at a dose of 53.3 kGy (Steris, Synergy Health Allershausen GmbH, Allershausen, Germany). To determine the coating and sealing lacquer weight, the samples were compared to the untreated samples at the same conditions (20 °C/65% RH).

2.2. Surface Characterization

The chemical surface composition was investigated u sing XPS and FTIR measurements.

The XPS data were obtained on a PHI 5000 Versa Probe II (ULVAC-PHI, Chigasaki, Japan) using monochromatic Al-Kα radiation with a photon energy of 1486.6 eV. The detector resolution measured at the Ag $3d_{5/2}$ peak was 0.6 eV at a pass energy of 23.5 eV. Detailed spectra of carbon (C 1*s*), oxygen (O 1*s*), bismuth (Bi 4*f*), silver (Ag 3*d*5) and aluminum (Al 2*p*) were recorded with a spot size of 200 μm, a pass energy of 46.95 eV and a step size of 0.1 eV. To avoid charging effects, the measurements were carried out by neutralizing sample charging. All spectra were shifted to the main C *1s* peak at 284.8 eV.

The FTIR measurements were performed using a PerkinElmer Frontier (PerkinElmer LAS (Deutschland) GmbH, Rodgau Jügesheim, Germany) with a diamond attenuated total reflection (ATR) (Specac Golden Gate GS 10515, Specac Ltd., Orpington, UK). The spectra were recorded at the range of 400–4000 cm^{-1}, averaged over 64 scans and a resolution of 4 cm^{-1}.

The UV-Vis absorbance spectra were recorded using a PerkinElmer 650 (PerkinElmer, Inc., Shelton, CT, USA) with an integrating sphere module in the range of 300–900 nm with a velocity of 10.5 nm/s and a slit width of 5 nm.

LSM (VK-X100, KEYENCE Deutschland GmbH, Neu-Isenburg, Germany) was used to determine the layer thickness. The measurement was conducted using a 20× objective for the polyester and CS coating and a 100× objective for the MOL. In comparison to the mean height of coated to uncoated area, the coating thickness could be determined. Before the analysis, a tilt correction was performed, and the image noise was reduced.

SEM images were performed on an EVO LS 15 (Carl Zeiss AG, Oberkochen, Germany) with an accelerating voltage of 15 kV. To avoid charging effects, the sample was pre-sputtered with a 20 nm gold layer.

2.3. Water Uptake and Release Tests

The water uptake and release tests were carried out in accordance with the procedure of Meyer et al. [28]. The uptake of liquid water, absorption and desorption of water vapor of the wood specimens were investigated after a treatment time of 24 h. A balance with an accuracy of 0.001 g was used to determine all masses (Kern KB 360-3N, KERN & Sohn GmbH, Balingen, Germany). For the water uptake and release tests, it is assumed that the coating and sealing lacquer do not absorb water.

2.3.1. Absorption

The samples were dried in an oven at 103 °C to constant mass and were weighed afterwards. Then, the samples were placed in a closed plastic box on a stainless-steel perforated plate over deionized water with the coated side up. The box was stored under constant conditions of 20 °C/65% RH. After 24 h, the samples were weighted and the moisture content of the samples was determined using the following equation:

$$WV\,A\,24_{100\%\mathrm{RH}} = \frac{(m_{100\%\mathrm{RH}} - m_0)}{(m_0 - m_c)} \times 100, \quad (1)$$

where $WV\,A\,24_{100\%\mathrm{Rh}}$ is the water vapor uptake after 24 h at 100% RH, $m_{100\%\mathrm{RH}}$ the mass after 24 h of vapor suspended, m_0 the mass after drying at 100 °C and m_c the mass of the coating and sealing lacquer.

2.3.2. Desorption

The samples were conditioned at 20 °C/100% RH until fiber saturation (after 28 days) and the mass was determined. Subsequently, the specimens were placed with the coated side on silica gel for 24 h in a closed plastic box. The box was stored under constant conditions of 20 °C/65% RH. After 24 h, the samples were weighted and the moisture content of the samples was determined using the following equation:

$$WV\,D\,24_{0\%\mathrm{RH}} = \frac{(m_{FS} - m_{0\%\mathrm{RH}})}{(m_{FS} - m_c)} \times 100, \quad (2)$$

where $WV\,D\,24_{0\%\mathrm{Rh}}$ is the water vapor desorption after 24 h at 0% RH, $m_{0\%\mathrm{RH}}$ the mass after 24 h suspended on silica gel, m_{FS} the mass at fiber saturation and m_c the mass of the coating and sealing lacquer.

2.3.3. Liquid Water Uptake

The specimens were dried in an oven at 103 °C to constant mass and were weighed afterwards. The samples were submersed in deionized water in a closed plastic box. The box was stored under constant conditions of 20 °C/65% RH. After 24 h, the samples were weighted and the moisture content of the samples was determined using the following equation:

$$W\,24_{SUB} = \frac{(m_{SUB} - m_O)}{(m_{SUB} - m_c)} \times 100, \tag{3}$$

where $W\,24_{SUB}$ is the water submersion after 24 h, m_{SUB} the mass after 24 h submersed in water, m_O the mass after drying at 103 °C and m_c the mass of the coating and sealing lacquer.

3. Results and Discussion

3.1. *Average Thickness and Surface Morphology*

LSM measurements were performed to determine the average layer thickness and are shown in Figure 1. The CSs consist of polyester and bismuth oxide or polyester and eCon and are abbreviated as PBi and PeCon in the following. The produced layer thickness of the Bi and eCon layers is approximately 3.1 µm and 4.4 µm, whereas the thickness of the polyester layer and the CSs is approximately 18.8 µm. The CSs did not show significant differences (Fischer's exact test) from the polyester coatings. Due to the constant layer thickness of the CSs and the polyester layer, it can be assumed that the metal oxide particles are embedded in the polyester coating.

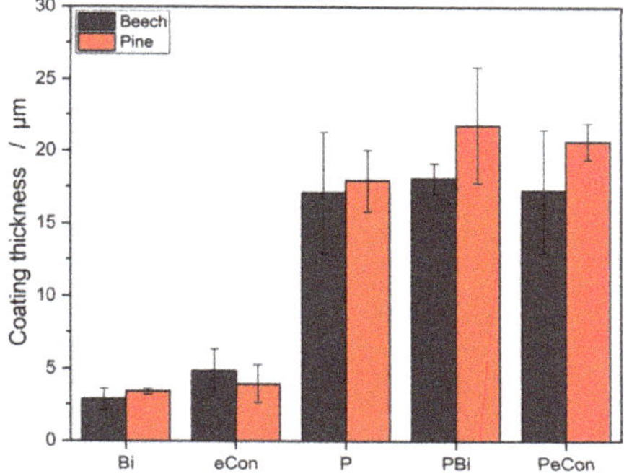

Figure 1. Coating thickness of metal oxide, polyester and coating system (CS) layer (n = 5).

The morphology of the coatings was assessed with SEM measurements and is shown in Figure 2. Pine and beech samples show the same layer morphology; therefore, only pine is shown here as an example. The pine reference shows the characteristic tracheides of the pine wood (a). Due to the impact on the wood surface the particles melted in the plasma form lenticular shapes which overlap each other and form a homogeneous surface. In the MOLs and the polyester coating (b)–(d) some isolated spots of the wood surface are apparent. The CSs (e) and (f) showed scattered holes on the surface. The metal oxide particles in the CSs seem to be under the surface of the polyester. In comparison to the MOLs and the CSs, the MOLs have a rather smooth surface. This could be due to the two-step coating process. In the second process step, in addition to the melting of the metal particles by the plasma, the surface of the previously deposited polyester is melted, too. It seems that by introducing the metal drops into the melt of the polyester, the metal drops are not transformed into lenticular structures but rather into spherical shapes.

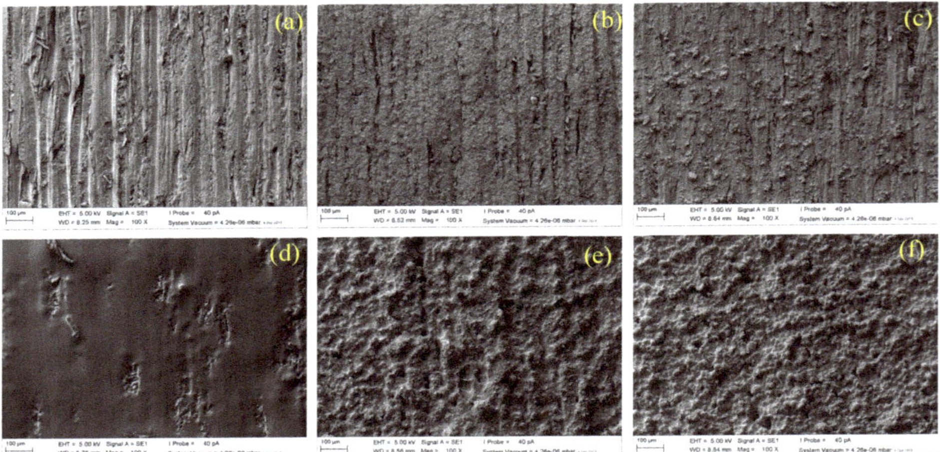

Figure 2. Scanning electron microscopy (SEM) images of pine wood (**a**) with different coatings: (**b**) Bi, (**c**) eCon, (**d**) polyester, (**e**) polyester and bismuth (PBi) and (**f**) polyester and eCon (PeCon).

3.2. Chemical Surface Characterisation

XPS measurements were performed in order to determine the elemental composition and binding states of the samples. Depending on the reference and the coating, different elements were detected:

- Wood reference and polyester layer: carbon (C 1*s*) and oxygen (O 1*s*),
- Bi coating and PBi: C 1*s*, O 1*s* and bismuth (Bi 4*f*)
- eCon and PeCon: C 1*s*, O 1*s*, aluminum (Al 2*p*) and silver (Ag 3*d*5).

A small amount of fluorine was detected in the layers with a polyester content. The atomic concentration of the samples is shown in Table 2.

Table 2. Atomic concentration of reference and coatings in atom—%.

Sample	C 1*s*	O 1*s*	Al 2*p*	Ag 3*d*5	Bi 4*f*	F 1*s*
Beech reference	68.14	31.86	–	–	–	–
Pine reference	71.22	28.78	–	–	–	–
Beech P	77.81	21.93	–	–	–	0.26
Pine P	77.98	21.7	–	–	–	0.32
Beech Bi	50.58	32.76	–	–	16.66	–
Pine Bi	63.63	24.75	–	–	11.62	–
Beech PBi	77.18	22.39	–	–	0.18	0.25
Pine PBi	77.3	22.29	–	–	0.15	0.26
Beech eCon	58.07	34.92	3.76	3.25	–	–
Pine eCon	58.24	32.82	5.09	3.85	–	–
Beech PeCon	77.13	22.17	0	0.21	–	0.49
Pine PeCon	77.05	22.51	0	0.22	–	0.22

In comparison to the reference of the polyester coating, the amount of carbon increased, and that of oxygen decreased by approximately 7–10 atom-%, respectively. The MOLs show a decrease of oxygen and an increase of the respective deposited metal. The chemical composition of the CSs was similar to the polyester coating with a small amount of the used metal. To confirm the assumption from the LSM and SEM measurements that the metal particles largely are embedded into the polyester, the detail spectra of carbon and oxygen were recorded, too. For a clearer presentation only the beech samples are shown. Figure 3 shows the C1s and O1s detail spectra of the reference, the polyester and CSs coated beech. The carbon and oxygen peek of the polyester coating have a similar structure to the CS—by the same structure and the low percentage of metal oxide at the surface it could confirm the embedding.

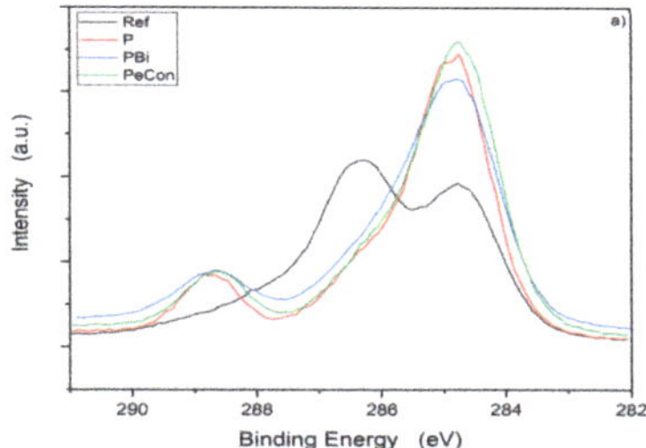

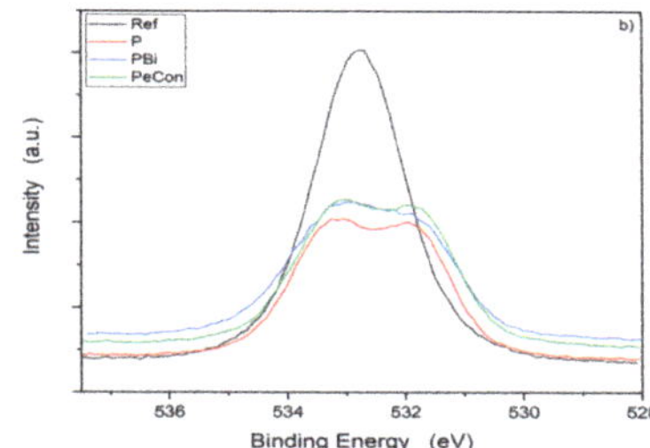

Figure 3. X-ray photoelectron spectroscopy (XPS) detail spectra of reference, polyester and CSs coated beech carbon (**a**) and oxygen (**b**).

To determine the oxidation state of the metal particle by the plasma torch, a detailed spectrum of bismuth, aluminum and silver was recorded and the peak position was analyzed. All peaks were normalized to the maximum of each spectrum. The bismuth detail spectra are shown in Figure 4a and consist of two peaks at the binding energies of 158.7 eV and 164.0 eV, which correspond to the Bi $4f_{7/2}$ and Bi $4f_{5/2}$. The peak position indicates that bismuth (III) oxide [29,30] was applied at the surface. The peak separation of the Bi $4f_{7/2}$ and Bi $4f_{5/2}$ is 5.3 eV. Similarly to bismuth, the silver spectrum consists of two peaks: the Ag $3d_{5/2}$ and Ag $3d_{3/2}$, with a peak separation of 6 eV (Figure 4b). The Ag $3d_{5/2}$ peak for the MOLs and the CS is located at 367.9 eV and 367.6 eV, respectively. The peak at 367.9 eV corresponds to the Ag(0) oxidation state [31,32]. The peak position at 367.6 eV is shifted to lower binding energies, which indicates the existence of silver in oxidation states [33,34]. Hsu et al. [35] explain the shift to a lower binding energy by a slight oxidation of the silver surface.

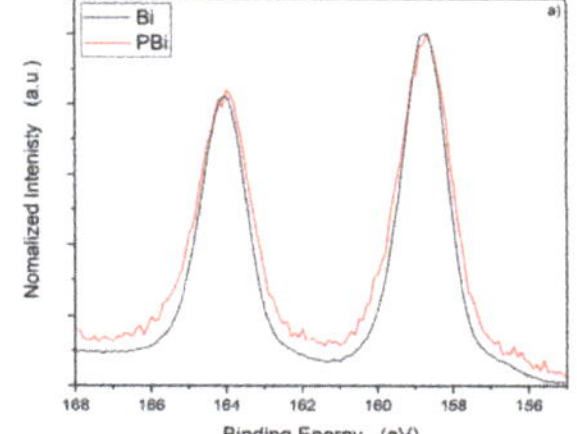

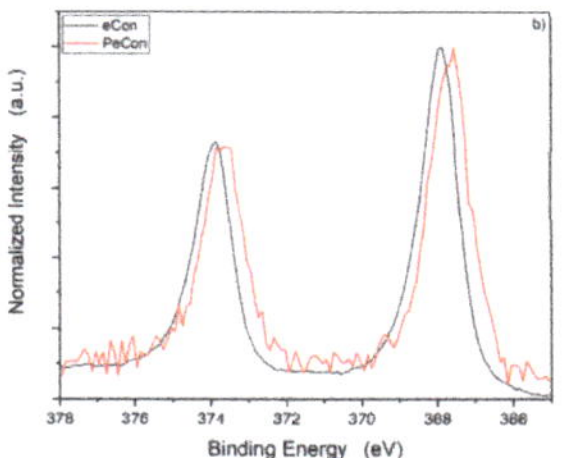

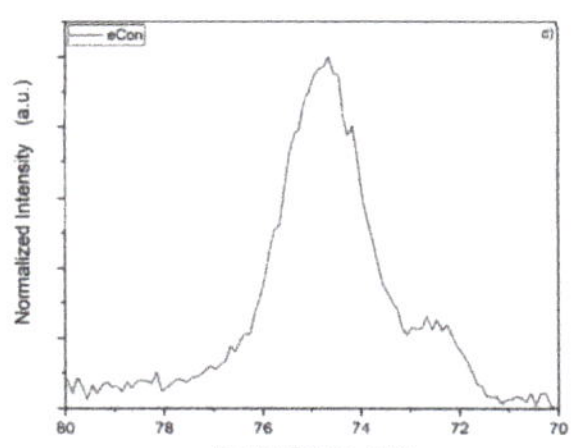

Figure 4. XPS detail spectra of (**a**) bismuth, (**b**) silver and (**c**) aluminum of beech coating.

Figure 4c shows the high-resolution spectra of aluminum. Only one spectrum could be shown because no aluminum was detected in the CS. The main peak at 74.7 eV is characteristic of Al_2O_3 [36] or $Al(OH)_3$ [37], whereas the small shoulder at 72.4 eV could be identified as Al(0) [38].

The results of the FTIR measurements are illustrated in Figure 5a–c. Figure 5a clearly shows the broad OH stretch band at 3300 cm^{-1} for the untreated reference specimens. The peak is pronounced since the main components of wood (lignin, cellulose, hemicellulose) all contain hydroxyl groups. After coating with polyester, a new spectrum is clearly evident—the OH band is no longer visible, but a characteristic carbonyl peak at about 1750 cm^{-1} is apparent. Due to the penetration depth of the ATR-IR measurement in the µm–range, a complete and area-wide coating can be assumed.

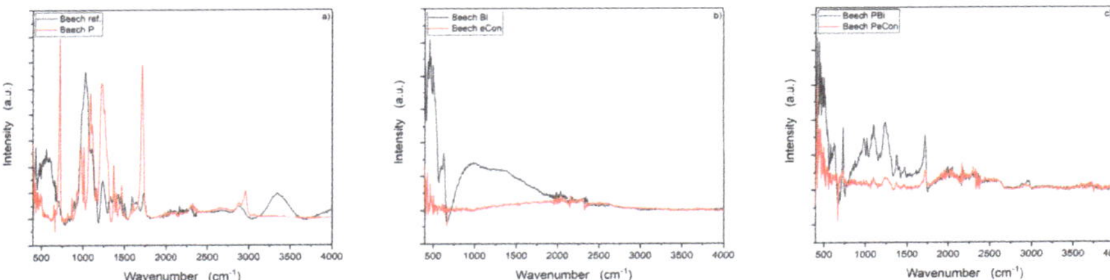

Figure 5. Coated and uncoated European beech with (**a**) reference and polyester coating, (**b**) Bi and eCon coating and (**c**) PBi and PeCon ($n = 3$).

Figure 5b compares Bi and eCon coated beech FTIR spectra. The reflective characteristic of the eCon particles in IR wavelength results in a mostly peak-free spectrum for eCon. Only for bismuth MOL a characteristic peak is visible in the range of 400–700 cm^{-1}, which is assigned to the Bi-O-Bi vibration band [39]. It should be emphasized, however, that also no peaks specific to wood are recognizable—a complete and area-wide deposited layer can again be assumed. Finally, PBi and PeCon deposits on beech are shown in Figure 5c. Again, no wood-typical peaks can be found. Characteristic polyester bands are clearly visible in case of PBi but are far less pronounced than in the pure polyester spectra shown in Figure 5a. This supports the conclusion of the XPS measurements that the metal particles are embedded. An attenuation of the characteristic polyester peaks can be explained by near-surface or not fully covering metal particles, whereby the IR radiation can indeed react with the polyester, but a significant proportion is reflected in advance by near-surface or on-surface of the metal particles.

3.3. UV/Vis Spectroscopy

Figure 6 shows the absorption spectra of the coatings and the wood reference measured at room temperature using UV-vis spectroscopy. The uncoated wood and the polyester coating show an absorbance in the range of 300–400 nm (Figure 6a). The measured spectra can be divided into two coating groups: with bismuth (Figure 6a) and with eCon (Figure 6b). All bismuth-based layers show an absorption behavior of the visible light with a wavelength <500–550 nm. The absorption behavior of the semiconductor allows the conclusion that bismuth oxide is present in a crystalline phase. The band gap of the crystalline semiconductor could be calculated using the Tauc plot method with the following equation [40]:

$$(\alpha hv)^{\frac{1}{n}} = A\left(hv - E_g\right), \tag{4}$$

where Eg, h, α, v, and A are the energy of the band gap, Planck's constant, absorption coefficient, light frequency and a constant. The use of n has been reported in former research [41]. The band gap energy is identified by extrapolating the straight line to the x-axis [42] in the plot $(\alpha hv)^2$ vs. (hv) [43] (Figure 6c). The thus determined band gap of the layers was at approximately 2.3 eV.

The eCon layers show no absorption behavior while the PeCon layers appear at a wide absorption peak at approximately 450 nm. These peaks could be associated with the plasmon resonance (PR) of the coated silver particles [44,45]. The PR can be influenced by different material parameters like size, shape, distance [46,47], composition of the nanostructures and the dielectric properties of the surrounding environment [48]. The effect is described in the literature for nanoparticles or thin stacked metal layers [49,50], which are not present in the layers produced here. A possible explanation for the PR could be that the particles were embedded close enough to each other in the polyester that PR would occur. The distance between the metal particles is sufficient for PR to occur, compared to Aslan et al. [51]. This effect cannot be observed in the eCon layers, since the particles are not embedded and no structures can be created by the impact of the particles on the wood surface, which allows PR. The PR is visible despite the slight formation of an oxide layer from the XPS results.

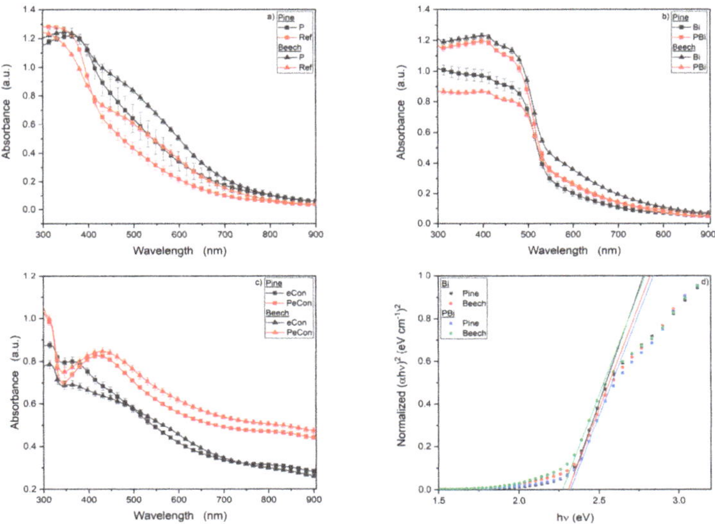

Figure 6. Absorbance spectra of coated and uncoated pine and beech wood: (**a**) reference and polyester coating, (**b**) coatings with bismuth oxide, (**c**) coatings with eCon (**d**) normalized band gap of the bismuth oxide existing layers ($n = 5$).

3.4. Water Uptake and Release Test

The absorption and desorption of water vapor and the absorption of liquid water were investigated in terms of the moisture behavior of the coated and uncoated wood samples. In general, the moisture content of the pine samples is higher than the beech samples. In Figure 7a, the water vapor absorption after 24 h is presented. In both wood species, the uncoated reference and the MOLs show the same absorption behavior, whereas the layers consisting of polyester can reduce the water vapor uptake by 53.5% and 66.0% in the beech and pine samples, respectively. The desorption is shown in Figure 7b. The MOLs exhibit a higher desorption than the uncoated references, while the polyester layers have the lowest desorption behavior. The CSs show the same desorption level as the uncoated wood references. The water submersion test shows no positive characteristics of the applied coatings (Figure 7c). All coated samples and the reference are at the same moisture content. The samples showed a concave form after 24 h. The concave forms could be due to the holes in the layer surface, which were detected in the SEM images (Figure 2) as the water could only enter the sample from this side unhindered. The bend of the samples leads to the failure of the coating and partly to cracks in the front edges. This means that more water can penetrate the samples, which is reflected in the equality of the measured values. After the water submersion, the coatings were investigated with the SEM measurements (Figure 8). In the reference and the MOLs, the roughness of the surface increased after the water treatment, individual fibers are visible, and the coating acts washed out. The images of the polyester layer and the CSs reveal cracks in the surface compared to the untreated samples.

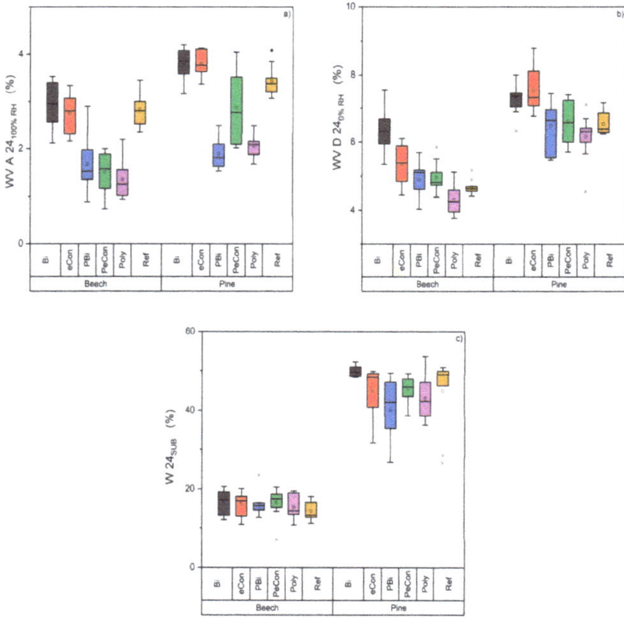

Figure 7. (**a**) Absorption (**b**) desorption of water vapor and (**c**) liquid water submersion of coated and uncoated beech and pine wood (n = 10).

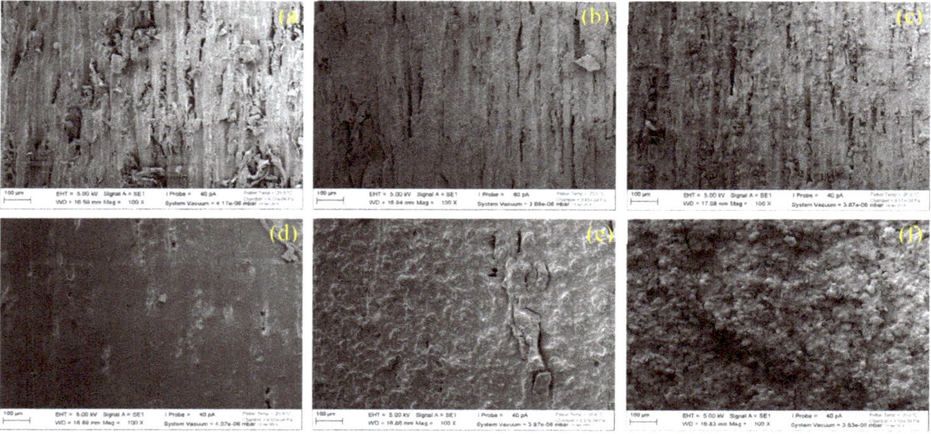

Figure 8. SEM images of pine wood (**a**) with different coatings: (**b**) Bi, (**c**) eCon, (**d**) polyester, (**e**) PBi and (**f**) PeCon after water uptake and release test.

4. Conclusions

In this study, three different layer types of beech- and pinewood were deposited. The results of the investigations allow the following conclusions:

- In the CSs, the metal oxide particles are largely embedded in the polyester.
- The layers containing bismuth oxide show a band gap in the visible spectral range of light at 540 nm, which allows its use as a potential UV protection of wood.
- The eCon MOLs show no protection against UV radiation but could reflect infrared light (<15 μm).
- The CSs with eCon show plasmon resonance probably due to their structure, which also serves as UV protection for wood.
- The bismuth MOLs show good reflective properties in long-wavelength infrared.
- The polyester-containing layers show a lower absorption behavior than the metal oxide layers.
- The metal oxide layers have a higher desorption behavior than the polyester-containing layers.
- The good behavior of the layers against water vapor could not be shown in the submersion test.
- The combination of metal oxide and polyester in the one-layer system combines the positive properties of the polyester layer in absorption and the metal oxide layer in desorption. The layers created in this way protect the wood from water vapor and exhibit good desorption properties.
- The combination of metal oxide and polyester in the one-layer system combines the protection properties of the single coatings against UV radiation and water vapor.

The results discussed here indicate that plasma powder coatings offer a promising way to produce protective wood layers. In further experiments, the UV-protective properties of the generated layers will be the focus. A possible lignin degradation by UV light should be evaluated with the aid of the zero-span test. Furthermore, the parameters such as layer thickness or cover ratio should be evaluated and undergo natural weather conditions.

Author Contributions: Conceptualization, R.K.; methodology, R.K., P.S.; validation, R.K., P.S., G.O., W.V., H.M.; investigation, R.K., P.S.; resources, G.O., W.V., H.M.; writing—original draft, R.K., P.S.; writing—review and editing, R.K., P.S., G.O., W.V., H.M.

Funding: This research was funded by Bundesministerium für Bildung und Forschung (03XP0015A/B).

Acknowledgments: The authors thank the German Research Foundation (DFG) for the provision of the XPS (INST 196/8-1), Akzo Nobel Powder Coatings GmbH and ECKART GmbH for providing the used powder material. We thank Michaela Zauner and Maximilian Wentzel for the support with the SEM measurements.

Conflicts of Interest: The authors declare no conflicts of interest.

References

1. Ishimaru, Y.; Arai, K.; Mizutani, M.; Oshima, K.; Iida, I. Physical and mechanical properties of wood after moisture conditioning. *J. Wood Sci.* **2001**, *47*, 185–191. [CrossRef]
2. Weichelt, F.; Emmler, R.; Flyunt, R.; Beyer, E.; Buchmeiser, M.R.; Beyer, M. ZnO-Based UV Nanocomposites for Wood Coatings in Outdoor Applications. *Macromol. Mater. Eng.* **2009**, *295*, 130–136. [CrossRef]
3. Gezici-Koç, Ö.; Erich, S.J.F.; Huinink, H.P.; van der Ven, L.G.J.; Adan, O.C.G. Bound and free water distribution in wood during water uptake and drying as measured by 1D magnetic resonance imaging. *Cellulose* **2017**, *24*, 535–553. [CrossRef]
4. Pearnchob, N.; Bodmeier, R. Dry polymer powder coating and comparison with conventional liquid-based coatings for Eudragit®RS, ethylcellulose and shellac. *Eur. J. Pharm. Biopharm.* **2003**, *56*, 363–369. [CrossRef]
5. Wallenhorst, L.; Rerich, R.; Vovk, M.; Dahle, S.; Militz, H.; Ohms, G.; Viöl, W. Morphologic and Chemical Properties of PMMA/ATH Layers with Enhanced Abrasion Resistance Realised by Cold Plasma Spraying at Atmospheric Pressure. *Adv. Condens. Matter Phys.* **2018**, *2018*, 1–11. [CrossRef]
6. Wallenhorst, L.M.; Dahle, S.; Vovk, M.; Wurlitzer, L.; Loewenthal, L.; Mainusch, N.; Gerhard, C.; Viöl, W. Characterisation of PMMA/ATH Layers Realised by Means of Atmospheric Pressure Plasma Powder Deposition. *Adv. Condens. Matter Phys.* **2015**, *2015*, 1–12. [CrossRef]
7. Köhler, R.; Sauerbier, P.; Militz, H.; Viöl, W. Atmospheric Pressure Plasma Coating of Wood and MDF with Polyester Powder. *Coatings* **2017**, *7*, 171. [CrossRef]
8. Sakakibara, A.; Samo, Y. Chemistry of Lignin. In *Wood and Cellulosic Chemistry*, 2nd ed.; Hon, D.N.-S., Shiraishi, N., Eds.; Marcel Dekker: New York, NY, USA, 2001.

9. Kiguchi, M.; Evans, P.D.; Ekstedt, J.; Williams, R.S.; Kataoka, Y. Improvement of the durability of clear coatings by grafting of UV-absorbers on to wood. *Surf. Coat. Int. Part B Coat. Trans.* **2001**, *84*, 263–270. [CrossRef]
10. Aloui, F.; Ahajji, A.; Irmouli, Y.; George, B.; Charrier, B.; Merlin, A. Inorganic UV absorbers for the photostabilisation of wood-clearcoating systems: Comparison with organic UV absorbers. *Appl. Surf. Sci.* **2007**, *253*, 3737–3745. [CrossRef]
11. Linsebigler, A.L.; Lu, G.; Yates, J.T. Photocatalysis on TiO_2 Surfaces: Principles, Mechanisms, and Selected Results. *Chem. Rev.* **1995**, *95*, 735–758. [CrossRef]
12. Prakash, J.; Sun, S.; Swart, H.C.; Gupta, R.K. Noble metals-TiO_2 nanocomposites: From fundamental mechanisms to photocatalysis, surface enhanced Raman scattering and antibacterial applications. *Appl. Mater. Today* **2018**, *11*, 82–135. [CrossRef]
13. Link, S.; Wang, Z.L.; El-Sayed, M.A. Alloy Formation of Gold–Silver Nanoparticles and the Dependence of the Plasmon Absorption on Their Composition. *J. Phys. Chem. B* **1999**, *103*, 3529–3533. [CrossRef]
14. George, B.; Suttie, E.; Merlin, A.; Deglise, X. Photodegradation and photostabilisation of wood–the state of the art. *Polym. Degrad. Stab.* **2005**, *88*, 268–274. [CrossRef]
15. Becheri, A.; Dürr, M.; Lo Nostro, P.; Baglioni, P. Synthesis and characterization of zinc oxide nanoparticles: Application to textiles as UV-absorbers. *J. Nanopart. Res.* **2008**, *10*, 679–689. [CrossRef]
16. Jnido, G.; Ohms, G.; Viöl, W. Deposition of TiO_2 Thin Films on Wood Substrate by an Air Atmospheric Pressure Plasma Jet. *Coatings* **2019**, *9*, 441. [CrossRef]
17. Raza, W.; Haque, M.M.; Muneer, M.; Harada, T.; Matsumura, M. Synthesis, characterization and photocatalytic performance of visible light induced bismuth oxide nanoparticle. *J. Alloys Compd.* **2015**, *648*, 641–650. [CrossRef]
18. Blaber, M.G.; Arnold, M.D.; Harris, N.; Ford, M.J.; Cortie, M.B. Plasmon absorption in nanospheres: A comparison of sodium, potassium, aluminum, silver and gold. *Phys. B Condens. Matter* **2007**, *394*, 184–187. [CrossRef]
19. Wallenhorst, L.; Gurău, L.; Gellerich, A.; Militz, H.; Ohms, G.; Viöl, W. UV-blocking properties of Zn/ZnO coatings on wood deposited by cold plasma spraying at atmospheric pressure. *Appl. Surf. Sci.* **2018**, *434*, 1183–1192. [CrossRef]
20. Wallenhorst, L.M.; Loewenthal, L.; Avramidis, G.; Gerhard, C.; Militz, H.; Ohms, G.; Viöl, W. Topographic, optical and chemical properties of zinc particle coatings deposited by means of atmospheric pressure plasma. *Appl. Surf. Sci.* **2017**, *410*, 485–493. [CrossRef]
21. Gascón-Garrido, P.; Mainusch, N.; Militz, H.; Viöl, W.; Mai, C. Effects of copper-plasma deposition on weathering properties of wood surfaces. *Appl. Surf. Sci.* **2016**, *366*, 112–119. [CrossRef]
22. Nejad, M.; Shafaghi, R.; Pershin, L.; Mostaghimi, J.; Cooper, P. Thermal Spray Coating: A New Way of Protecting Wood. *BioResources* **2016**, *12*. [CrossRef]
23. Gascón-Garrido, P.; Thévenon, M.F.; Mainusch, N.; Militz, H.; Viöl, W.; Mai, C. Siloxane-treated and copper-plasma-coated wood: Resistance to the blue stain fungus Aureobasidium pullulans and the termite Reticulitermes flavipes. *Int. Biodeterior. Biodegrad.* **2017**, *120*, 84–90. [CrossRef]
24. Beier, O.; Pfuch, A.; Horn, K.; Weisser, J.; Schnabelrauch, M.; Schimanski, A. Low Temperature Deposition of Antibacterially Active Silicon Oxide Layers Containing Silver Nanoparticles, Prepared by Atmospheric Pressure Plasma Chemical Vapor Deposition. *Plasma Process. Polym.* **2013**, *10*, 77–87. [CrossRef]
25. Gerullis, S.; Pfuch, A.; Spange, S.; Kettner, F.; Plaschkies, K.; Küzün, B.; Kosmachev, P.V.; Volokitin, G.G.; Grünler, B. Thin antimicrobial silver, copper or zinc containing SiOx films on wood polymer composites (WPC) applied by atmospheric pressure plasma chemical vapour deposition (APCVD) and sol–gel technology. *Eur. J. Wood Prod.* **2018**, *76*, 229–241. [CrossRef]
26. Tshabalala, M.A.; Sung, L.-P. Wood surface modification by in-situ sol-gel deposition of hybrid inorganic–organic thin films. *J. Coat. Technol. Res.* **2007**, *4*, 483–490. [CrossRef]
27. Köhler, R.; Siebert, D.; Kochanneck, L.; Ohms, G.; Viöl, W. Bismuth Oxide Faceted Structures as a Photocatalyst Produced Using an Atmospheric Pressure Plasma Jet. *Catalysts* **2019**, *9*, 533. [CrossRef]
28. Meyer-Veltrup, L.; Brischke, C.; Alfredsen, G.; Humar, M.; Flæte, P.-O.; Isaksson, T.; Brelid, P.L.; Westin, M.; Jermer, J. The combined effect of wetting ability and durability on outdoor performance of wood: Development and verification of a new prediction approach. *Wood Sci. Technol.* **2017**, *51*, 615–637. [CrossRef]

29. Meng, L.; Xu, W.; Zhang, Q.; Yang, T.; Shi, S. Study of nanostructural bismuth oxide films prepared by radio frequency reactive magnetron sputtering. *Appl. Surf. Sci.* **2019**, *472*, 165–171. [CrossRef]
30. Chang, B.; Liu, Q.; Chen, N.; Yang, Y. A Flower-like Bismuth Oxide as an Efficient, Durable and Selective Electrocatalyst for Artificial N 2 Fixation in Ambient Condition. *ChemCatChem* **2019**, *11*, 1884–1888. [CrossRef]
31. Kowalska, E.; Wei, Z.; Karabiyik, B.; Herissan, A.; Janczarek, M.; Endo, M.; Markowska-Szczupak, A.; Remita, H.; Ohtani, B. Silver-modified titania with enhanced photocatalytic and antimicrobial properties under UV and visible light irradiation. *Catal. Today* **2015**, *252*, 136–142. [CrossRef]
32. Maiti, N.; Thomas, S.; Debnath, A.; Kapoor, S. Raman and XPS study on the interaction of taurine with silver nanoparticles. *RSC Adv.* **2016**, *6*, 56406–56411. [CrossRef]
33. Prieto, P.; Nistor, V.; Nouneh, K.; Oyama, M.; Abd-Lefdil, M.; Díaz, R. XPS study of silver, nickel and bimetallic silver–nickel nanoparticles prepared by seed-mediated growth. *Appl. Surf. Sci.* **2012**, *258*, 8807–8813. [CrossRef]
34. Ivanova, T.; Homola, T.; Bryukvin, A.; Cameron, D. Catalytic Performance of Ag_2O and Ag Doped CeO_2 Prepared by Atomic Layer Deposition for Diesel Soot Oxidation. *Coatings* **2018**, *8*, 237. [CrossRef]
35. Hsu, K.-C.; Chen, D.-H. Microwave-assisted green synthesis of Ag/reduced graphene oxide nanocomposite as a surface-enhanced Raman scattering substrate with high uniformity. *Nanoscale Res. Lett.* **2014**, *9*, 193. [CrossRef] [PubMed]
36. Potter, D.B.; Powell, M.J.; Parkin, I.P.; Carmalt, C.J. Aluminum /gallium, indium/gallium, and aluminum /indium co-doped ZnO thin films deposited via aerosol assisted CVD. *J. Mater. Chem. C* **2018**, *6*, 588–597. [CrossRef]
37. Kloprogge, J.T.; Duong, L.V.; Wood, B.J.; Frost, R.L. XPS study of the major minerals in bauxite: Gibbsite, bayerite and (pseudo-)boehmite. *J. Colloid Interface Sci.* **2006**, *296*, 572–576. [CrossRef]
38. Strohmeier, B.R. An ESCA method for determining the oxide thickness on aluminum alloys. *Surf. Interface Anal.* **1990**, *15*, 51–56. [CrossRef]
39. Astuti, Y.; Fauziyah, A.; Nurhayati, S.; Wulansari, A.D.; Andianingrum, R.; Hakim, A.R.; Bhaduri, G. Synthesis of α-Bismuth oxide using solution combustion method and its photocatalytic properties. *IOP Conf. Ser. Mater. Sci. Eng.* **2016**, *107*, 12006. [CrossRef]
40. Viezbicke, B.D.; Patel, S.; Davis, B.E.; Birnie, D.P. Evaluation of the Tauc method for optical absorption edge determination: ZnO thin films as a model system. *Phys. Status Solidi B* **2015**, *252*, 1700–1710. [CrossRef]
41. Köhler, R.; Ohms, G.; Militz, H.; Viöl, W. Atmospheric Pressure Plasma Coating of Bismuth Oxide Circular Droplets. *Coatings* **2018**, *8*, 312. [CrossRef]
42. Li, X.; Sun, Y.; Xiong, T.; Jiang, G.; Zhang, Y.; Wu, Z.; Dong, F. Activation of amorphous bismuth oxide via plasmonic Bi metal for efficient visible-light photocatalysis. *J. Catal.* **2017**, *352*, 102–112. [CrossRef]
43. Zhang, L.; Wang, W.; Yang, J.; Chen, Z.; Zhang, W.; Zhou, L.; Liu, S. Sonochemical synthesis of nanocrystallite Bi2O3 as a visible-light-driven photocatalyst. *Appl. Catal. A Gen.* **2006**, *308*, 105–110. [CrossRef]
44. Jiang, M.-M.; Chen, H.-Y.; Li, B.-H.; Liu, K.-W.; Shan, C.-X.; Shen, D.-Z. Hybrid quadrupolar resonances stimulated at short wavelengths using coupled plasmonic silver nanoparticle aggregation. *J. Mater. Chem. C* **2014**, *2*, 56–63. [CrossRef]
45. Pawar, O.; Deshpande, N.; Dagade, S.; Waghmode, S.; Nigam Joshi, P. Green synthesis of silver nanoparticles from purple acid phosphatase apoenzyme isolated from a new source Limonia acidissima. *J. Exp. Nanosci.* **2016**, *11*, 28–37. [CrossRef]
46. Klantsataya, E.; François, A.; Ebendorff-Heidepriem, H.; Sciacca, B.; Zuber, A.; Monro, T.M. Effect of surface roughness on metal enhanced fluorescence in planar substrates and optical fibers. *Opt. Mater. Express* **2016**, *6*, 2128. [CrossRef]
47. Feng, A.L.; You, M.L.; Tian, L.; Singamaneni, S.; Liu, M.; Duan, Z.; Lu, T.J.; Xu, F.; Lin, M. Distance-dependent plasmon-enhanced fluorescence of upconversion nanoparticles using polyelectrolyte multilayers as tunable spacers. *Sci. Rep.* **2015**, *5*, 7779. [CrossRef] [PubMed]
48. Li, M.; Cushing, S.K.; Wu, N. Plasmon-enhanced optical sensors: A review. *Analyst* **2015**, *140*, 386–406. [CrossRef]
49. Rifat, A.A.; Mahdiraji, G.A.; Chow, D.M.; Shee, Y.G.; Ahmed, R.; Adikan, F.R.M. Photonic crystal fiber-based surface plasmon resonance sensor with selective analyte channels and graphene-silver deposited core. *Sensors (Basel)* **2015**, *15*, 11499–11510. [CrossRef]

50. Jeong, S.-H.; Choi, H.; Kim, J.Y.; Lee, T.-W. Silver-Based Nanoparticles for Surface Plasmon Resonance in Organic Optoelectronics. *Part. Part. Syst. Charact.* **2015**, *32*, 164–175. [CrossRef]
51. Aslan, K.; Leonenko, Z.; Lakowicz, J.R.; Geddes, C.D. Annealed silver-island films for applications in metal-enhanced fluorescence: Interpretation in terms of radiating plasmons. *J. Fluoresc.* **2005**, *15*, 643–654. [CrossRef]

Article

Wetting Behavior of Alder (*Alnus cordata* (Loisel) Duby) Wood Surface: Effect of Thermo-Treatment and Alkyl Ketene Dimer (AKD)

Teresa Lovaglio [1], Wolfgang Gindl-Altmutter [2], Tillmann Meints [3], Nicola Moretti [1] and Luigi Todaro [1,*]

1 School of Agricultural, Forestry, Food and Environmental Science, University of Basilicata, V. le Ateneo Lucano 10, 85100 Potenza, Italy
2 Department of Materials Science and Process Engineering, University of Natural Resources and Life Science Vienna, Konrad Lorenzstrasse 24, 3430 Tulln, Austria
3 Kompetenzzentrum Holz GmbH, Altenbergerstrasse 69, 4040 Linz, Austria
* Correspondence: luigi.todaro@unibas.it; Tel.: +39-0971-205340

Received: 26 July 2019; Accepted: 3 September 2019; Published: 5 September 2019

Abstract: The main purpose of this study was to investigate the hydrophobic effect and chemical changes induced by thermo-treatment and alkyl ketene dimer (AKD) on the surface properties of Alder (*Alnus cordata* (Loisel) Duby) wood before and after an artificial weathering test. Thermal treatment was conducted at a temperature of 200 °C for 4 h in a thermo-vacuum cylinder. Then, the paper sizing agent, AKD at different concentrations of a solution of 0.1%, 0.5% and 10% was used as a potential hydrophobizing reagent for untreated and thermally treated alder wood surfaces. The contact angle measurement, ATR-FTIR analysis and colour variation were carried out for the samples. The preliminary results revealed that the contact angle values of the wood materials increased with thermal modification. However, the influence of the thermal treatment on hydrophobicity was small when compared to the substantial effect of the AKD application in this respect, and also after the artificial weathering test. The FTIR analysis supported the hypothesis that AKD could make bonds chemically stable even when using a small concentration of AKD. The findings acquired in this work provide important information for future research and the utilization of the AKD on lesser-used wood species.

Keywords: underused wood species; thermo-vacuum treatment; surface degradation; hydrophobic effect

1. Introduction

It is known fact that wood undergoes degradation processes when exposed to exterior conditions. In particular, the surface is affected by exposure to light and water. The hygroscopic nature of wood produces some undesirable properties [1]. Due to water absorption and desorption, it is easily subjected to the biological attack of fungi and insects as well as dimensional instability. All of these adverse factors influence the overall properties of wood in outdoor use.

To enhance the performance of wood under critical environmental conditions, different approaches may be applied. There are several chemical treatments to preserve wood surfaces, but these could cause toxic emissions during the production, service life and after use [2]. An important alternative to preserve wood is thermal modification.

As noted by Hill [3], the heating temperature used during the treatment provoked changes in the wood. Thermo-treatment removes water from wood and causes changes in the chemical composition of wood and ultrastructure through autocatalytic reactions and cross-linking of the cell-wall constituents [4,5]. The wood became less hygroscopic, and an improvement in dimensional stability could be observed [6]. However, thermally modified wood shows a general decrease in mechanical properties. The colour change is not permanent in time when the wood is exposed to exterior situations. Decay resistance does show some improvement, however is not recommended to use thermally treated wood in ground contact situations [3,6].

Hemicellulose degradation starts with the hydrolytic cleavage of acetyl followed by carboxyl groups from the side chains, which in the presence of temperature and water, leads to the formation of acetic and formic acids. Afterwards, organic acids catalyze the further hydrolysis of the cell wall polysaccharides. The corresponding sugars can be further dehydrated to aldehydes, with furfural and hydroxymethylfurfural being typical products originating from pentoses and hexoses [7,8].

The degradation of amorphous cellulose also occurs which implies the inaccessibility of hydroxyl groups to water molecules. This is one of the factors that contribute to a decrease of equilibrium moisture content. However, the lignin degrades and its content increases and new methylene bridges arise which are responsible for crosslinking [9–11]. At the same time, heat treatments modify the wettability of wood, which is directly related to the chemical changes of the wood structure, especially to the degradation of the hemicelluloses.

The effectiveness of water-repellency treatment can be defined as the ability of a treatment to prevent or control the rate of liquid water uptake [12]. Several direct and indirect analytical approaches have been adopted to wood samples, including the measurement of wettability through the water contact angle (WCA) [12,13]. The relevance of the determination of the wettability of wood and its significance in wood science was critically discussed by Petrič and Oven [13].

Numerous investigations were carried out on the wettability changes of wood during heat treatment. General statement thermo-treatment has induced a reduced wettability [3]. However, Huang et al. [14] stated that induced weathering increased the wettability of heat-treated wood.

Weathering is a complex process that modifies and degrades the overall molecular structure of wood and wood-based products. The ultraviolet (UV) radiation is one of the main responsible of the photodegradation, causing the fastest and the strongest degradation of the wood components. The quantum energies of UV light break many of the chemical bonds presented in the wood constituents [15].

Thus, a certain degree of surface degradation of the treated wood used in outdoor conditions would still be expected. Therefore, surface protection of thermo-treated wood products would be necessary to enhance the overall effective service life of the final product [15].

While thermal modification of wood is an important alternative for improving technological performance, numerous studies have proposed making wood materials water-repellent by using various derivatives of anhydride, epoxide, alkyl chloride, aldehyde and chlorosilane [16], such as octadecyltrichlorosilane, which according to Hui et al. [17], have exhibited hydrophobic properties. Other chemical treatments of esterification, such as acetylation of cellulose [18] or natural fibers [19], can also be used.

An interesting treatment regarding the use of fluorine, which is a chemical element with the highest electronegativity, has led to hydrophobic characteristics in carbon fiber [20]. For wood, fluorination converts the bond between the carbon and hydroxyl groups into C–F bonds, thus making it hydrophobic [21].

Alkyl ketene dimer (AKD) can be also used to make wood surfaces hydrophobic [22] and is a widely used hydrophobization chemical in the papermaking industry [23]. Compared with other chemical compounds, it exhibits low toxicity, requires no special equipment, and the solvents involved are recyclable. Each AKD molecule is composed of two alkyl chains and one hetero four-member ring that reacts to hydroxyl groups to bind β-ketoester linkages [23,24].

There were three different hypotheses of the mechanisms of the reaction of AKD with cellulose materials. The first hypothesis states that the substance could react directly with cellulose hydroxyl (principally with C6 hydroxyl) as it is less sterically hindered when compared to the hydroxyl group on C2 or C3 [25–28]. The second one indicates that the mechanism of the reaction of AKD and water molecules is faster than cellulose even if it produces an unstable β-keto acid [29,30]. The third hypothesis takes into account whether the reaction could occur between AKD molecules forming an oligomer [31]. Other authors [32,33] have argued that the reaction between the hydroxyl groups in cellulose and AKD is essential to repel water.

However, although many studies were carried out related to the use of AKD on the wood surface, the investigation of the chemical reaction after an accelerating weathering test remains poorly investigated.

During the last decades, the demand for sustainable material removing the use of toxic chemicals obtaining materials with great potential and durability has been increasing. For this reason, it has raised the necessity to modify the wood products. High-value wood resources might be obtained with different processes, including impregnation, chemical modification or thermal treatments, among others. The potential economic gain is more attractive when considering lesser-used or minor wood species of relatively low market value.

The effects of thermal treatments on properties of lesser-used hardwood species, such as Alder wood, have been understudied [15]. Italian Alder (*Alnus cordata* (Loisel) Duby) can grow very rapidly (up to 30 m tall and reaching 70–80 cm in diameter) in the soil at increased humidity along watercourses [34–36]. Italian alder grows along the watershed, even at high altitudes mainly in Southern Italy. These species have been introduced in Sardinia and the Southern Alps. Recent plantations were also established in other European countries including France, Spain, Portugal, England, and the Netherlands. It has also been recently introduced in New Zealand and Chile [15]. It shows a density of 0.560 g/cm^3 at 12% moisture content, while the strength is variable with density. Normally, the compression value is approximately 36 N/mm^2, and the bending approximately 80 N/mm^2 [37]. It is mainly used for energy, but also for furniture, paneling, plywood, and paper pulp. Italian Alder wood, similarly to Black alder, for its machinability, colour homogeneity, and pleasant appearance, could be an alternative for different industrial destinations [38,39].

At present, this wood material has low commercial value for different reasons, mainly related to the economic competition caused by wood species from abroad. To reduce the importation of wood material, an environmentally friendly improvement of the properties of native or adapted minor species is one of the new challenges for wood researchers.

Our research assumes a gradual increase of hydrophobicity of untreated and thermo-treated Alder woods using AKD. The choice of a water-repellent agent, such as AKD, for the superficial processing of Alder wood is conditioned by the need for the application of a low toxicity chemical agent able to maintain the properties of natural wood by improving its water-repellency.

This study combined thermal bulk-modification alder with additional surface treatment using AKD at different concentrations. The main purpose of the present study was to investigate the hydrophobic effect and chemical changes induced by thermo-treatment and AKD on the surface properties of Alder wood before and after an artificial weathering test. The changes in surface hydrophilicity were investigated utilizing WCA measurements. The attenuated total reflectance-Fourier transform infrared spectroscopy (ATR-FTIR) was used to investigate the chemical changes of the wood surface.

2. Materials and Methods

2.1. Sample Preparation

Four Alder (A) trees were selected and harvested in the Apennine forest (Calvello municipality, Basilicata Region, Southern Italy, 40°28′12.1′′N 15°51′59.3′′E). The logs were then sawn in a commercial sawmill in boards of 40 mm (tangential) × 200 mm (radial) × 2200 mm (longitudinal). For this study, 1 m^3 of boards were randomly selected. Half of the boards were initially dried to 0% moisture content in a vacuum ranging from 185–200 mbar at a temperature of 90 °C for 4 h. The drying process started at an initial temperature level of 30 °C and increased to 90 °C at 5 °C per hour intervals. In the next step, the thermal treatment was applied to the boards by gradually increasing to a final temperature of 200 °C, under 220–320 mbar of pressure conditions for 4 h in a thermo-vacuum cylinder (WDE-Maspell s.r.l., Terni, Italy). Further information on this technology has been described by Todaro et al. [40].

After the thermal-vacuum treatment, 24 defect-free specimens (5 mm × 50 mm × 50 mm tangential, radial and longitudinal) were cut and randomly assigned to groups according to Table 1.

Table 1. Experimental design of untreated and thermo-treated alder samples.

Treatment	Pre UV Test	Post UV Test
Untreated	A	A post UV
Thermo-treated	ATH	ATH post UV
Untreated + 0.1% AKD	A + 0.1%AKD	A + 0.1%AKD post UV
Thermo-treated + 0.1% AKD	ATH + 0.1%AKD	ATH + 0.1%AKD post UV
Untreated + 5% AKD	A + 5%AKD	A + 5%AKD post UV
Thermo-treated + 5% AKD	ATH + 5%AKD	ATH + 5%AKD post UV
Untreated + 10% AKD	A + 10%AKD	A + 10%AKD post UV
Thermo-treated + 10% AKD	ATH + 10%AKD	ATH + 10%AKD post UV

(UV: Ultraviolet; A: Alder; ATH: Thermo-treated wood; AKD: Alkyl ketene dimer).

The untreated and the thermo-treated wood samples were then exposed for one month to a climatic chamber in a controlled environment (20 °C and 65% of humidity). The samples were immersed in three different AKD/toluene solutions at an ambient temperature for 10 s each.

The AKD solutions were obtained by dissolving AKD wax C18 in the toluene solvent to achieve: 0.1%, 5% and 10% concentration (w/w). Then, the samples were put under a fume hood where the toluene was allowed to evaporate overnight. Finally, the samples were kept at the standard climate of 20 °C and 65% of the relative humidity.

2.2. Water Contact Angle (WCA)

The angle formed between the wood surface and the line tangent to the droplet radius from the point of contact with the wood surface was determined using digital image analysis software (Drop Shape Analysis, Krüss GmbH, Germany). Demineralized water of 6 μL was employed for the measurements. The WCA was recorded every 0.5 s for a total of 120 s, thus for each drop, a total number of 240 WCA values were recorded. Each measurement was undertaken 3 times on each of the 24 wood specimens before and after the weathering test.

2.3. ATR-FTIR Analysis

Each of the 24 samples was analyzed, before and after the weathering test, with an ATR-FTIR spectrometer (PerkinElmer, FT-IR Spectrometer Frontier, Perkin-Elmer Inc., Norwalk, CT, USA.) with Spectrum™ Software operating on 4 scans and 2 cm^{-1} resolution. The spectra were recorded within a wavelength from 650 cm^{-1} to 4000 cm^{-1} and were baseline corrected and normalized by the software. To normalize the infrared spectra, the band referred to 1025 cm^{-1} in the fingerprint region was used. For the samples with AKD, the band usually used was at 2917 cm^{-1}

2.4. Artificial Weathering Test of the Samples

The wood samples were tested in artificial weathering using the fluorescent UV lamp and water test. The test was carried out according to the European standard (EN 927-6) and the experiment lasted one week. The cycle started with 24 h conditioning, followed by 2.5 h UV and then 0.5 h water spraying. These two phases were repeated for 6 days. All the samples were put in a chamber test for 116 h having 24 h conditioning, 15 h spraying water, and 77 h UV irradiation.

The apparatus was equipped with ultraviolet A (UVA) 340 nm lamps maintaining the constant temperature with a value of 45 °C during the conditional phase and 60 °C during the UV irradiation.

The irradiation set point was 89 W/m^2 at 340 nm and the conditions during the spray phase were from 6 L/min to 7 L/min during UV off.

2.5. Colour Variation

The colour changes during aging were constantly monitored on the specimens based on the method of the Commission International de l'Eclairage (CIE) using color parameters, L*, a* and b* [15] and their average values were calculated for each treatment.

Colorimetric measurements, by using Minolta CM-2002 spectrophotometer (Minolta Corp., Ramsey, NJ, USA), were also performed on three separated locations and repeated three times before thermo-treatment, after thermo treatment and after AKD. In addition, the measures were repeated after weathering test.

3. Results and Discussion

3.1. Water Contact Angle (WCA)

The results of the WCA measurements of untreated and thermo-treated alder wood, before and after UV, were displayed in Figures 1 and 2. As expected, water was absorbed faster by the untreated wood (A) compared to the thermo-treated wood (ATH) (Figures 1a and 2a). The lower wettability on the wood surface after thermal treatment could be ascribed to several events taking place in the wood. One of the probable causes are the migration of non-polar extractives to the wood surface during heating [7,14,41].

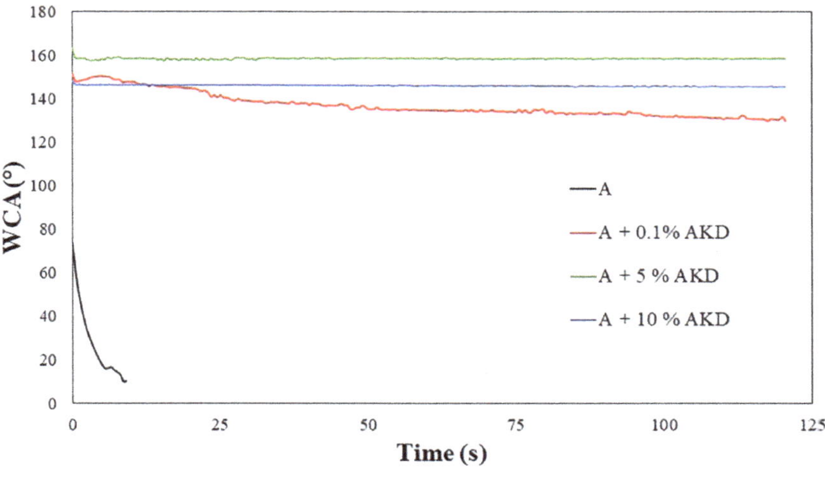

(a)

Figure 1. *Cont.*

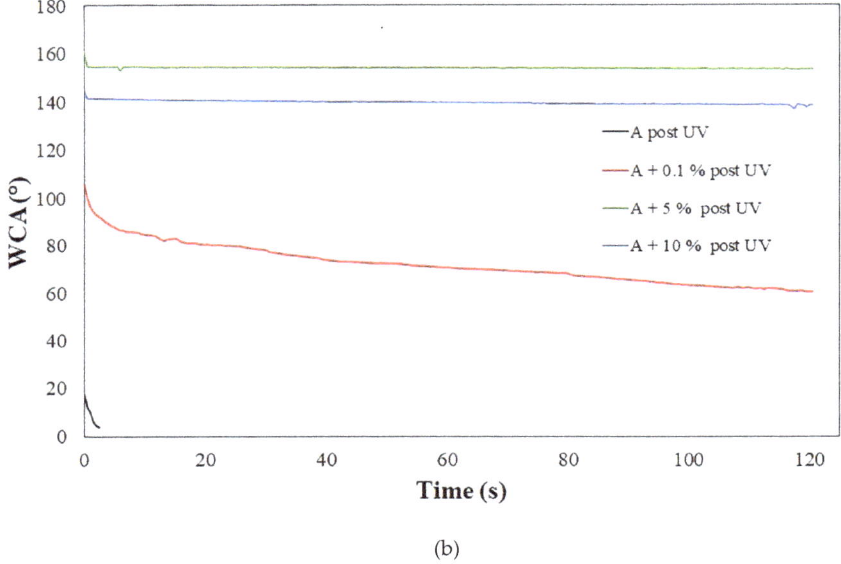

(b)

Figure 1. The water contact angle (WCA) during time on untreated alder samples (A), before (**a**) and after UV exposure (**b**).

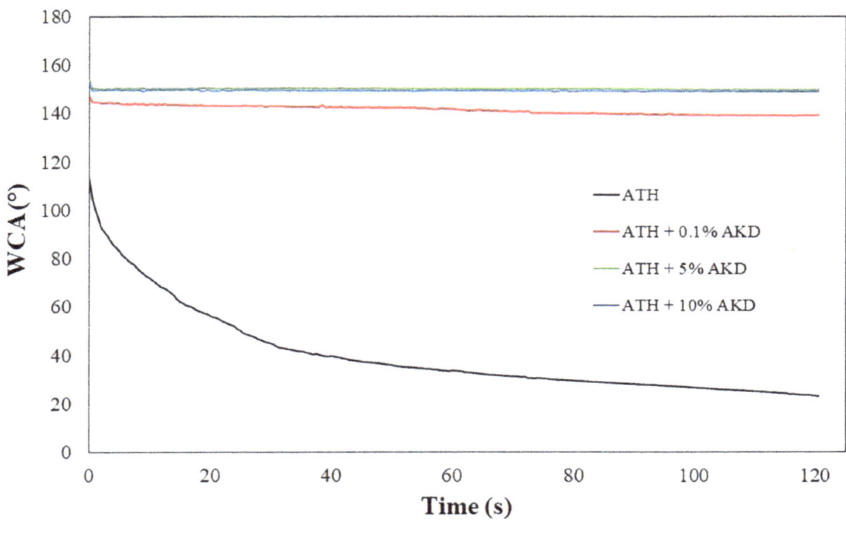

(a)

Figure 2. *Cont.*

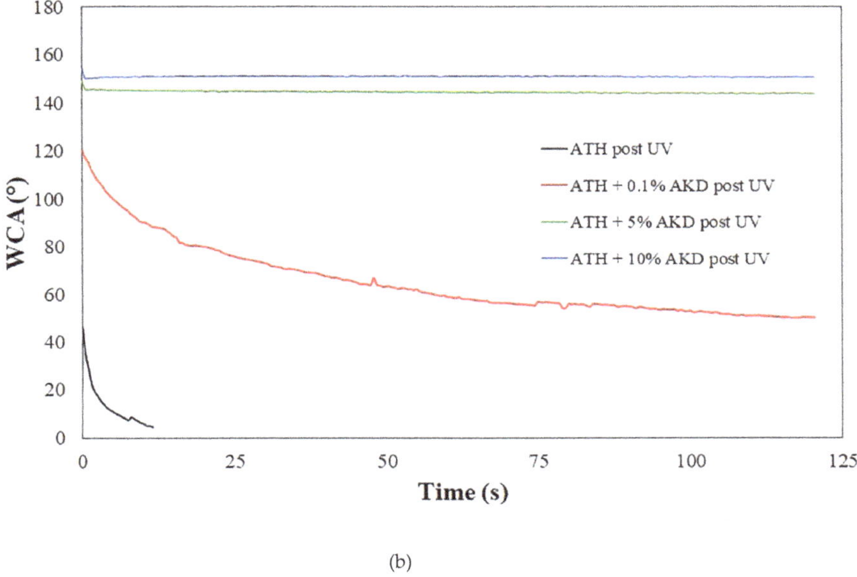

(b)

Figure 2. WCA during time on thermo-treated alder samples (ATH), before (**a**) and after UV exposure (**b**).

Several other parameters also play an important role in the reduced hygroscopicity of wood surfaces [4,42]. For example, the different types of reactions between the cell-wall components occur during heating and they imply a cross-linking between the lignin and polysaccharides, and a reduction in free hydroxyl groups. At high temperatures, the hemicelluloses, as one of the first structural components of wood that changes during heating [6], may undergo oxidation and pyrolysis reactions. They can also cause changes in compounds, like the formation of furfural polymers, which are considered to be less hygroscopic compared to hydroxyl groups [43]. The hydrophobicity of both A (10.1 after 9 s) and ATH (23.1 after 120 s), before artificial weathering, were greatly improved after the application of AKD (Figures 1a and 2a). The hydrophobic effect of the AKD substance could be caused by the esterification reactions between AKD and wood compounds, like hemicellulose [23,24].

After artificial weathering, the WCA of A (3.9 after 2.5 s) and mostly ATH (4.6 after 11 s) decreased significantly (Figures 1b and 2b). This result is similar to other published studies. The initial WCA of the heat-treated specimens were higher than the untreated samples [44], but after the weathering test, the difference in WCA between the control and TH was limited. Some of the previous studies stated that these changes in the wettability were strictly due to the combination of structural and chemical changes of wood surfaces [45,46].

The Figures 1 and 2 suggest that after artificial weathering, the hydrophobicity remains unchanged for the A and ATH wood surface treated with 5% and 10% of AKD. However, for the wood sample A + 0.1% AKD (from 129.8 to 60.58) and ATH + 0.1% AKD (from 139.0 to 50.21), a decrease of WCA was observed.

3.2. ATR-FTIR Analysis

In Figure 3, the FTIR spectra before UV exposure, between the wavelengths of 4000 cm^{-1} and 600 cm^{-1} of both untreated (A) and thermo-treated alder wood (ATH), are displayed.

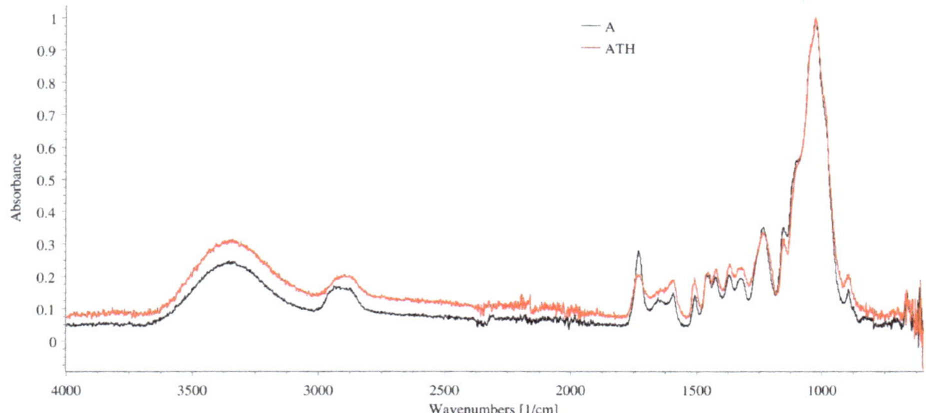

Figure 3. Comparison of FTIR spectra of alder wood (A and ATH, respectively).

As stated by Colom et al. [47], in the spectra, two different areas can be observed: The first was between 3600 cm^{-1} and 2700 cm^{-1} and it was referred to the stretching vibrations of O–H and C–H groups; the second between 1800 cm^{-1} and 800 cm^{-1}, is called the fingerprint region, assigned to stretching vibrations of wood structural compounds.

Clear differences could be noticed in the fingerprint region. The first important difference was the intensity of the band of 1734 cm^{-1}, assigned to the C=O stretching vibration of acetyl or carboxylic acid which was slightly smaller in ATH. As stated by Colom et al. [47], this variation indicated the cleavage of the acetyl side chains in the hemicelluloses.

After thermo-treatment, a band at 1335 cm^{-1} appeared. A previous study has used this doublet to monitor the described degradation process. Colom et al. [47] reported that a decrease in the ratio between the absorption band of 1335 cm^{-1} and 1316 cm^{-1} signifies an increase in crystallinity. This indicated that the degradation process was more intense in the amorphous cellulose component than in the crystallized portion.

Another significant difference was in the doublet at 1610–1595 cm^{-1} which occurred only in spectra A. The presence of this doublet before the heat treatment showed that the lignin was a mixture of the guaiacyl and the syringyl components [48]. After thermo-treatment, the strong decrease of a band at 1610 cm^{-1} meant that only the degradation of the the guaiacyl occurred, most likely due to its easier degradation compared to the syringyl [49]. This behavior was confirmed by bands at 1595 cm^{-1} and 1510 cm^{-1}, based on the aromatic skeletal vibration in the lignin having a similar intensity before and after the treatment. This result can be attributed to the predominant syringyl units, which remained unchanged after the heating process [47].

There were other signals that did not change after thermo-treatment but were equally important—1377 cm^{-1}, 1240 cm^{-1}, 1165 cm^{-1}, and 1030 cm^{-1} were assigned to C–H, C–O–C and C–O stretching vibrations of the different groups in carbohydrates, and 898 cm^{-1} was connected to the amorphous region of cellulose.

In Figure 4, the FTIR spectra of A and ATH, pre and post UV, are displayed showing only the fingerprint region.

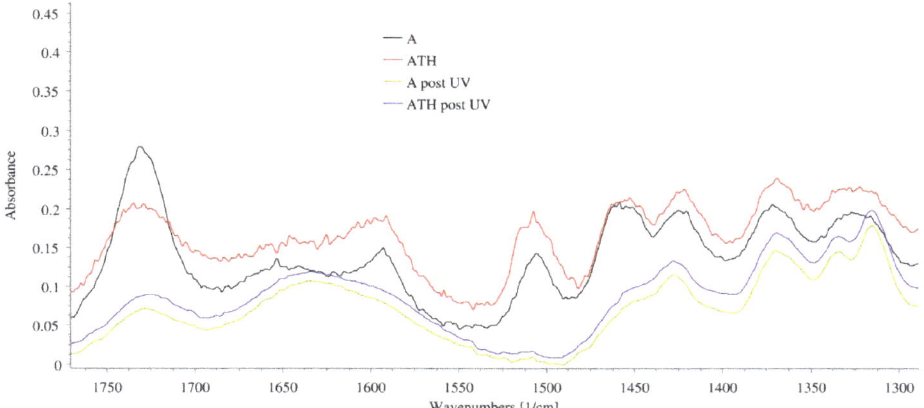

Figure 4. Comparison of FTIR spectra of A and ATH before and after UV of the fingerprint region.

After UV treatment, many changes take place in the fingerprint region (Figure 4). The bands 1595 cm^{-1}, 1510 cm^{-1}, and 1465 cm^{-1}, assigned to the lignin component, decreased significantly in both A and ATH as a result of the irradiation process. This means that the structure of lignin polymer was degraded to a significant extent. The decrease of bands at 1595–1510 cm^{-1} showed that the guaiacyl nuclei (1510 cm^{-1}) are more sensitive to the artificial aging degradation process that the syringyl nuclei (1595 cm^{-1}). In contrast to what was claimed by [50], the absorbance to stretching vibration of acetyl or carboxylic acid at 1730 cm^{-1} decreased greatly. After UV, the doublet at 1335–1316 cm^{-1} assigned to the cellulosic component developed, resulting in a high crystallized cellulose content. During the weathering test, the degradation of the amorphous cellulose component was more intense. The band at 1426 cm^{-1} related to CH_2 stretching, shifted at 1430 cm^{-1} after the degradation of amorphous cellulose [51].

For a better understanding of the effect of the hydrophobizing reagent on wood surfaces, it is essential to analyze, in advance, the ATR-FTIR spectra of AKD (Figure 5).

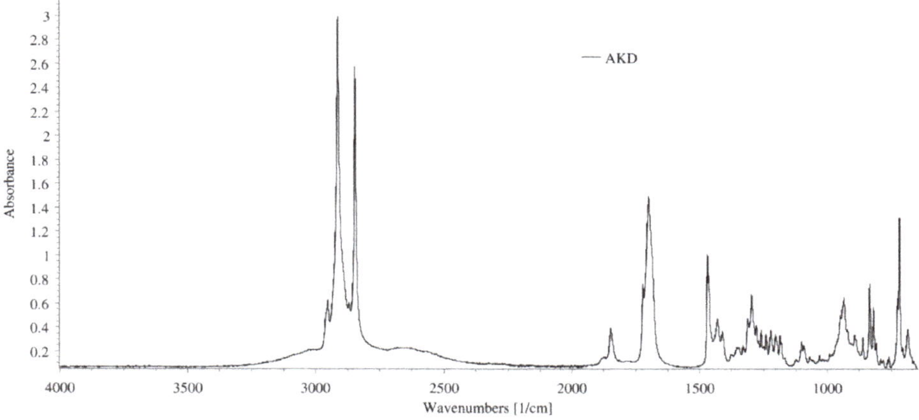

Figure 5. FTIR spectra of alkyl ketene dimer (AKD).

The specification bands of AKD are at 2915 cm^{-1} and 2848 cm^{-1} which characterize the C–H stretching vibration representing methyl and methylene groups of the alkyl chain in the AKD molecule.

There are other relevant signals, 1848 cm^{-1} connected to the stretching of C=C and the band at 1700 cm^{-1}, related to the group C=O. These characteristic signals related to AKD can be observed also in Figure 6. The FTIR spectra of untreated (A) and thermo-treated alder (ATH) with different concentration of AKD on their surfaces showed high intensity bands ranging from 2850–2950 cm^{-1}. This is characteristic of the AKD molecule and the sharp peaks located at 1467 cm^{-1} and 1845 cm^{-1} that represent the C=C and C=O groups in the lactone ring of the AKD molecule.

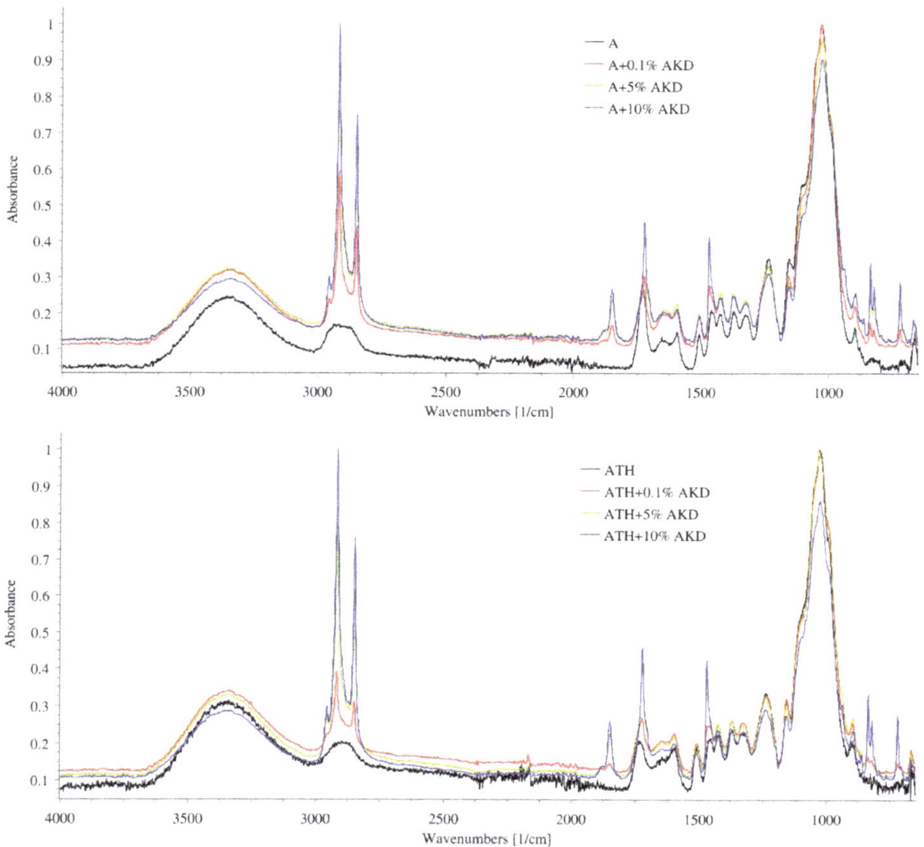

Figure 6. Comparison of FTIR spectra of A (**top**) and ATH (**bottom**) with AKD.

The reaction mechanisms of AKD with wood fiber hydroxyl groups of cellulose were considered as an esterification reaction by Hubbe [23] and Seppänen [24]. The evidence of this reaction was the presence of the band at 1720 cm^{-1}, indicating that the carbonyl groups originated from both the AKD and wood compounds. This band would underline the reaction between the AKD and wood surface since the band relative the stretching C=O in wood and in pure AKD was present at different wavenumbers.

As expected, the significant differences in intensity of the bands among a different concentration of AKD were recorded. Furthermore, there were differences between A and ATH samples immersed in AKD solutions, in particular at the concentration of 0.1% AKD. This was correlated to the thermo-treatment that destroyed and changed some chemical components of the cell walls that may act as binding sites [4,5]. Therefore, it could be assumed that the effect of AKD was significant, even at a low concentration of the solution.

After the weathering test, the analysis of ATR-FTIR showed some relevant differences (Figure 7). As previously discussed for Figure 4, the main changes occurred in the fingerprint area of the wood.

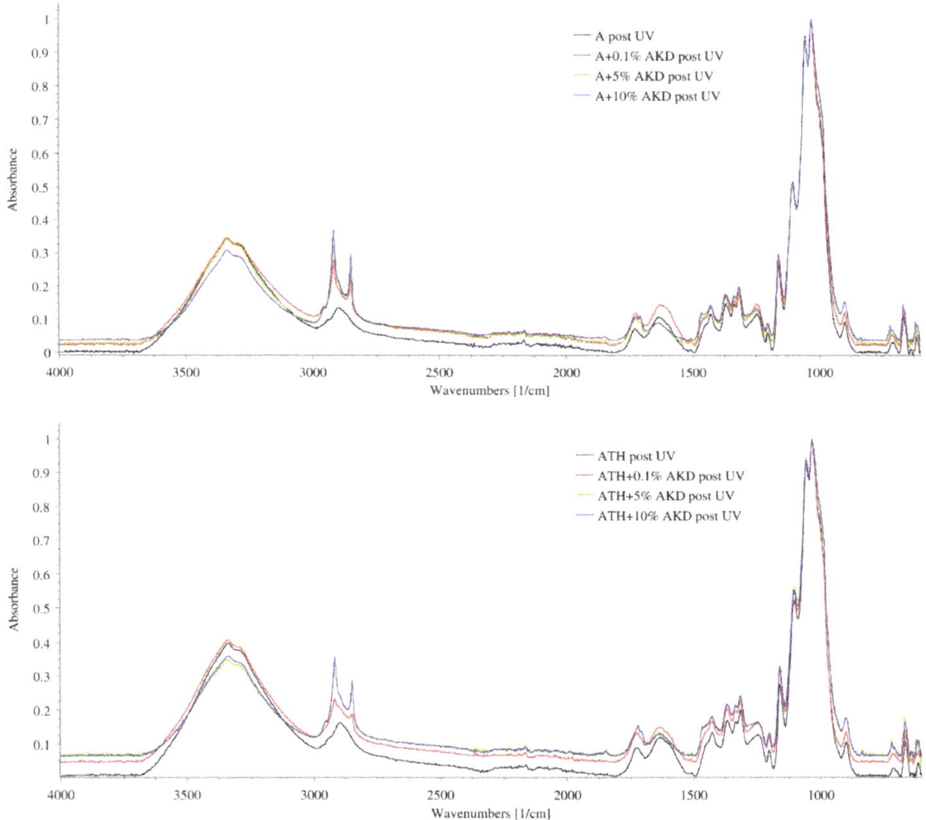

Figure 7. Comparison of FTIR spectra of A (**top**) and ATH (**bottom**) with AKD after UV exposure.

The other relevant changes were displayed in the spectra: The bands ranging from 2850–2950 cm^{-1}, which is typical of a AKD molecule; and the bands at approximately 1730 cm^{-1}.

The 2915 cm^{-1} and 2848 cm^{-1} peaks which were C–H stretching vibration representing methyl and methylene groups of the alkyl chain in the AKD had a minor intensity in the spectra ATH + AKD post UV compared to A + AKD post UV. This could be dependent on the degradation of the methyl group of AKD. In the case of ATH samples, the C–H wood groups already modified by the heat treatment continue to alter further by lowering the intensity of this area of the FTIR spectrum. The band at 1730 cm^{-1} almost disappeared. The ester group, and thus the principal interaction between AKD and wood compounds, had been degraded by UV irradiation. Considering that the effect of waterproofing of AKD did not vary significantly, according to the WCA analysis, it could be assumed that there were other interactions more or less strong between the wood and wax AKD.

Several authors have demonstrated [26–28,32,33] that AKD has the potential to substitute other surface treatments to obtain a durable hydrophobic effect. Furthermore, AKD could be also used as an antifungal treatment [52]. In fact, the authors demonstrated that wood treated with AKD showed a promising resistance against G. trabeum, C. versicolor, P. placenta, but failed to prevent C. puteana attacks.

A previous study investigated the use of AKD to modify particleboard surfaces [53]. The authors stated that the hydrophobicity attained with AKD was due to esterification with wood hydroxyl groups and a subsequent orientation of the hydrocarbon chains [54]. Although FTIR-spectroscopy indicated ester formation, the strong absorption bands decreased after toluene extraction.

3.3. Colour Variation

The changing in natural coloration of wood and, in particular the darker tonality after heating (−26.39), can be attributed to the chemical reactions of oxidation or to the formation of a new compound by degradation of the wood component (Table 2). These results and the differences between the untreated and thermo treated samples before and after the weathering test, are largely discussed in a previous work [15].

Table 2. The colour change (mean value) for both the treated and untreated alder. The colour variation of ATH samples are referred to A control samples.

ALDER	ΔL	Δa	Δb	ΔE
Control samples				
A	0	0	0	0
A + 0.1%AKD	−0.14	0.03	0.36	0.42
A + 5%AKD	−0.62	0.77	0.42	1.34
A + 10%AKD	0.05	0.63	0.45	0.87
Thermo-treated				
ATH	−26.39	3.10	0.83	26.58
ATH + 0.1%AKD	−0.38	−0.44	−0.01	1.32
ATH + 5%AKD	0.57	−2.02	−8.27	8.58
ATH + 10%AKD	−0.04	−1.37	−8.78	8.90

After immersion of each sample in AKD solutions, the change in colour for the untreated samples are negligible. In case of thermo-treated wood, the total colour variation was due mainly to the modification of b* coordinate (−8.27 and −8.78).

The color change of the specimens in terms of ΔL, Δa, Δb and ΔE occurring after the weathering test are displayed in Table 3.

Table 3. The colour change (mean value) for both the untreated and treated alder after the weathering test.

ALDER Post UV	ΔL	Δa	Δb	ΔE
Control samples				
A	−5.02	−0.34	1.66	5.31
A + 0.1%AKD	−8.52	4.76	7.52	12.36
A + 5%AKD	−5.73	−1.23	−3.61	7.00
A + 10%AKD	−6.34	−0.73	−5.35	8.34
Thermo-treated				
ATH	1.11	−2.50	−2.58	3.90
ATH + 0.1%AKD	4.22	−2.72	−4.13	6.58
ATH + 5%AKD	6.24	−2.74	−1.32	7.28
ATH + 10%AKD	7.68	−3.04	0.59	8.47

Alder wood, in both untreated and thermo treated samples, showed that AKD did not create a layer to protect and cover the wood by a photo-degradation. In fact, in terms of the total colour variation, no significant difference was depicted when AKD was used. On the contrary, compared to A and ATH, the wood surface covered by AKD seemed to be less appropriate to protect wood by photodegradation. For the ATH sample at an increasing concentration of AKD solution, the ΔL^* increased and the surfaces were thus slightly lightened. The interaction of photon energy with the wood samples covered with AKD solution involves complex physical and chemical reactions only partially explained through ATR-FTIR analysis [15].

Our preliminary research presented an assessment of the effectiveness of the chemical modification of underused, untreated, and thermo-treated woods using AKD in improving its hydrophobicity. The AKD/water dispersions are commercially available and more environmentally friendly than toluene. However, the scope of this research was to highlight the main evidence of the relationship between AKD and alder wood surface.

4. Conclusions

- The untreated and thermo-treated wood samples treated with 0.1% of AKD had a significant decrease in WCA after the weathering test.
- The FTIR analysis showed that guaiacyl nuclei in the alder wood was more sensitive to heat degradation than syringyl nuclei.
- The presence of the band between 1720 cm^{-1} in the FTIR spectra before the use of UV could indicate the formation of the carbonyl groups originated from the reaction between AKD and hydroxyl groups of cellulose, making the surface water repellent.
- After the weathering test, the band between 1700 cm^{-1} and 1750 cm^{-1} in FTIR spectra disappeared, thus it could be assumed that there were other interactions, more or less, strong between the wood and AKD.
- The analysis of WCA showed that the investigated range of AKD, at 5% and 10%, did not differ in wettability also after weathering test. This evidence suggests that the optimal solution could be found in this range.
- In terms of colour variation, the results revealed that AKD in outdoor conditions was not very effective in restricting UV light induced degradation in the wood specimens.

Author Contributions: T.L. wrote most of the paper and performed the experimental testing; T.M. and N.M. performed part of the experimental testing; L.T. and W.G.-A. conceived and designed the study and revised the paper contributing to its drafting; all authors participated to the discussion of the work.

Funding: This research was funded by the PRIN2015 project (No. 2015YW8JWA) "Short chain in the biomass-wood sector: Supply, traceability, certification and carbon sequestration. Innovations for bio-building and energy efficiency", coordinated by G. Scarascia-Mugnozza.

Acknowledgments: The authors would like to acknowledge the contribution of the COST Action FP1407. The Ph.D. program in "Agricultural, Forest and Food Science" coordinated by M. Borghetti, at the University of Basilicata, supported T. Lovaglio.

Conflicts of Interest: The authors declare no conflicts of interest. The funders had no role in the design of the study; in the collection, analyses, or interpretation of data; in the writing of the manuscript, or in the decision to publish the results.

References

1. Todaro, L.; D'Auria, M.; Langerame, F.; Salvi, A.M.; Scopa, A. Surface characterization of untreated and hydro-thermally pre-treated Turkey oak woods after UV-C irradiation. *Surf. Interface Anal.* **2015**, *47*, 206–215. [CrossRef]
2. Hingston, J.A.; Collins, C.D.; Murphy, R.J.; Lester, J.N. Leaching of chromated copper arsenate wood preservatives: A review. *Environ. Pollut.* **2001**, *111*, 53–66. [CrossRef]
3. Hill, C.A. *Wood Modification: Chemical, Thermal and Other Processes*; John Wiley & Sons: Hoboken, NJ, USA, 2007; p. 260.

4. Tjeerdsma, B.F.; Militz, H. Chemical changes in hydrothermal treated wood: FTIR analysis of combined hydrothermal and dry heat-treated wood. *Holz Roh-und Werkst.* **2005**, *63*, 102–111. [CrossRef]
5. Kocaefe, D.; Poncsak, S.; Boluk, Y. Effect of thermal treatment on the chemical composition and mechanical properties of birch and aspen. *BioResources* **2008**, *3*, 517–537.
6. Sandberg, D.; Haller, P.; Navi, P. Thermo-hydro and thermo-hydro-mechanical wood processing: An opportunity for future environmentally friendly wood products. *Wood Mater. Sci. Eng.* **2013**, *8*, 64–88. [CrossRef]
7. Fengel, D.; Wegener, G. *Wood: Chemistry, Ultrastructure, Reactions*; Walter de Gruyter: Berlin, Germany, 1984; Volume 613, pp. 1960–1982.
8. Altgen, M.; Willems, W.; Militz, H. Wood degradation affected by process conditions during thermal modification of European beech in a high-pressure reactor system. *Eur. J. Wood Wood Prod.* **2016**, *74*, 653–662. [CrossRef]
9. Wikberg, H.; Maunu, S.L. Characterisation of thermally modified hard-and softwoods by 13 C CPMAS NMR. *Carbohydr. Polym.* **2004**, *58*, 461–466. [CrossRef]
10. Boonstra, M.J.; Tjeerdsma, B. Chemical analysis of heat treated softwoods. *Eur. J. Wood Wood Prod.* **2006**, *64*, 204–211. [CrossRef]
11. Mahnert, K.C.; Adamopoulos, S.; Koch, G.; Militz, H. Topochemistry of heat-treated and N-methylol melamine-modified wood of koto (Pterygota macrocarpa K. Schum.) and limba (Terminalia superba Engl. et. Diels). *Holzforschung* **2013**, *67*, 137–146. [CrossRef]
12. Rowell, R.M.; Banks, W.B. *Water Repellency and Dimensional Stability of Wood*; US Department of Agriculture, Forest Service, Forest Products Laborator: Madison, WI, USA, 1985; p. 50.
13. Petrič, M.; Oven, P. Determination of wettability of wood and its significance in wood science and technology: A critical review. *Rev. Adhes. Adhes.* **2015**, *3*, 121–187. [CrossRef]
14. Huang, X.; Kocaefe, D.; Kocaefe, Y.; Boluk, Y.; Pichette, A. Changes in wettability of heat-treated wood due to artificial weathering. *Wood Sci. Technol.* **2012**, *46*, 1215–1237. [CrossRef]
15. D'Auria, M.; Lovaglio, T.; Rita, A.; Cetera, P.; Romani, A.; Hiziroglu, S.; Todaro, L. Integrate measurements allow the surface characterization of thermo-vacuum treated alder differentially coated. *Measurement* **2018**, *114*, 372–381. [CrossRef]
16. Rowell, R.M. Chemical modification of wood. *Handb. Wood Chem. Wood Compos.* **2005**, *14*, 381.
17. Hui, B.; Li, Y.; Huang, Q.; Li, G.; Li, J.; Cai, L.; Yu, H. Fabrication of smart coatings based on wood substrates with photoresponsive behavior and hydrophobic performance. *Mater. Des.* **2015**, *84*, 277–284. [CrossRef]
18. Ashori, A.; Babaee, M.; Jonoobi, M.; Hamzeh, Y. Solvent-free acetylation of cellulose nanofibers for improving compatibility and dispersion. *Carbohydr. Polym.* **2014**, *102*, 369–375. [CrossRef]
19. Tserki, V.; Zafeiropoulos, N.E.; Simon, F.; Panayiotou, C. A study of the effect of acetylation and propionylation surface treatments on natural fibres. *Compos. Part A Appl. Sci. Manuf.* **2005**, *36*, 1110–1118. [CrossRef]
20. Bismarck, A.; Tahhan, R.; Springer, J.; Schulz, A.; Klapötke, T.M.; Zeil, H.; Michaeli, W. Influence of fluorination on the properties of carbon fibres. *J. Fluor. Chem.* **1997**, *84*, 127–134. [CrossRef]
21. Pouzet, M.; Gautier, D.; Charlet, K.; Dubois, M.; Beakou, A. How to decrease the hydrophilicity of wood flour to process efficient composite materials. *Appl. Surf. Sci.* **2015**, *353*, 1234–1241. [CrossRef]
22. Yan, Y.; Amer, H.; Rosenau, T.; Zollfrank, C.; Dörrstein, J.; Jobst, C.; Li, J. Dry, hydrophobic microfibrillated cellulose powder obtained in a simple procedure using alkyl ketene dimer. *Cellulose* **2016**, *23*, 1189–1197. [CrossRef]
23. Hubbe, M.A. Paper's resistance to wetting—A review of internal sizing chemicals and their effects. *BioResources* **2007**, *2*, 106–145.
24. Seppänen, R. On the Internal Sizing Mechanisms of Paper with AKD and ASA Related to Surface Chemistry, Wettability and Friction. Ph.D. Thesis, KTH Royal Institute of Technology, Stockholm, Sweden, 2007.
25. Yoshida, Y.; Isogai, A. Preparation and characterization of cellulose β-ketoesters prepared by homogeneous reaction with alkylketene dimers: Comparison with cellulose/fatty acid esters. *Cellulose* **2007**, *14*, 481. [CrossRef]
26. Yoshida, Y.; Heux, L.; Isogai, A. Heterogeneous reaction between cellulose and alkyl ketene dimer under solvent-free conditions. *Cellulose* **2012**, *19*, 1667–1676. [CrossRef]
27. Song, X.; Chen, F.; Liu, F. Study on the reaction of alkyl ketene dimer (akd) and cellulose fiber. *BioResources* **2012**, *7*, 652–662.

28. Yang, Q.; Takeuchi, M.; Saito, T.; Isogai, A. Formation of nanosized islands of dialkyl β-ketoester bonds for efficient hydrophobization of a cellulose film surface. *Langmuir* **2014**, *30*, 8109–8118. [CrossRef]
29. Roberts, J.C.; Garner, D.N. The mechanism of alkyl ketene dimer sizing of paper, part I. *Tappi J.* **1985**, *68*, 118–121.
30. Seo, W.S.; Cho, N.S. Effect of water content on cellulose/AKD reaction. *Appita J.* **2005**, *58*, 122.
31. Bottorff, K.J.; Sullivan, M.J. New insights into the AKD sizing mechanism. *Nordic Pulp Pap. Res. J.* **1993**, *8*, 86–95. [CrossRef]
32. Kumar, S.; Chauhan, V.S.; Chakrabarti, S.K. Separation and analysis techniques for bound and unbound alkyl ketene dimer (AKD) in paper: A review. *Arabian J. Chem.* **2012**, *23*, 1–7. [CrossRef]
33. Shi, Z.; Fu, F.; Wang, S.; He, S.; Yang, R. Modification of Chinese fir with alkyl ketene dimer (AKD): Processing and characterization. *BioResources* **2012**, *8*, 581–591. [CrossRef]
34. Loewe, V.; Álvarez, A.; Barrales, L. Growth development of hardwood high value timber species in central south Chile, South America. In Proceedings of the 4th International Scientific Conference on Hardwood Processing, Delft, The Netherlands, 28–30 August 2013; p. 50.
35. Praciak, A.; Pasiecznik, N.; Sheil, D.; van Heist, M.; Sassen, M.; Correia, C.S.; Dixon, C.; Fyson, G.; Rushford, K.; Teeling, C. *The CABI Encyclopedia of Forest Trees*; CABI: Oxfordshire, UK, 2013.
36. Caudullo, G.; Mauri, A. Alnus cordata in Europe: Distribution, habitat, usage and threats. In *European Atlas of Forest Tree Species*; San-Miguel-Ayanz, J., de Rigo, D., Caudullo, G., Eds.; EU: Brussels, Belgium, 2016.
37. Giordano, G. *Tecnologia del Legno*; I.N.A.P.L.I.: Roma, Italy, 1981; pp. 938–939.
38. Salca, E.A. Black alder (Alnus glutinosa L.)—A resource for value-added products in furniture industry under European screening. *Curr. For. Rep.* **2019**, *5*, 41–54. [CrossRef]
39. Salca, E.A.; Krystofiak, T.; Lis, B. Evaluation of selected properties of alder wood as functions of sanding and coating. *Coatings* **2017**, *7*, 176. [CrossRef]
40. Todaro, L.; Rita, A.; Moretti, N.; Cuccui, I.; Pellerano, A. Assessment of Thermo-treated Bonded Wood Performance: Comparisons among Norway Spruce, Common Ash, and Turkey Oak. *BioResources* **2014**, *10*, 772–781. [CrossRef]
41. Christiansen, A.W. How overdrying wood reduces its bonding to phenol-formaldehyde adhesives: A critical review of the literature. Part I. Physical responses. *Wood Fiber Sci.* **2007**, *22*, 441–459.
42. Repellin, V.; Guyonnet, R. Evaluation of heat-treated wood swelling by differential scanning calorimetry in relation to chemical composition. *Holzforschung* **2005**, *59*, 28–34. [CrossRef]
43. Hillis, W.E. High temperature and chemical effects on wood stability. *Wood Sci. Technol.* **1984**, *18*, 281–293. [CrossRef]
44. Kocaefe, D.; Poncsak, S.; Doré, G.; Younsi, R. Effect of heat treatment on the wettability of white ash and soft maple by water. *Holz Roh-und Werkst.* **2008**, *66*, 355–361. [CrossRef]
45. Windeisen, E.; Strobel, C.; Wegener, G. Chemical changes during the production of thermo-treated beech wood. *Wood Sci. Technol.* **2007**, *41*, 523–536. [CrossRef]
46. Aydemir, D.; Gunduz, G.; Altuntas, E.; Ertas, M.; Sahin, H.T.; Alma, M.H. Investigating changes in the chemical constituents and dimensional stability of heat-treated hornbeam and uludag fir wood. *BioResources* **2011**, *6*, 1308–1321.
47. Colom, X.; Carrillo, F.; Nogués, F.; Garriga, P. Structural analysis of photodegraded wood by means of FTIR spectroscopy. *Polym. Degrad. Stab.* **2003**, *80*, 543–549. [CrossRef]
48. Carrillo, F.; Colom, X.; Valldeperas, J.; Evans, D.; Huson, M.; Church, J. Structural characterization and properties of lyocell fibers after fibrillation and enzymatic defibrillation finishing treatments. *Text. Res. J.* **2003**, *73*, 1024–1030. [CrossRef]
49. Evans, P.A. Differentiating "hard" from "soft" woods using Fourier transform infrared and Fourier transform spectroscopy. *Spectrochim. Acta Part A Mol. Spectrosc.* **1991**, *47*, 1441–1447. [CrossRef]
50. Hon, N.S. Surface chemistry of oxidized wood. In *Cellulose and Wood Chemistry and Technology*; Schuerch, C., Ed.; John Wiley and Sons: New York, NY, USA, 1989; p. 1401.
51. Liang, C.Y.; Marchessault, R.H. Infrared spectra of crystalline polysaccharides. II. Native celluloses in the region from 640 to 1700 cm^{-1}. *J. Polym. Sci. Part A Polym. Chem.* **1959**, *39*, 269–278. [CrossRef]
52. Suttie, E.D.; Hill, C.A.; Jones, D.; Orsler, R.J. Chemically modified solid wood. I. Resistance to fungal attack. *Mater. Org.* **1998**, *32*, 159–182.

53. Hundhausen, U.; Militz, H.; Mai, C. Use of alkyl ketene dimer (AKD) for surface modification of particleboard chips. *Eur. J. Wood Wood Prod.* **2009**, *67*, 37–45. [CrossRef]
54. Odberg, L.; Lindstrom, T.; Liedberg, B.; Gustavsson, J. Evidence for β-ketoester formation during the sizing of paper with alkylketene dimers. *Tappi J.* **1987**, *70*, 135–139.

Article

Wood Surface Changes of Heat-Treated *Cunninghamia lanceolate* Following Natural Weathering

Xinjie Cui [1] and Junji Matsumura [2,*]

[1] Graduate School of Bioresource and Bioenvironmental Sciences, Faculty of Agriculture, Kyushu University, 744 Motooka, Nishi-ku, Fukuoka 819-0395, Japan; Cuixinjie0731@outlook.com

[2] Laboratory of Wood Science, Faculty of Agriculture, Kyushu University, 744 Motooka, Nishi-ku, Fukuoka 819-0395, Japan

* Correspondence: matumura@agr.kyushu-u.ac.jp; Tel.: +81-092-802-4656

Received: 21 August 2019; Accepted: 9 September 2019; Published: 11 September 2019

Abstract: To quickly clarify the effect of heat treatment on weatherability of *Cunninghamia lanceolate* (Lamb.) Hook., we investigated the surface degradation under natural exposure. A comparison between heat-treated and untreated samples was taken based on surface color changes and structural decay at each interval. Over four weeks of natural exposure, multiple measurements were carried out. Results show that color change decreased in the order of 220 °C heat-treated > untreated > 190 °C heat-treated. The results also indicate that the wood surface color stability was improved via the proper temperature of thermal modification. Low vacuum scanning electron microscopy (LVSEM) results expressed that thermal modification itself had caused shrinking in the wood surface structure. From the beginning of the weathering process, the heat treatment affected the surface structural stability. After natural exposure, the degree of wood structure decay followed the pattern 220 °C heat-treated > 190 °C heat-treated > untreated. Therefore, when considering the impact on the structure, thermal modification treatment as a protective measure to prevent weathering was not an ideal approach and requires further improvement.

Keywords: natural weathering; heat-treated; color change; wood anatomical; *Cunninghamia lanceolate*

---

1. Introduction

In recent years, the wood protection industry has been paying greater attention to environmentally friendly substitutes for traditional wood protection treatments. One existing environmentally conscious method is the thermal modification of wood. Modified wood does not impregnate the material with any harmful substances or chemicals and the finished material does not produce environmental pollution. In the heat-treatment process, the wood is heated to high temperatures, ranging from 160 °C to 260 °C, for various standing times based on the species and the desired material properties [1–4]. Thermal modification offers particular benefits. In general, it reduces the equilibrium moisture content, improves hydrophobicity, enhances dimensional stability, maintains uniform wood color, and offers improved protection, especially against damage caused by micro-organisms and insects. However, brittleness, cracking, and other forms of mechanical strength loss are the main disadvantages of heat-treated wood [1,3,5]. Thermally-modified wood has many applications for exterior structures, including terraces, fences, decks, cladding, garden furniture, doors, and windows; as well as interior uses, e.g., decorative panels, parquet, kitchen furniture, and saunas [3,6,7].

Weathering is the general term used to refer to the slow degradation of materials when exposed to the weather. It can be influenced by several factors, including sunlight, moisture, heat/cold, wind, air pollutants, and biological agents [8]. Weathering of wood surfaces can cause color, chemical, physical, mechanical, and microscopic changes. These changes occur in the wood surface at a depth of 0.05–2.5 mm during the initial weathering period [9].

Earlier research generally examined the resistance of heat-modified wood to artificial weather [10–16]. Heat-treated wood exhibits high physical characteristics and low mechanical properties during artificial weathering. Artificial weathering tests are generally considered to be a simulation of outdoor conditions, but this method includes only UV light and moisture cycles. There are many other degradation factors in the natural environment, such as microbes, air pollutants, chemicals, and biological agents [8]. Therefore, and artificial weathering experiment cannot fully substitute for a study of natural weathering, and it is still necessary to test the service life of wood. Research of natural weathering always takes many years. In fact, the surface changes of weathering characteristics begin from the moment the wood is exposed to the outdoor environment. It has been reported that the surface characteristics of wood can change significantly during a short exposure period [17]. Therefore, surface change analysis is a method that can evaluate the weatherability of wood over a short period of time.

The purpose of this study is to clarify the effect of heat treatment on the weatherability of *Cunninghamia lanceolate* (Lamb.) Hook. wood. We conducted an outdoor exposure experiment for one month and examined the anatomical and physical changes of wood surface. In order to attain these objectives, some techniques and methods for the study of wood surfaces were used such as a colorimeter for physical color changes and low vacuum scanning electron microscopy (LVSEM) for anatomical changes. These analytical tools provided accurate insight into the degradation process, which allowed a comparison between heat-treated and untreated Chinese fir exposed to natural conditions.

2. Materials and Methods

2.1. Preparation of Wood Samples

All the samples used in this study were taken from Chinese fir (*Cunninghamia lanceolate*) [18]. Samples were obtained from sapwood straight off the grain, and each was free of knots and visible defects. Samples were 10 (R) × 10 (T) × 10 (L) mm in size. Under drying conditions, three parts of the samples were cut with a hammer microtome. Lumber was heat-treated at 190 °C and 220 °C for 120 min, respectively, under steam. This range was chosen because 220 °C is a significant critical point for the heat treatment of Chinese fir [19]. From previous studies of Chinese fir [20], heat treatment at 190 °C and 220 °C was determined to be optimal temperatures for strength and corrosion resistance, respectively. Timbers were dried to a moisture content of 8% prior to the steam-heat treatments. About 4% moisture content was achieved after the steam-heat treatments. Samples were classified into three groups as follows: (1) untreated controls, (2) 190 °C heat-treatment, and (3) 220 °C heat-treatment.

2.2. Natural Weathering Conditions

Untreated and heat-treated Chinese fir samples were placed in an outdoor environment for four weeks. The experiment was conducted during the summer season of 2018 in Fukuoka, Japan. The statistics mean temperature, humidity, duration of sunshine, total rainfall and wind speed were 30.01 ± 1.41 °C, 67.81% ± 6.31%, 9.13 ± 3.57 h, 1.68 mm, and 2.92 ± 0.80 m/s, respectively. They were orientated in a south-facing position at 45° relative to the horizontal (latitude 33° north) according to JIS K 5600-7-6 [21]. After one day of exposure, all the specimens were retrieved from the weathering racks, and measurements were taken in a laboratory, after which they were put back on the weathering racks. This was repeated after 2, 4, 7, 14, 21, and 28 days [22]. On each occasion, a puff of nitrogen was used to gently clear the surface of dust.

2.3. Color Measurements

Color measurements were used to evaluate the degree of color change on the sample surfaces after each period of natural weathering [22]. There were 10 replicates for each treatment group.

A mathematical average was established based on 30 measurements for every data point of color measurement. For each sample, the color coordinates L*, a*, and b* were established both before and after exposure to weather. A handy colorimeter (Model NR-3000, Nippon Denshoku, Tokyo, Japan) was used to measure the color change, with a D65 standard illuminant and the 2° standard observer. The color difference values of ΔC*, ΔE*, and ΔH* were determined before and after the natural conditions. C* indicates chroma, while ΔC* is the chroma difference. The parameter ΔE* indicates the total difference value. ΔH* is the difference in hue [23,24]. Average values and standard deviation were computed for each sample. These parameters were then employed to establish ΔC*, ΔE*, and ΔH* based on the following formulae:

$$C^* = [(a^*)^2 + (b^*)^2]^{1/2} \tag{1}$$

$$\Delta C^* = C^* - C_o^* \tag{2}$$

$$\Delta E^* = [(\Delta L^*)^2 + (\Delta a^*)^2 + (\Delta b^*)^2]^{1/2} \tag{3}$$

$$\Delta H^* = [(\Delta E^*)^2 - (\Delta L^*)^2 - (\Delta C^*)^2]^{1/2} \tag{4}$$

Positive values of ΔC* indicate greater vividness, whereas a negative value indicates a more faded hue in comparison to the initial color. A low ΔE* indicates that the color was not altered very much or remained the same, while a high value indicates obvious color change. ΔH* records the extent of the hue change.

2.4. Statistical Analysis

By statistical analysis of the color data before and after natural weathering, the color difference values of different exposure times did not conform to normal distribution. The rank sum test method of paired samples was selected in this experiment. With the change of exposure time, the color difference of 190 °C and 220 °C heat-treated wood was compared separately with that of untreated samples. The measurement index is the color difference values of ΔC*, ΔE*, and ΔH*, which were the quantitative data of the paired design. The rank of the contrast difference was reported in the results. The results of the Wilcoxon signed ranks test indicated the Z value, the approximate method was used to calculate the p value, and the statistical significance of the difference was determined.

2.5. Low-Vacuum Scanning Electron Microscopy (LVSEM)

During the natural weathering experiment, samples that had no coating were observed via low-vacuum scanning electron microscopy (LV-SEM, Model JSM-5600LV, JEOL, Tokyo, Japan). The microscopy conditions were as follows: a voltage of 15 kV, a pressure of approximately 10–30 Pa, and a working distance of 10–20 mm. This LVSEM method was deployed over the exact same area of wood surface before and after exposure. LVSEM was a valuable tool to observe the anatomical changes in the process of weathering [25,26].

3. Results

3.1. Discoloration of Wood Surfaces

Wood absorbs light, and polymeric compounds inside wood interact with photons. These factors lead to the deterioration and discoloration of wood. When wood is exposed to natural conditions, other factors such as moisture and microorganism growth also contribute to the degradation. The color of the exposed wood is influenced by light radiation, rain water, and temperature. UV light causes particularly remarkable color changes [24,27–29]. Color stability is an important parameter among the physical properties of wood.

After heat treatment, the color exhibited a notable change, and the colors visible to the naked eye became darker. L* (lightness) decreased, while a* (redness) increased. Different temperatures had effects on the color of wood. B* (yellowness) and C* (chroma) increased after 190 °C heat treatment, while these variables decreased after 220 °C heat treatment. These changes might be due to hydroxyl groups that were oxidized to carbonyl groups and carboxyl groups during heat treatment. If this is the case, then the wood color was darkened because the carbonyl groups belong to chromophric groups and the carboxyl groups to auxochromic groups [30].

Figure 1 shows the color modifications (L*, a*, b*, and c*) of untreated and heat-treated wood surfaces while exposed to natural conditions. Increases in a* (redness) were observed with the increase in exposure time during the first stages. Redness of heat-treated samples underwent an especially noticeable increase after the first weathering day. However, a slow increase in redness was also shown during the first 4 exposure days for untreated samples. After the first 4 days, the value of a* decreased quickly for both untreated samples. After the first 4 days, the value of a* decreased quickly for both untreated and heat-treated wood. Increases in b* (yellowness) and c* (chroma) were observed with increases in the number of natural weathering days. In particular, b* and c* exhibited rapid increases during the first week. Nevertheless, a slow increase in these values was also shown in the natural weathering experiment after the first week. This implies that the first week is a critical period for changes in yellow and intense chroma during the natural weathering. The L* (lightness) value of untreated wood demonstrated a decrease in each treatment, but the L* values increased as the number of days increased for wood that received heat treatment at 190 °C and 220 °C. As time increased during the first week, the value of L* changed rapidly. After one week, however, the value of L* changed incrementally. The darkening of untreated wood might be caused by high temperature inducing the migration of extractives to the wood surface. Several previous studies have reported similar results [11,12,31–33]. The surface of heat-treated wood became lighter during weathering as the extractives were degraded and removed. This lightening was mainly due to the photodegradation of lignin [12].

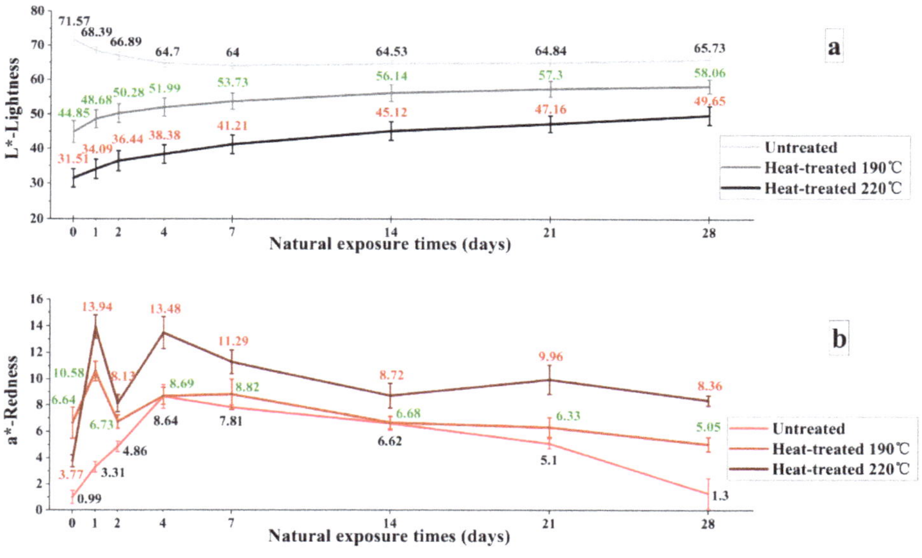

Figure 1. *Cont.*

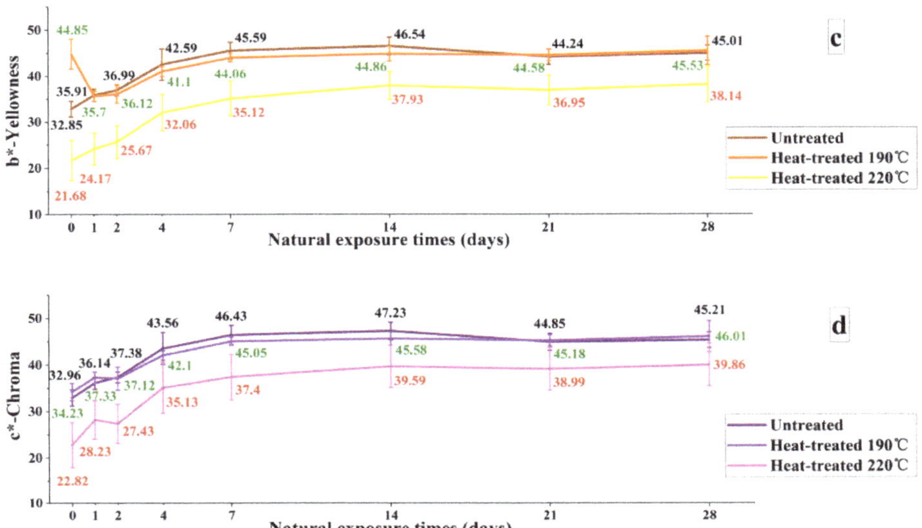

Figure 1. Color modifications for each of the four weeks of natural exposure: (**a**) Lightness L*; (**b**) redness a*; (**c**) yellowness b*; (**d**) chroma c*.

Figure 2 shows the effect of natural exposure on ΔC*, ΔE*, and ΔH*, revealing the color changes as a result of the effects of natural weathering on wood samples. The ΔC* (chroma difference) values in untreated and in heat-treated wood increased significantly in the first seven days, and the values then stabilized after the first week. The ΔC* value for 190 °C heat-treated wood was lower than other treatments. This result indicates that 190 °C heat-treatment could attenuate the chroma difference during the natural weathering process. The values of ΔH* (hue difference) exhibited insignificant changes during the four weeks of the experiment for both untreated and heated-treated wood (Figure 2c). Meanwhile, ΔE* (total color difference) values in untreated wood increased quickly over the initial seven-day period, and the values then started decreasing after two weeks. However, this phenomenon was not found in heat-treated wood. This implies that untreated and heat-treated samples exhibited different color change responses during the four weeks of natural weathering.

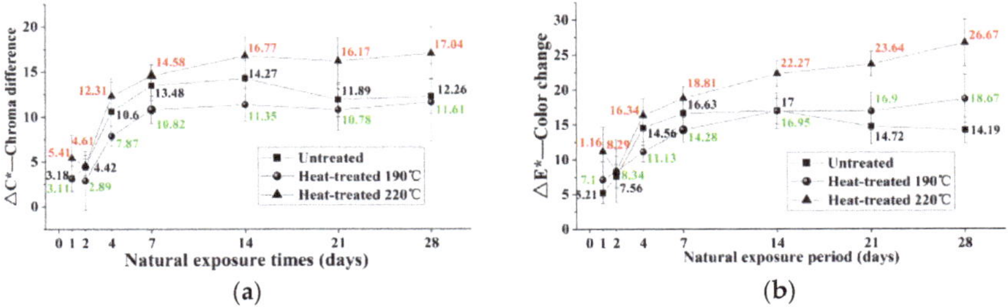

Figure 2. *Cont.*

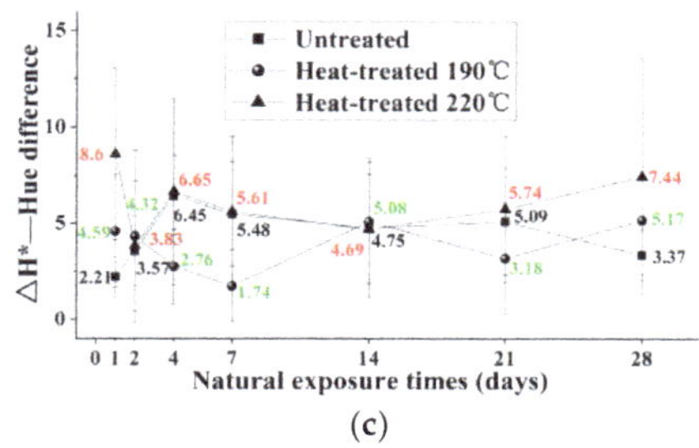

(c)

Figure 2. Color changes of wood samples subjected to natural conditions: (**a**) Chroma difference Δc*; (**b**) color change ΔE*; (**c**) hue difference ΔH*.

Wilcoxon signed ranks test statistics demonstrated significant differences in ΔC*, ΔE*, and ΔH* values during the four weeks of natural exposure for the untreated, 190 °C heat-treated, and 220 °C heat-treated wood samples (Table 1). The results illustrated that the surface chroma exhibited a significant negative difference for both the 190 °C heat-treated and untreated in the initial stage of natural weathering (Table 2). The 190 °C heat treatment clearly promoted the stability of chromaticity. The same results were also obtained for ΔE* (total color difference value) and ΔH* (hue difference), but the differences were not so significant. However, the 220 °C heat-treated wood exhibited a significant positive difference in ΔC*, ΔE*, and ΔH* when compared to untreated wood (Table 2). This shows that 220 °C heat treatment could not stabilize the color during the process of weathering. On the contrary, this treatment brings about greater color change (Table 1).

Table 1. Ranks heat-treated–untreated.

			N	Mean Rank	Sum of Ranks
ΔC*	190 °C heat-treated–untreated	Negative ranks	48 [a]	37.11	1781.50
		Positive ranks	20 [b]	28.23	564.50
		Ties	0 [c]		
		Total	68		
	220 °C heat-treated–untreated	Negative ranks	17 [a]	22.12	376.00
		Positive ranks	53 [b]	39.79	2109.00
		Ties	0 [c]		
		Total	70		
ΔE*	190 °C heat-treated–untreated	Negative ranks	39 [a]	30.74	1199.00
		Positive ranks	29 [b]	39.55	1147.00
		Ties	0 [c]		
		Total	68		
	220 °C heat-treated–untreated	Negative ranks	4 [a]	14.50	203.00
		Positive ranks	56 [b]	40.75	2282.00
		Ties	0 [c]		
		Total	70		
ΔH*	190 °C heat-treated–untreated	Negative ranks	41 [a]	35.30	1447.50
		Positive ranks	27 [b]	33.28	898.50
		Ties	0 [c]		
		Total	68		
	220 °C heat-treated–untreated	Negative ranks	28 [a]	30.20	845.50
		Positive ranks	42 [b]	39.04	1639.50
		Ties	0 [c]		
		Total	70		

[a] Heat-treated < untreated; [b] Heat-treated>untreated; [c] Heat-treated = untreated.

Table 2. Test statistics; [a] heat-treated–untreated.

	ΔC*		ΔE*		ΔH*	
	190 °C	220 °C	190 °C	220 °C	190 °C	220 °C
Z	−3.718 [b]	−5.071 [b]	−0.159 [b]	−6.083 [b]	−1.677 [b]	−2.323 [b]
Asymp. Sig. (2-talled)	**	**	0.874	**	0.093	*

[a] Wilcoxon signed ranks test; [b] Based on negative ranks. * Statistically significant $p < 0.05$; ** Statistically significant $p < 0.01$.

Other researchers [34,35] reported that cellulose remained relatively stable after heat treatment of Chinese fir. In hemicelluloses, the acetyl group was decomposed from molecular chains to acetic acid and degraded pyranose. A cross-linking occurrence had been formed among the aromatic units in the lignin. The observed that higher heat treatment temperatures intensified the reaction [19]. In the present study, there are two potential reasons for the favorable color stability under natural weathering of wood heat-treated at 190 °C. First, the heat treatment temperature of 190 °C may not have been enough to greatly change the chemical composition of wood because hemicellulose degradation was less 3% [34]. Second, color deepening after heat treatment could inhibit the color change to a certain extent [36,37]. As a result, the color change of wood during natural exposure was weakened and the color stability was improved after 190 °C heat treatment. The relative content of lignin in wood was significantly increased after heat treatment at 220 °C because of the degradation of hemicellulose [19,34]. The most important change in the process of weathering was the decomposition of lignin [24,27,38–41]. This directly results in the aggravation of weathering, and the color change was very significant. Therefore, selecting the appropriate thermal modification temperature played a substantial role in color stability during the weathering process.

3.2. Wood Structure Changes

The changes of wood surface structure accompany other physical changes taking place during the natural weathering. Heat treatment changes the characteristics of wood, including microscopic structural changes. In this study, low vacuum scanning electron microscopy (LVSEM) was used to investigate the wood structure degradation of heat-treated Chinese fir (*Cunninghamia lanceolate*) subjected to natural exposure. Both heat-treated and untreated wood surfaces were surveyed for comparison. LVSEM analysis of the three sections of heat-treated Chinese fir sapwood showed obvious microscopic structural changes that took place during the short-term exposure to conditions (Figures 3 and 4). It was found that microstructure did not change in the first two days of natural weathering, until after four days had a little change (shown in Figure 4g).

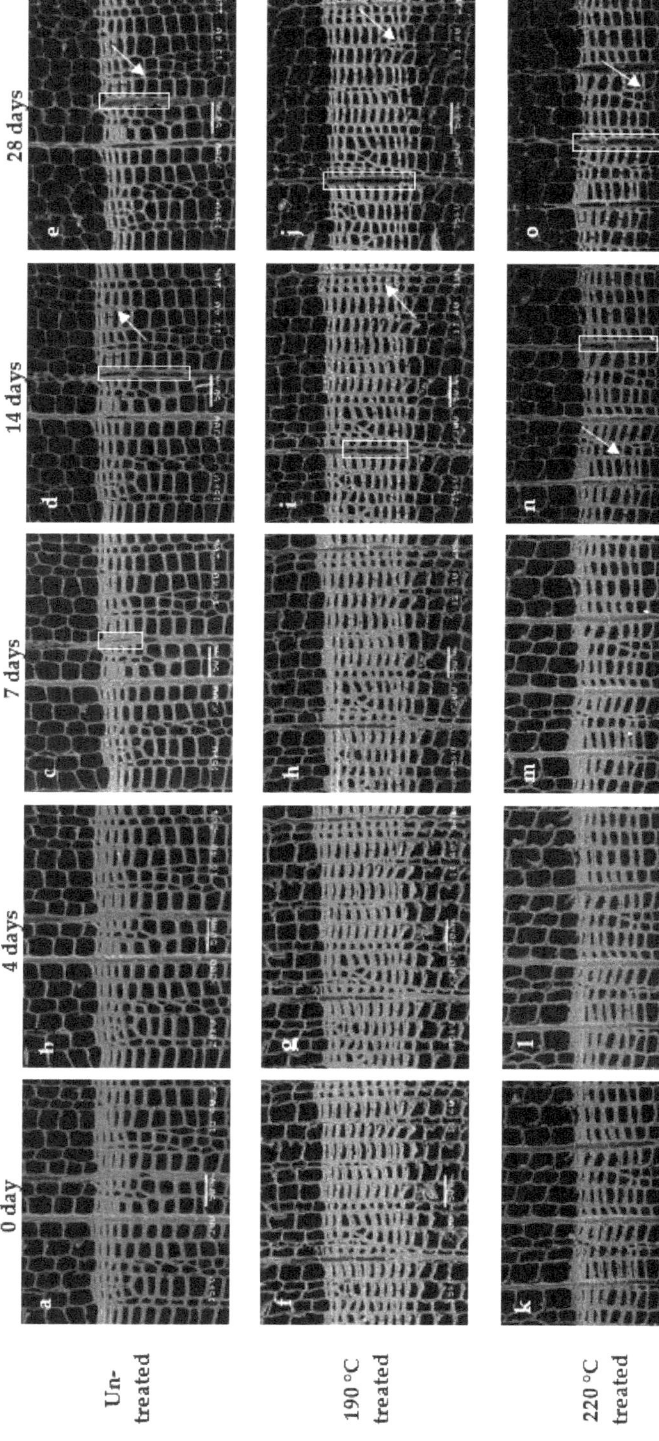

Figure 3. Scanning electron micrographs of cross sections of untreated and heat-treated wood before and after natural weathering.

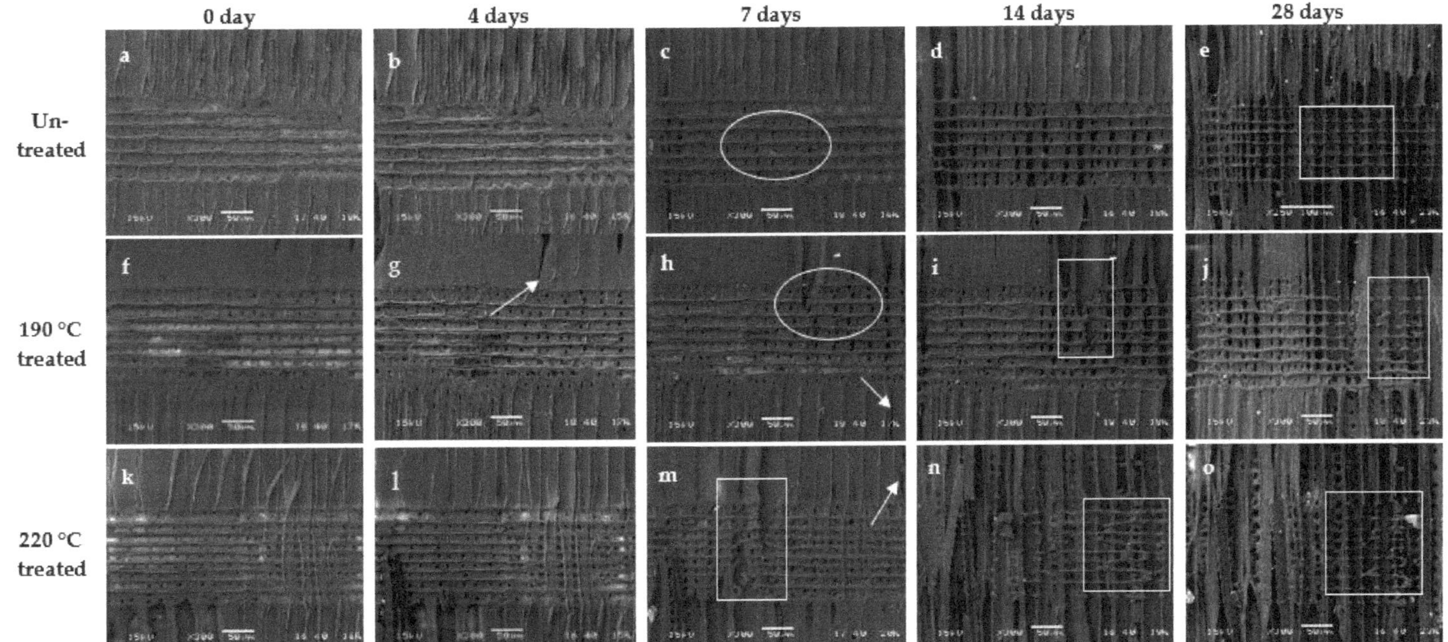

Figure 4. Scanning electron micrographs of radial sections of untreated and heat-treated wood before and after natural weathering.

3.2.1. Cell Wall

As shown in Figure 3a,f,k the microstructure of the heat-treated wood was rougher than the microstructure of untreated wood. Heat treatment caused the cell wall to become slightly plasticized [42]. The surface of the heat-treated Chinese fir wood was more brittle than the surface of untreated wood. In contrast, the 220 °C heat treatment had greater effects than the 190 °C heat treatment. After four weeks of natural weathering, the middle lamella and primary cell wall had substantially disappeared (as shown by the arrows in Figure 3e,j,o). Compared with the unexposed condition, the secondary cell wall became obviously thinner. The degradation of the secondary cell wall in heat-treated wood was more severe than untreated wood under natural weathering conditions. The comparison between the two heat treatment groups found that higher heat treatment temperatures produced more severe weathering erosion of the secondary cell wall (Figure 3).

The S1 layer comprises 30.0% cellulose, 51.7% lignin, and 18.3% hemicellulose. The S2 layer comprises 54.3% cellulose, 15.1% lignin, and 30.6% hemicelluloses. The S3 layer is composed of 13% cellulose, no or only negligible amounts of lignin, and 87% hemicelluloses [43]. The lignin content in wood exposed for four weeks is somewhat lower than the content in wood exposed for seven days. Compared with the untreated wood, the heat treatment caused significant degradation of hemicellulose and cellulose in the cell wall. Therefore, the lignin percentage was relatively increased. Chemical composition changes caused differences in the structure. As a result, the change in the material from exposure was more significant in heat treatment groups. However, there is a rupture phenomenon in the weathering stage due to the high level of lignin in the cell wall of the wood exposure for seven days. Previous studies on the heat treatment of Chinese fir showed that chemical changes increased with high temperature. The degree of wood plasticization increased with the proportion of hemicellulose and cellulose [7,19]. The results of this experiment are consistent with previous research conclusions that reported that heat treatment produced microstructural changes [12].

In the case of Chinese fir, the untreated and the heat-treated wood tended to show degradation in the primary cell wall and middle lamella when subjected to natural weathering after two weeks (as shown by the arrows in Figure 3d,i,n). According to the literature [43], the primary cell wall and middle lamella comprised primarily lignin (84%), a smaller percentage of hemicelluloses (13.3%), and a very small percentage of cellulose (0.7%). The disappearance of the primary cell wall and lamella in wood after four weeks of weathering evidenced that the most photosensitive component of the cell wall was lignin. This conclusion conformed to previous studies [12]. The lignin content in the cell walls of the heat-treated Chinese fir was much higher than that of the untreated control group, and higher heat treatment temperatures produced higher proportions of lignin in the cell wall [19,43]. The cell wall lignin content exhibited the following pattern: 220 °C treated > 190 °C treated > untreated. Structures with high lignin content were more seriously eroded during weathering due to the photodegradation of lignin. The degree of cell wall damage after weathering followed the pattern: 220 °C treated > 190 °C treated > untreated.

3.2.2. Rays

Figure 3a,f,k show that the heat treatment apparently did not alter the microstructure of the wood rays. After natural weathering over 28 days, ray parenchyma cells exhibited a much higher the degree of degradation in the heat treatment groups and untreated groups than surrounding tracheids (as shown by the square area in Figure 3e,j,o). It is well known that parenchyma cells only possess very thin primary cell walls and no secondary walls. The chemical constituents of these cell walls are mostly lignin [14]. After heat treatment, these cell walls are still mainly composed of lignin. The attenuation of the cell wall in the parenchyma cells during the weathering process was obvious. This provides more evidence that the photosensitivity of lignin was particularly prominent.

3.2.3. Cross-Field

SEM pictures of the cross-field structure on the radial section of untreated and heat-treated samples displayed certain differences (Figure 4). Serious structural damage was found at the cross-field after heat treatment at 220 °C (shown in Figure 4m). Longer exposure times produced the more obvious damage. After 28 days of weathering, the cell walls of the ray parenchyma at the cross-field position were significantly degraded (as shown by the square area in Figure 4e,j,o). Compared with the thickness of cross-field tracheid walls after weathering, the degradation of ray parenchyma cells is obvious. These cells only possess primary walls and do not contain secondary walls. Therefore, their main components were lignin [14,43], and lignin was highly degraded during weathering. The degradation degree of tracheid cell walls also increased with higher heat treatment temperatures.

In SEM micrographs of the cross-field pits of Chinese fir, it can be seen that the pits did not suffer damage due to the heat treatment process (Figure 4f,k). The edges of the bordered pits were destroyed after 1 week of weathering and cracked along the direction of the microfibril (as shown by the circle area in Figure 4c,h,m). Higher heat treatment temperatures produced higher plasticity. During the process of weathering, cracking was more likely to occur under the influence of shrinkage and swelling.

3.2.4. Tracheids

Micrographs of the radial surface of untreated and heat-treated wood demonstrated that the natural weathering process caused degradation of the wood surface. After only a week, severe longitudinal micro-cracks appeared on the tangential surface tracheids of the heat-treated wood (as shown by the arrows in Figure 4h,m). Cellulose is responsible for wood strength, hemicelluloses and lignin compose the matrix system in wood, and lignin provides wood with rigidity or stiffness. During the process of weathering, the change of humidity and the loss of lignin in the secondary cell wall caused serious surface shrinkage and made the wood crack more easily. After heat treatment, plasticization exacerbated the cracking.

4. Discussion

In summary, it can be seen that the more serious deterioration during the natural weathering process was caused by the anatomical position of high lignin content. Lignin was always the most sensitive chemical component in weathering, and this is consistent with the results of previous studies [24,27,38–41]. Relative lignin content depended more on the degradation of hemicellulose at higher treatment temperatures [30,34]. The more obvious the decay phenomenon happened in the high lignin content anatomical position under natural exposure. We conclude that heat treatment of wood could not improve structural stability; on the contrary, it would aggravate the decline of the anatomical structural structure. Both 190 °C and 220 °C heat treatments accelerated the aging phenomenon, and these results differ only in degree. These trends were contrary to the results of color stability.

Heat-treated wood, considered comprehensively, can enhance the dimensional stability of wood, maintain uniform wood color, and, most importantly, prevent degradation caused by microorganisms and insects [1,3,5]. These advantages play a very positive role in the natural weathering of wood. In particular, reducing the damage caused by microorganisms and insects greatly improves the degradation resistance of the wood itself, and it mitigates the effects of natural weathering caused by decay. However, because heat treatment reduces the mechanical strength and structural stability of wood, heat treatment as a means of wood weatherproofing should be comprehensively evaluated first. In the process of heat treatment, the changes of mechanical strength and structural stability should be considered and serve as the focus of future research. Making full use of the advantages of improving corrosion resistance and reducing the loss of mechanical strength and structural stability, heat-treatment technology is better applied to the environmentally friendly modification of wood weathering prevention, and the utilization rate of wood is thus improved.

5. Conclusions

Heat-treated and untreated *Cunninghamia lanceolate* samples were exposed to natural conditions for one month. The physical and anatomical changes were examined on wood surfaces by different analysis methods.

After 190 °C heat treatment, the color change of the Chinese fir surface was significantly restrained in the process of natural weathering. Heat treatment played a positive role in maintaining the physical properties of wood.

The LVSEM study indicated that heat treatment aggravated the degradation of the wood structure in Chinese fir subjected to natural conditions. Thermal modification was not conducive to the maintenance of structural stability.

The effect of heat treatment on the weatherability of Chinese fir wood was beneficial for aesthetic properties and harmful for structural properties. If heat treatment is deployed as an anti-weathering treatment method, practitioners should comprehensively consider the advantages and disadvantages of thermally modified wood.

Author Contributions: Data curation: X.C.; supervision: J.M.; writing—original draft preparation: X.C.; writing—review and editing: J.M.

Funding: This research received no external funding.

Conflicts of Interest: The authors declare no conflict of interest.

References

1. Alfred, J.S.; Horace, K.B.; Albert, A.K. Staybwood—Heat-stabilized wood. *Ind. Eng. Chem.* **1946**, *38*, 630–634.
2. Alfred, J.S. Thermal degradation of wood and cellulose. *Ind. Eng. Chem.* **1956**, *48*, 413–417.
3. Esteves, B.M.; Pereira, H.M. Wood modification by heat treatment: A review. *Bioresources* **2009**, *4*, 370–404.
4. Militz, H. Heat Treatment of Wood: European Processes and Their Background. In *International Research Group Wood Pre, Section 4-Processes*; N° IRG/WP; International Research Group on Wood Protection: Stockholm, Sweden, 2002; p. 40241.
5. Kocaefe, D.; Poncsak, S.; Doré, G.; Younsi, R. Effect of heat treatment on the wettability of white ash and soft maple by water. *Holz Roh. Werkst.* **2008**, *66*, 355–361. [CrossRef]
6. Bengtsson, C.; Jermer, J.; Brem, F. Bending Strength of Heat-Treated Spruce and Pine Timber. In *International Research Group Wood Pre, Section 4-Processes*; N° IRG/WP; International Research Group on Wood Protection: Stockholm, Sweden, 2002; p. 40242.
7. Syrjänen, T.; Kangas, E. Heat-Treated Timber in Finland. In *International Research Group Wood Pre, Section 4-Processes*; N° IRG/WP; International Research Group on Wood Protection: Stockholm, Sweden, 2000; p. 40158.
8. Williams, R.S. Weathering of wood. In *Handbook of Wood Chemistry and Wood Composites*, 2nd ed.; Rowell, R.M., Ed.; CRC Press: Boca Raton, FL, USA, 2005; pp. 139–185.
9. William, C.F.; David, N.S.H. Chemistry of weathering and protection. In *The Chemistry of Solid Wood, Advances in Chemistry Series 207*; Rowell, R.M., Ed.; American Chemical Society: Washington, DC, USA, 1984; pp. 401–451.
10. Ayadi, N.; Lejeune, F.; Charrier, F.; Charrier, B.; Merlin, A. Color stability of heat–treated wood during artificial weathering. *Holz Roh. Werkst.* **2003**, *61*, 221–226. [CrossRef]
11. Deka, M.; Humar, M.; Rep, G.; Kričej, B.; Šentjurc, M.; Petrič, M. Effects of UV light irradiation on colour stability of thermally modified, copper ethanolamine treated and non-modified wood: EPR and DRIFT spectroscopic studies. *Wood Sci. Technol.* **2008**, *42*, 5–20. [CrossRef]
12. Huang, X.; Kocaefe, D.; Kocaefe, Y.; Boluk, Y.; Pichette, A. Study of the degradation behavior of heat-treated jack pine (*Pinus banksiana*) under artificial sunlight irradiation. *Polym. Degrad. Stabil.* **2012**, *97*, 1197–1214. [CrossRef]
13. Yildiz, S.; Tomak, E.D.; Yildiz, U.C.; Ustaomer, D. Effect of artificial weathering on the properties of heat reated wood. *Polym. Degrad. Stabil.* **2013**, *98*, 1419–1427. [CrossRef]

14. Todaro, L.; D'Auria, M.; Langerame, F.; Salvi, A.M.; Scopa, A. Surface characterization of untreated and hydro-thermally pre-treated Turkey oak woods after UV-C irradiation. *Surf. Interface Anal.* **2015**, *47*, 206–215. [CrossRef]
15. Xing, D.; Wang, S.; Li, J. Effect of Artificial weathering on the properties of industrial-scale thermally modified wood. *BioResources* **2015**, *10*, 8238–8252. [CrossRef]
16. Shen, H.; Cao, J.; Sun, W.; Peng, Y. Influence of post-extraction on photostability of thermally modified scots pine wood during artificial weathering. *BioResources* **2016**, *11*, 4512–4525. [CrossRef]
17. Evans, P.D.; Thay, P.D.; Schmalzl, K.J. Degradation of wood surfaces during natural weathering. Effects on lignin and cellulose and on the adhesion of acrylic latex primers. *Wood Sci. Technol.* **1996**, *30*, 411–422. [CrossRef]
18. Li, K.F.; Yong, F.Y.; Robert, R.M. Taxodiaceae. *Flora China* **1999**, *4*, 54–61.
19. Cao, Y.; Jiang, J.; Lu, J.; Huang, R.; Jiang, J.; Wu, Y. Color change of Chinese fir through steam-heat treatment. *Bioresources* **2012**, *7*, 2809–2819.
20. Cheng, D.L. The Study on Technics and Properties of Heat Treated Fir Wood. Master's Thesis, Nanjing Forest University, Nanjing, China, June 2007.
21. Japan Meteorological Agency. Available online: https://www.jma.go.jp/jma/index.html (accessed on 1–31 August 2018).
22. Kataoka, Y.; Kiguchi, M. Weatherability of water-borne wood preservation semi-transparent coatings (I)—Coating performance during 24 months of natural weathering. *Wood Preserv.* **2009**, *35*, 204–214. [CrossRef]
23. Robertson, A.R. The CIE 1976 color-difference formulae. *Color Res. Appl.* **1977**, *2*, 7–11. [CrossRef]
24. Tolvaj, L.; Faix, O. Artificial ageing of wood monitored by DRIFT spectroscopy and CIE L*a*b* color measurements. I. Effect of UV light. *Holzforschung* **1995**, *49*, 397–404. [CrossRef]
25. Hatae, F.; Kataoka, Y.; Kiguchi, M.; Matsunaga, H.; Matsumura, J. In Site Visualization of Wood Degradation during Artificial Weathering by Variable Pressure Scanning Electron Microscopy. In Proceedings of the 11th Pacific Rim Bio-Based Composites Symposium (BIOCOMP 2012), Shizuoka, Japan, 27–30 November 2012; pp. 226–232.
26. Hatae, F.; Kataoka, Y.; Kiguchi, M.; Matsunaga, H.; Matsumura, J. Use of Variable Pressure Scanning Electron Microscopy for in Situ Observation of Degradation of Wood Surfaces during Artificial Weathering. In Proceedings of the International Research Group on Wood Protection, IRG/WP 12-20489, Kuala Lumpur, Malaysia, 6–10 May 2012.
27. Pandey, K.K. Study of the effect of photo-irradiation on the surface chemistry of wood. *Polym. Degrad. Stabil.* **2005**, *90*, 9–20. [CrossRef]
28. Mitsui, K.; Tsuchikawa, S. Low atmospheric temperature dependence on photodegradation of wood. *J. Photochem. Photobiol. B* **2005**, *81*, 84–88. [CrossRef]
29. Miklecic, J.; Jirouš-Rajković, V.; Antonović, A.; Španić, N. Discoloration of thermally modified wood during simulated indoor sunlight exposure. *BioResources* **2011**, *6*, 434–446.
30. Funaoka, M.; Kako, T.; Abe, I. Condensation of lignin during heating of wood. *Wood Sci. Technol.* **1990**, *24*, 277–288. [CrossRef]
31. Masanori, K.; Nakano, T. Artificial weathering of tropical woods. Part 2: Color change. *Holzforschung* **2004**, *58*, 558–565.
32. Dubey, M.K.; Pang, S.; Walker, J. Color and dimensional stability of oil heat-treated radiata pinewood after accelerated UV weathering. *Forest Prod. J.* **2010**, *60*, 453–459. [CrossRef]
33. Temiz, A.; Terziev, N.; Jacobsen, B.; Eikenes, M. Weathering, water absorption, and durability of silicon, acetylated, and heat-treated wood. *J. Appl. Polym. Sci.* **2006**, *102*, 4506–4513. [CrossRef]
34. Cheng, S.; Huang, A.; Wang, S.; Zhang, Q. Effect of different heat treatment temperatures on the chemical composition and structure of Chinese fir wood. *BioResources* **2016**, *11*, 4006–4016. [CrossRef]
35. Xing, D.; Li, J. Effects of heat treatment on thermal decomposition and combustion performance of *Larix* spp. Wood. *BioResources* **2014**, *9*, 4274–4287. [CrossRef]
36. Svetlana, B.; Marko, H.; Timo, K. Weathering of wood-polypropylene composites containing pigments. *Eur. J. Wood Prod.* **2012**, *70*, 719–726.
37. Hyvärinen, M.; Butylina, S.; Kärki, T. Accelerated and natural weathering of wood-polypropylene composites containing pigments. *Adv. Mater. Res.* **2015**, *1077*, 139–145. [CrossRef]

38. Tolvaj, L.; Faix, O. Artificial ageing Colom, X.; Carrillo, F.; Nogue's, F.; Garriga, P. Structural analysis of photodegraded wood by means of FTIR spectroscopy. *Polym. Degrad. Stabil.* **2003**, *80*, 543–549.
39. Pandey, K.K.; Vuorinen, T. Comparative study of photodegradation of wood by a UV laser and a xenon light source. *Polym. Degrad. Stabil.* **2008**, *93*, 2138–2146. [CrossRef]
40. Preklet, E.; Papp, G.; Barta, E.; Tolvaj, L.; Berkesi, O.; Bohus, J.; Szatmári, S. Changes in DRIFT spectra of wood irradiated by lasers of different wavelength. *J. Photochem. Photobiol. B* **2012**, *112*, 43–47. [CrossRef] [PubMed]
41. Calienno, L.; Monaco, A.L.; Pelosi, C.; Picchio, R. Colour and chemical changes on photodegraded beech wood with or without red heartwood. *Wood Sci. Technol.* **2014**, *48*, 1167–1180. [CrossRef]
42. Boonstra, M.J.; Rijsdijk, J.F.; Sander, C.; Kegel, E.; Tjeerdsma, B.; Militz, H.; Acker, J.V.; Stevens, M. Microstructural and physical aspects of heat treated wood. Part 1. Softwoods. *Maderas Cienc. Tecnol.* **2006**, *8*, 193–208. [CrossRef]
43. Rowell, R.M.; Roger, P.; James, S.H.; Jeffrey, S.R.; Mandla, A.T. Cell wall chemistry. In *Handbook of Wood Chemistry and Wood Composites*, 2nd ed.; Rowell, R.M., Ed.; CRC Press: Boca Raton, FL, USA, 2005; pp. 35–74.

Article

The Performance of Wood Decking after Five Years of Exposure: Verification of the Combined Effect of Wetting Ability and Durability

Miha Humar [1,*], Davor Kržišnik [1], Boštjan Lesar [1] and Christian Brischke [2]

1. Biotechnical Faculty, Department for Wood Science and Technology, University of Ljubljana, Ljubljana SI1000, Slovenia; davor.krzisnik@bf.uni-lj.si (D.K.); Bostjan.lesar@bf.uni-lj.si (B.L.)
2. Wood Biology and Wood Products, University of Goettingen, D-37077 Goettingen, Germany; christian.brischke@uni-goettingen.de

* Correspondence: Miha.Humar@bf.uni-lj.si; Tel.: +386-31843724

Received: 31 August 2019; Accepted: 10 October 2019; Published: 14 October 2019

Abstract: Wood is one of the most important construction materials, and its use in building applications has increased in recent decades. In order to enable even more extensive and reliable use of wood, we need to understand the factors affecting wood's service life. A new concept for characterizing the durability of wood-based materials and for predicting the service life of wood has recently been proposed, based on material-inherent protective properties, moisture performance, and the climate- and design-induced exposure dose of wooden structures. This approach was validated on the decking of a model house in Ljubljana that was constructed in October 2013. The decay and moisture content of decking elements were regularly monitored. In addition, the resistance dose D_{Rd}, as the product of the critical dose D_{crit}, and two factors taking into account the wetting ability of wood (k_{wa}) and its inherent durability (k_{inh}), were determined in the laboratory. D_{Rd} correlated well with the decay rates of the decking of the model house. Furthermore, the positive effect of thermal modification and water-repellent treatments on the outdoor performance of the examined materials was evident, as well as the synergistic effects between moisture performance and inherent durability.

Keywords: decay; decking; inherent durability; moisture performance; resistance model; service life

---

1. Introduction

Wood is one of the most important building materials. It is frequently used outdoors, where it is exposed to weathering and degradation. In Europe, wood-degrading fungi are the predominant reason for failures of wood used in outdoor applications [1,2]. Various solutions are used to prevent fungal decay and to achieve the desired service life, namely the use of biocides, wood modification, proper design and the use of domestic or imported durable wood species [3]. More recently, consumers are avoiding tropical wood species, so the importance of domestic wood species is increasing [4]. Unfortunately, the majority of European wood species do not provide a sufficiently high durability [5]. Particular emphasis is therefore placed on the utilization of domestic wood species [4].

Service life prediction of wooden objects is challenging, because the time during which a particular wooden structure will fulfil its function depends on a variety of factors, including the wood material used, the protection applied and various climate-related parameters [6]. In addition to the material-inherent durability, the moisture and temperature conditions inside the wood, i.e., the material climate, are the most important factors influencing the ability of fungi to decompose wood [2,7,8]. These two factors are influenced by the design of the construction, the exposure conditions and local climatic conditions (microclimate).

Building information modelling (BIM) software packages nowadays require information about service life and maintenance intervals for key materials used in the planning phase [9]. This information is required by the European Construction Products Regulations [10] and is needed for performance-based design [11]. The Eurocode [12] provides indicative design working lives of 10 years for temporary structures, 50 years for building structures and 100 years for monumental building structures and bridges. These values are set for objects regardless of the materials used. The expected service life of wooden structures under the given exposure conditions is a key parameter in the selection of materials for construction. Unfortunately, the current European standardization system provides neither information about expected service lives nor a methodology for the service life prediction of wood-based materials and components.

A new concept for characterizing the durability of wood-based materials and for predicting the service life of wood was recently proposed by Meyer-Veltrup et al. [13,14], based on the material-inherent protective properties, the moisture performance and the climate- and design-induced exposure dose of wooden structures. This approach has been successfully applied to untreated wood [15–17]. In this study, this modelling approach will be expanded to variously modified and preservative-treated woods.

2. Materials and Methods

2.1. Materials

This study investigated the performance of 19 different wood species and wood-based materials used in a decking application (Table 1). The selected materials were eight untreated wood species, Norway spruce (*Picea abies* (L.) H. Karst), European larch heartwood (*Larix decidua* Mill.), European beech (*Fagus sylvatica* L.), European ash (*Fraxinus excelsior* L.), Scots pine heartwood and sapwood (*Pinus sylvestris* L.), sweet chestnut (*Castanea sativa* Mill.) and European oak heartwood (*Quercus* sp.), and 11 materials that had been treated or modified in different ways. Although the authors are aware that not all these materials are traditionally used in decking, we have included them as reference wood species. In addition, the objective of this paper was not only to determine the performance of decking per se, but to validate the model [13]. In order to address this objective, materials of various durability have to be investigated.

Thermal modification (TM) was performed according to a commercial process (Silvapro®, Silvaprodukt, Ljubljana, Slovenia), with an initial vacuum in the first step of the treatment [18,19]. The modification was performed for 3 h at the target temperature (ranging between 210 °C and 230 °C, depending on the wood species). Impregnation was performed with a commercial copper–ethanolamine solution Silvanolin® (Silvaprodukt, Ljubljana, Slovenia), which consisted of copper, ethanolamine, boric acid and quaternary ammonium compounds [20]. The concentration of active ingredients and consequent retention met use class 3 (UC 3) [21] requirements. Impregnation was performed according to the full cell process in a laboratory impregnation setup. It consisted of 30 min vacuum (80 kPa), 180 min pressure (1 MPa) and 20 min vacuum (80 kPa). The same procedure was applied for the impregnation of wood with 5% commercially available natural wax dispersion with a solid content up to 50% by weight (Montax 50, Romonta, Germany) [22]. The acrylic surface coating Silvanol® Lazura B (Silvaprodukt, Ljubljana, Slovenia) was manually applied on the wood by brushing in two layers, with a 24-h drying time between them.

Table 1. Nineteen different investigated wood species and wood-based materials.

	Wood Species								Treatment			
	Norway Spruce (*Picea abies*)	Scots Pine – Sapwood (*Pinus sylvestris*)	Scots Pine – Heartwood (*Pinus sylvestris*)	European Larch (*Larix decidua*)	European Ash (*Fraxinus excelsior*)	European Beech (*Fagus sylvatica*)	Sweet Chestnut (*Castanea sativa*)	English Oak (*Quercus* sp.)	Thermal Modification	Impregnation with Natural Wax	Copper–Ethanolamine Impregnation	Water Borne Acrylic Surface Coating
Abbreviation	PA	PS	PH	LD	FE	FS	CS	Q	TM	NW	CE	AC
PA	**PA**											
PA–NW	**PA**									**NW**		
PA–AC	**PA**											**AC**
PA–CE	**PA**										**CE**	
PA–CE–NW	**PA**									**NW**	**CE**	
PA–TM	**PA**								**TM**			
PA–TM–NW	**PA**								**TM**	**NW**		
PA–TM–CE	**PA**								**TM**		**CE**	
PS		**PS**										
PH			**PH**									
LD				**LD**								
LD–TM				**LD**					**TM**			
FE					**FE**							
FE–TM					**FE**				**TM**			
FS						**FS**						
FS–TM						**FS**			**TM**			
FS–TM–NW						**FS**			**TM**	**NW**		
CS							**CS**					
Q								**Q**				

2.2. Outdoor Exposure

Figure 1 shows the wooden model house unit at the Department of Wood Science and Technology in Ljubljana, Slovenia (46°02′55.7″ N, 14°28′47.3″ E, elevation above sea level 293 m), where the in-service performance of decking elements was tested. This object has been used in several studies. The objective of this model house was to comprehensively assess the technical and aesthetic service life of wood. The aesthetic service life has been reported already [23], so this manuscript focusses on the technical service life only. The average temperature in Ljubljana is 10.4 °C, the annual precipitation is 1290 mm and the Scheffer Climate Index is 55.3. The test specimens, with a cross section of 2.5×5.0 cm^2, were exposed on the decking of the model house. At least seven samples of the wood material were exposed on the decking. The in-service testing started in October 2013 and the prime objective was to monitor the occurrence and development of decay (functional service life) and the moisture performance. Decay was visually evaluated annually and rated (0—no attack; 1—slight attack; 2—moderate attack; 3—severe attack; 4—failure) as prescribed by EN 252 [24]. Only the decking specimens were considered within this study.

Figure 1. Wooden model house unit in October 2013 at the beginning of the exposure. Yellow arrow is pointing north.

Moisture content (MC) during service life is one of the indicators of moisture performance. For MC measurements, resistance sensors were applied at 19 positions, with one pair of sensors for each wood material. They were linked to a signal amplifier (Gigamodule, Scanntronik, Zorneding, Germany) that enabled wood MC measurements between 6% and 60%. Pairs of stainless-steel screws with a diameter of 3.9 mm and length of 25 mm served as resistance electrodes, fastened in the middle of the tangential surface with a longitudinal distance of 32 mm between them. The screws were insulated with a universal heat-shrinking tube, except for the tip, which served as the point of measurement. Hence, the measurements take place approximately 10 mm below the surface. Sensors were located at least 20 cm from the cross section. The electrical resistance of the wood was measured every 12 h, and these data were used for calculating the wood's MC. Resistance characteristics for each material were determined as reported by Kržišnik et al. (in press) [25], using the methodology described by Brischke and Lampen [26] and Otten et al. [27].

2.3. Determination of Factors Describing Inherent Durability (k_{inh})

Agar block tests with pure fungal culture were used for the assessment of inherent durability. A decay test was performed according to a modified CEN/TS 15083–1 procedure [28]. Specimens (1.5 × 2.5 × 5.0 cm^3) made of 19 wood materials (Table 1) were conditioned in a standard laboratory climate (T = 25 ± 1; RH = 65 ± 2) and steam-sterilized in an autoclave before incubation with decay fungi; 350-mL experimental glass jars with aluminium covers and cotton wool with 50 mL of potato dextrose agar (DIFCO, Fisher Scientific, Franklin Lakes, NJ, USA) were prepared and inoculated with white rot fungi *Trametes versicolor* (L.) Lloyd (ZIM L057) and two brown rot fungi *Gloeophyllum trabeum* (Pers.) Murrill (ZIM L018) and *Fibroporia vaillantii* (DC.) Parmasto (ZIM L037). The fungal isolates originated from the fungal collection of the Biotechnical Faculty, University of Ljubljana, and are available to research institutions on demand [29]. Information regarding the origin of the fungal isolates and details about identification are available in the appropriate catalogue. One week after inoculation, two random specimens per jar were positioned on a plastic high-density polyethylene (HDPE) mesh, which was used to avoid direct contact between the samples and the medium. The assembled test glasses were then incubated at 25 °C and 80% relative humidity (RH) for 16 weeks, as prescribed by the standard. After incubation, specimens were cleaned from adhering fungal mycelium, weighed to the nearest 0.0001 g, oven-dried at 103 ± 2 °C, and weighed again to the nearest 0.0001 g to determine mass loss through wood-destroying basidiomycetes. Five replicate specimens for each of the selected materials/wood species were used in this test.

2.4. Determination of Factors Describing Wetting Ability (k_{wa})

For the assessment of wetting ability, a range of laboratory tests were performed. Tests was performed on five replicate samples (1.5 × 2.5 × 5.0 cm^3) of each material. One set of specimens was used for sorption tests, and the other for various immersion tests. The average relative values of the multiple test were combined to calculate the wetting ability factor.

Short-term capillary water uptake was carried out at 20 °C and 50 ± 5% RH, on a Tensiometer K100MK2 device (Krüss, Hamburg, Germany), according to a modified EN 1609 [30] standard, after conditioning at 20 °C and 65% RH until constant mass. The axial surfaces of the specimens were positioned to be in contact with the test liquid (distilled water), and their masses were subsequently measured continuously every 2 s for up to 200 s. Other parameters used were: velocity before contact with water 6 mm/min, sensitivity of contact 0.005 g, and depth of immersion 1 mm. The uptake of water was calculated in g/cm^2 on the basis of the final mass change of the immersed sample and the surface in contact with water.

Long-term water uptake was based on the ENV 1250–2 [31] leaching procedure. Before the test, specimens were oven-dried at 60 ± 2 °C until constant mass and weighed to determine the oven-dried mass. The dry wood blocks were placed in a glass jar and weighted down to prevent them from floating; 100 g of distilled water was then added per specimen. The mass of the specimens was determined after 24 h, and the MC of five replicate specimens was calculated. MC was determined gravimetrically, as a ratio between the retained water and the oven-dried mass of the specimens.

To determine the sorption properties of the samples, a water vapour uptake test in a water-saturated atmosphere with a drying process above freshly activated silica gel was performed. Specimens were oven-dried at 103 ± 2 °C until constant mass and weighed. The specimens were stacked in a glass climate chamber with a ventilator above distilled water. Specimens were positioned on mesh above the water using thin spacers [13]. After 24 h of exposure, they were weighed again and the MC was calculated. Specimens were then left in the same chamber for an additional three weeks until a constant mass was achieved. In addition to wetting, outdoor performance is also influenced by drying. After three weeks of conditioning, most specimens were positioned above freshly activated silica gel for 24 h in a closed container, and the MC of the specimens was calculated according to the procedure described by Meyer-Veltrup et al. [13]. Five replicate specimens were used for this analysis.

2.5. Factor Approach for Quantifying the Resistance Dose D_{Rd}

A modelling approach was applied according to Meyer-eltrup et al. [13] and Isaksson et al. [32] in order to predict the field performance of the examined materials. The model describes climatic exposure and the resistance of the material. The acceptability of the chosen design and material is expressed as follows:

$$\text{Exposure} \leq \text{Resistance}. \tag{1}$$

The exposure can be expressed as an exposure dose (D_{Ed}), determined by daily averages of temperature and MC. The material property is expressed as the resistance dose (D_{Rd}) in days [d], with optimum wood MC and wood temperature conditions for fungal decay [33]:

$$D_{Ed} \leq D_{Rd}, \tag{2}$$

where D_{Ed} is the exposure dose [d] and D_{Rd} is the resistance dose [d].

The exposure dose D_{Ed} depends on the annual dose at a specific geographical location and several factors describing the effect of driving rain, local climate, sheltering, distance from the ground and detail design. Isaksson et al. [32] give a detailed description of the development of the corresponding exposure model. The present study focussed on the counterpart of the exposure dose, which is the resistance, expressed as resistance dose D_{Rd}. This is considered to be the product of the critical dose D_{crit} and two factors expressing the wetting ability of wood (k_{wa}) and its inherent durability (k_{inh}). The approach is given by Equation (3), according to Isaksson et al. [32] (Table 2):

$$D_{Rd} = D_{crit} \times k_{wa} \times k_{inh}, \tag{3}$$

where D_{crit} is the critical dose corresponding to decay rating 1 (slight decay), according to EN 252 [24], k_{wa} is a factor accounting for the wetting ability of the tested materials [-], relative to the reference Norway spruce, and k_{inh} is a factor accounting for the inherent protective properties of the tested materials against decay [-], relative to the reference Norway spruce. Namely, the wetting ability and inherent durability of the Norway spruce were set to 1. Materials with either of these values better than the one determined for Norway spruce have higher values overall, but limited to a value of 5.

Table 2. Description of key terms addressed in respective article.

Term	Description
k_{wa}	Factor describing the wetting ability of wood-based materials. Factor is expressed in relative values, relative to the wetting ability of the spruce.
k_{inh}	Factor describing the inherent durability of wood-based materials. Factor is expressed in relative values, relative to the inherent durability of the spruce.
D_{Rd}	Resistance dose reflects the material property and is expressed in days (d), with optimum wood MC and wood temperature conditions for fungal decay, before the first evidence of decay.
Rel. D_{Rd}	Relative resistance dose. Usually spruce is used as the normalisation factor.

Based on the results of the various moisture tests presented in this paper, the wetting ability factor k_{wa} was calculated. The methodology for the calculation of k_{wa} followed the Meyer-Veltrup procedure [13], except that the size of the specimens differed. The original model prescribes specimens (0.5 × 1.0 × 10.0 cm^3) that are of a different shape from that used in the present study (1.5 × 2.5 × 5.0 cm^3). Since the methodology is based on relative values, the sample size has a minor influence on the outcome. Results from durability tests were used to evaluate the inherent resistance factor k_{inh}, and both factors were used to determine the resistance dose D_{Rd} of the 19 wood materials examined in this study. Only basidiomycetes were applied to determine k_{inh} in this research. Terrestrial microcosm tests and in-ground durability tests were not performed, as prescribed by the original Meyer-Veltrup approach [13].

2.6. Dose–Response Model

A dose–response model for the fungal decay of wood in aboveground situations, as described in detail by Brischke and Meyer-Veltrup [34], was applied to the recorded material–climatic data (***MC, T***). For comparative analysis, the total dose D (= cumulative daily dose d over time) was determined. A moisture-induced dose component $\boldsymbol{d_{MC}}$ and a temperature-induced dose component d_T were therefore calculated based on the physiological needs of decay fungus and optimized on the basis of long-term field tests at several climatically different locations in Europe by Brischke and Rapp [7]. The model considers wood MC and T as the key parameters for fungal growth and decay and allows a daily dose between 0 for adverse conditions and 1 for favourable conditions. The two-dose components $\boldsymbol{d_{MC}}$ and $\boldsymbol{d_T}$ are calculated separately, as follows:

$$d_{MC} = 6.75 \times 10^{-10} MC^5 - 3.50 \times 10^{-7} MC^4 + 7.18 \times 10^{-5} MC^3 - 7.22 \times 10^{-3} MC^2 + 0.34 MC - 4.98;\ \text{if MC} \geq 25\% \quad (4)$$

$$d_T = -1.8 \times 10^{-6} T^4 + 9.57 \times 10^{-5} T^3 - 1.55 \times 10^{-3} T^2 + 4.17 \times 10^{-2} T;\ \text{if } 40\ ^\circ\text{C} > \text{T} > -1\ ^\circ\text{C}, \quad (5)$$

where $\boldsymbol{d_{MC}}$ is the moisture-induced daily dose (*d*), $\boldsymbol{d_T}$ is the temperature-induced daily dose (*d*), ***MC*** is the daily wood ***MC*** (%), and *T* is the daily average wood temperature (°C). To consider the impact of ***MC*** and temperature on decay, a weighting factor (a) was added to calculate the daily dose (*d*), as follows. The following conditions were considered: the daily dose (d) of days with a temperature above 40 °C, with a temperature below −1 °C, or with an ***MC*** below 25% was set to 0:

$$d = ((a \times d_T) + d_{MC})/(a + 1);\ \text{if } d_T > 0 \text{ and } d_{MC} > 25\%, \quad (6)$$

where ***d*** is the daily dose (*d*) and a = 3.2 is the weighting factor of the temperature-induced daily dose component $\boldsymbol{d_T}$.

For n days of exposure, total exposure dose is given by:

$$d(n) = \sum_1^n d_i = \sum_1^n (f(d_T(T_i), d_{MC}(MC_i))), \quad (7)$$

where T_i is the average temperature (°C) and MC_i is the average moisture content for day (%).

Decay is initiated when the accumulated dose reaches a critical level. The dose is thus defined as a material–climate index and the response is considered to be the mean decay rating according to EN 252 [24], or the resulting decay rate (i.e., the decay rating per year). Expected service lives were estimated according to Equation (8) using a mean decay rating 2 (moderate decay) as a limit state. Any decay rating above this limit state means that serviceability is no longer given. A critical dose d_{crit} = 670 (for white and soft rot) and d_{crit} = 356 (for brown rot) was needed to reach the limit state.

$$ESL = \frac{d_{crit}}{d_a}, \quad (8)$$

3. Results and Discussion

3.1. Resistance Dose Based on Inherent Durability and Wetting Ability

Our data clearly confirm that the resistance of different wood species and treated or modified wood products in aboveground applications is primarily dependent on the degree of inherent material resistance against fungal decay (k_{inh}), but also on the wetting ability (k_{wa}) of the particular material. The material resistance dose D_{Rd} is a product of both factors and the respective critical dose D_{crit}, as summarized for all materials in Table 3. Since k_{inh} and k_{wa} are normalized to Norway spruce, the relative material resistance dose (rel. D_{Rd}) of Norway spruce is 1.0. The rel. D_{Rd} of Beech (FS) is 0.88, while the rel. D_{Rd} of Ash (FE; 1.22) and Scots pine sapwood (PS; 1.32) are slightly higher. The highest relative D_{Rd} among the nontreated wood species was determined for Oak (Q; 5.92) and Sweet chestnut (CS; 6.40) (Table 3). Rel. D_{Rd} of Oak (Q; 5.92) and Scots pine heartwood (PH; 2.97) were similar to those reported by Meyer-Veltrup et al. (2017) (Q; 5.10; PH; 2.75). This confirms the robustness and reliability of the approach. However, the main objective of the study reported by Meyer-Veltrup et al. [13] was to determine the rel. D_{Rd} of nontreated wood. In contrast, this study focused on the comparison of the performance and validation of the methodology with wood treated with biocides (copper–ethanolamine) or water repellents (wax) and (thermally) modified wood.

Table 3. Material resistance dose D_{Rd} and MC data for k_{inh} and k_{wa} calculated based on the Meyer-Veltrup et al. (2018) methodology [14].

Material	k_{inh}	k_{wa}	D_{Rd}	rel. D_{Rd}
PA	1.0	1.0	325	1.00
PA–NW	1.3	2.4	977	3.01
PA–AC	1.1	2.9	1009	3.10
PA–CE	5.0	1.5	2356	7.25
PA–CE–NW	5.0	2.9	4705	14.48
PA–TM	3.1	1.8	1763	5.43
PA–TM–NW	3.4	2.5	2698	8.30
PA–TM–CE	5.0	1.2	1978	6.09
PS	1.1	1.2	430	1.32
PH	2.5	1.2	966	2.97
LD	1.6	1.9	1002	3.08
LD–TM	2.7	3.2	2746	8.45
FE	1.2	1.0	396	1.22
FE–TM	2.9	1.9	1771	5.45
FS	0.9	1.0	284	0.88
FS–TM	2.6	2.1	1773	5.46
FS–TM–NW	3.3	2.6	2815	8.66
CS	5.0	1.3	2080	6.40
Q	3.9	1.5	1923	5.92

The treatment of Norway spruce wood with water repellents (wax) had a positive effect on the wetting ability. The factor k_{wa} increased from 1.0 (reference spruce) to 2.4 for wax-treated wood. A similar effect was noted for Norway spruce wood when applying an acrylic coating (PA–AC; 2.9). As a consequence, the rel. D_{Rd} of wax-treated Norway spruce (PA–NW; 3.01) was similar to that of Scots pine heartwood (PH; 2.97) or larch heartwood (LD; 3.08) (Table 3). As expected, the highest inherent durability against fungal decay was found for copper–ethanolamine-treated wood. Biocidal ingredients (copper, boron and quaternary ammonium compounds) effectively prevent decay [35,36]. The respective factor k_{inh} increased up to the defined maximum of 5.0. When the wetting ability of copper-treated wood was improved with a wax treatment (PA–CE–NW), rel. D_{Rd} reached the highest value (14.48). The positive effect of water repellents on the outdoor performance of preservative-treated wood has been shown previously by Obanda et al. [37] and Lesar et al. [38], who showed that hydrophobic treatments reduced the water uptake and limited leaching of active ingredients.

Thermal modification has a positive effect on both the inherent protective properties against fungal decay and the wetting ability. This is in line with findings from previous studies [39]. The highest improvement was observed for larch heartwood. The rel. D_{Rd} increased from 3.08 to 8.45 (Table 3). Larch was modified under fairly mild conditions, so the treatment presumably did not degrade biologically active extractives but had a positive effect on the wetting ability [40]. Thermal modification combined with a wax treatment resulted in the second highest rel. D_{Rd}, for wax-treated and thermally modified Beech (FS–TM–NW, 8.66). Apparently, wax treatment and thermal modification act synergistically. Thermal modification improves the durability and sorption properties of wood, while wax treatment improves its resistance against liquid water uptake [22].

3.2. Moisture Performance of Decking

Moisture dynamics are an essential parameter for the overall outdoor performance of wood, in addition to the inherent durability. In order to present the data more clearly, we have decided to summarize the data and present them in Table 4 and Figure 2. In addition to average and extreme data, the percentage of wet days was found to be an important parameter as well. The term "wet day" refers to days when the wood MC exceeds the predefined threshold of 25% while the temperature stays between 4 °C and 40 °C. Respective wood moisture measurements were performed in the centre of the wood samples, approximately 10 mm below the surface. It should be considered that the MC on the surface of the wood might have been even higher, as the surface is directly exposed to weathering. In addition, the authors are aware that different thresholds could be taken into account, depending on the wood sorption properties. For example, the threshold for thermally modified wood might be lower than that for beech. In general, the 25% wood moisture threshold is considered to be the minimum MC required for fungal decay on the majority of untreated woods from temperate regions. This value represents a conservative fibre saturation (FSP) value. However, it should be considered that lower threshold values are possible if fungi can transport water from a neighbouring moisture source to the wood [41,42]. In addition, it is generally accepted that fibre saturation is a range rather than a fixed threshold [43] and varies between 22% and 36% [42,43], depending on the wood species. In modified wood these values can be considerably lower.

Table 4. Measurements of moisture content (MC) of wood decking at the wooden model house unit. Calculated median and average values of all measurements, and the number and percentages of the measurements with MC equal to or higher than 25% are shown. Measurements were performed in the period between 11.4.2014 and 26.11.2018 (n = 3381).

Material	Average MC (%)	Median MC (%)	No. of meas. MC > 25%	% of meas. MC > 25%
PA	27.9	21.7	1.075	31.8%
PA–NW	20.1	17.8	838	24.8%
PA–AC	18.5	16.9	672	19.9%
PA–CE	16.9	15.9	247	7.3%
PA–CE–NW	20.1	18.9	242	7.2%
PA–TM	25.4	25.5	1.764	52.2%
PA–TM–NW	16.1	12.4	643	19.0%
PA–TM–CE	19.1	14.0	927	27.4%
PS	49.4	55.3	2.841	84.0%
PH	19.5	15.9	844	25.0%
LD	19.3	17.8	723	21.4%
LD–TM	13.3	12.0	6	0.2%
FE	14.6	14.0	94	2.8%
FE–TM	17.5	16.8	779	23.0%
FS	27.3	25.7	1.747	51.7%
FS–TM	18.9	19.5	811	24.0%
FS–TM–NW	17.7	14.0	771	22.8%
CS	17.3	15.6	608	18.0%
Q	16.6	15.8	370	10.9%

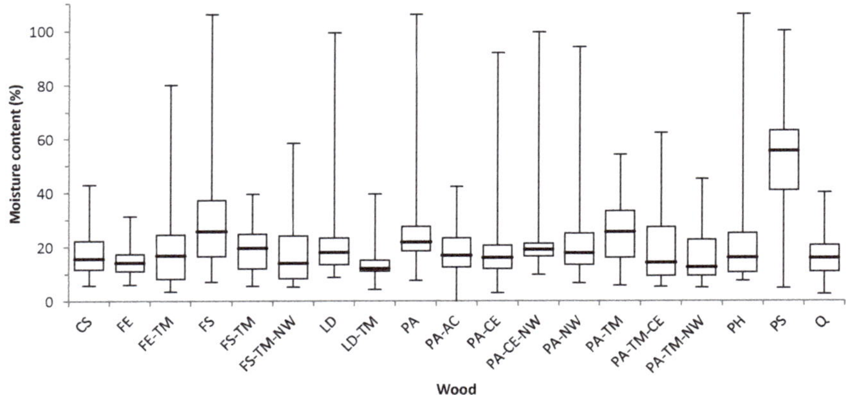

Figure 2. Distribution of moisture content in various wood-based materials exposed as decking of the model house in Ljubljana in the period between 11 April 2014 and 26 November 2018 (n = 3381).

Median values are more indicative than average values, since resistance-based measurements are fairly inaccurate at higher MC, above 50% to 60%, depending on the wood species. We will focus on median values below. The highest median value was reported for Scots pine sapwood. The median MC was 55.3%, with 84.0% of the measurements being above the threshold of 25%. The low moisture performance of Scots pine sapwood was expected and has been reported, for instance, by Žlahtič-Zupanc et al. [44]. The second-highest median MC was for beech wood decking elements. This coincides with its good permeability [4]. Surprisingly, thermally modified wood did not exhibit a good moisture performance. However, the moisture performance of freshly modified wood was fairly good (Table 3). The excellent moisture performance of freshly thermally modified wood has often been reported [39]. However, as can be seen from the data presented in Table 4, exposure under use class 3.2 [21] (above ground, exposed to weathering) conditions apparently led to an increased water uptake [44–46]. The drop in moisture performance can be ascribed to the formation of microcracks, bacterial degradation of pit membranes and blue staining [47]. The combination of thermal modification and wax treatment considerably improved the moisture performance of decking elements (Table 4, Figure 3, and Figure 4). Wax formed a hydrophobic layer on the surface that limited the penetration of liquid water into the wood [22]. Wax-treated, thermally modified Norway spruce wood thus exhibited the lowest median MC, 12.0%. A similar but less prominent effect was also observed for wood coated with acrylic coatings. Due to their anatomical features (tyloses, aspirated pits, etc.), heartwoods (PH, Q, and CS) revealed a fairly good moisture performance (Figure 2).

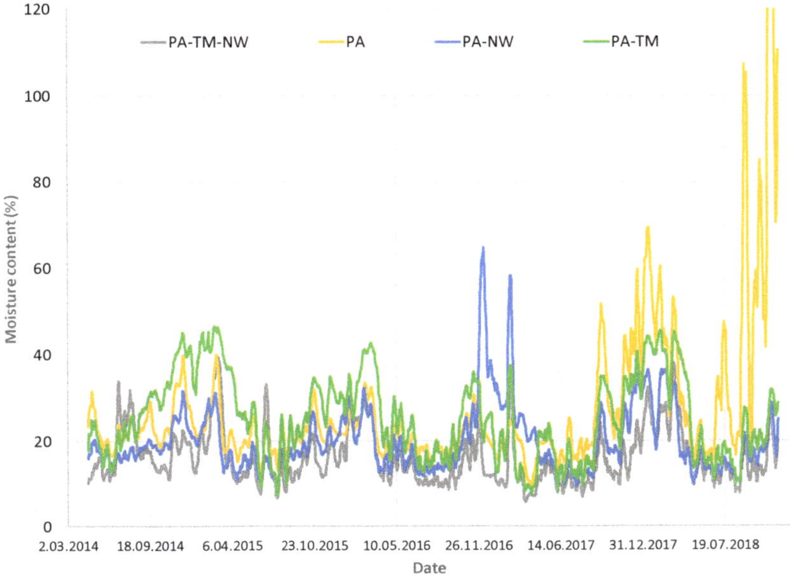

Figure 3. MC of spruce (PA), thermally modified spruce (PA–TM), wax-treated spruce (PA–NW) and wax-treated thermally modified spruce (PA–TM–NW) decking of model house in Ljubljana in the period between 11 April 2014 and 26 November 2018. Plots displayed are moving averages of 20 measurements.

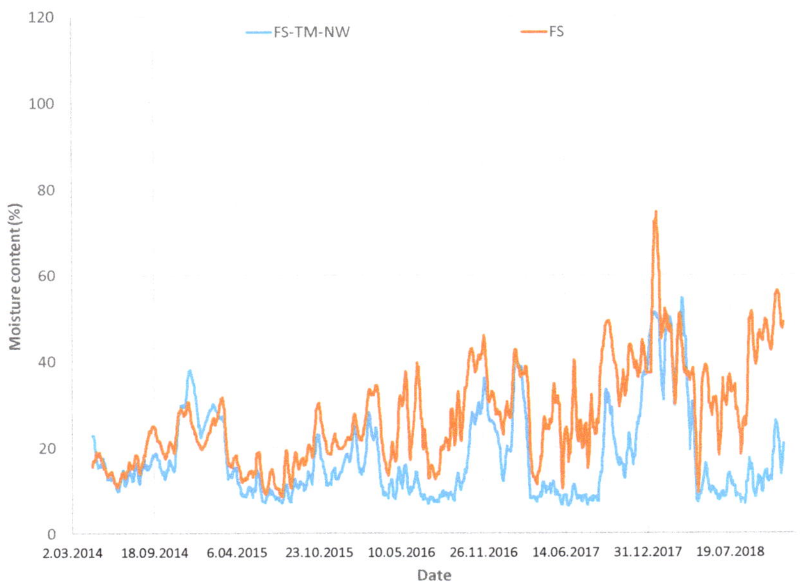

Figure 4. MC of beech (FS) and wax-treated, thermally modified beech (FS–TM–NW) decking of model house in Ljubljana in the period between 11 April 2014 and 26 November 2018. Plots displayed are moving averages of 20 measurements.

The high moisture performance of wax-treated, thermally modified wood (PA–TM–NW; FS–TM–NW) is evident from Figures 4 and 5. At almost any time, the MC of the thermally modified and wax-treated wood was significantly below that of the untreated reference. This is further evidence of the synergistic effect of wax and thermal modification. However, from Figures 3 and 4 it can be seen that the moisture performance of untreated spruce wood and untreated beech wood decreased after a certain period of exposure. We assume that the decreased moisture performance may be associated with fungal decay. Fungi open up new voids in the cell matrix, which results in better permeability [48]. One possible explanation for increased electrical conductivity (hence, increased MC) could be the consequence of fungal colonisation, due to the presence of electrolytes excreted by the fungi. However, recent results clearly indicate that the increased moisture content of decayed wood cannot be ascribed to the changed relationship between electrical resistance and MC, as reported by Brischke and coworkers [49].

Figure 5. Laser confocal image of the surface of the spruce wood coated with an acrylic coating after five years of exposure. The remaining coating on the decking element is brown, while parts where the coating was removed remain lighter. The surface was severely damaged. Field of view 5850 × 5882 μm.

In addition to the median MC, the number of measurements with MC equal to or higher than 25% is important, since it accounts for the time component (time of wetness). This value provides information about the MC of wood for which conditions are suitable for fungal decay. However, it should be noted that, although the 25% threshold might be suitable for nonmodified wood, recent studies have indicated that the minimum MC required for the decay of thermally modified wood is lower than that of untreated wood [50].

3.3. Decay Rate in the Decking of the Model House

During the first year of exposure, there was no decay to the decking of the model house in Ljubljana. In the second year, the first signs of decay developed on Norway spruce (PA), beech (FS) and Scots pine sapwood (PS). This is in line with findings from previous studies [13,51]. One of the possible reasons for the lesser decay of Scots pine sapwood could be associated with pinosylvin. This extract in pine sapwood causes a delay in spore germination [52]. Decay occurred in the third year. In addition, decay developed on coated Norway spruce (PA–AC), Scots pine heartwood (PH), Ash (FE), wax-treated Norway spruce (PA–NW) and larch (LD). In the fourth year, the first signs of decay also appeared on oak (Q). After four years of exposure, only sweet chestnut (CS), copper-treated spruce (PA–CE; PA–CE–NW) and thermally modified wood remained without visible signs of decay (Table 5). After five years of exposure, Norway spruce wood was completely degraded, followed by beech and

Scots pine sapwood. Fairly prominent decay was noted on Norway spruce coated with acrylic coatings. It must be noted that the acrylic coating was not maintained, so its initially positive effect turned negative, as previously reported by Isaksson et al. [53]. Coating limited liquid penetration in the first stages, but later on when cracks form coatings limit the drying of the wood, which enables fungal development below the acrylic coating. First, cracks formed; later, flakes of coating appeared as well (Figure 5). Brown rot fungi caused the majority of decay on softwood species, e.g., fruiting bodies of *Gloeophyllum* sp. were found. On hardwoods, white rot was more dominant. Fruiting bodies of *Trametes versicolor* were frequently found.

Table 5. Decay rating of the decking elements determined according to EN 252 [24].

Material	Average Decay Rating of the Decking Elements				
	2014	2015	2016	2017	2018
PA	0.0	1.0	2.4	3.7	4.0
PA–NW	0.0	0.0	0.2	1.1	1.9
PA–AC	0.0	0.0	0.8	1.6	2.8
PA–CE	0.0	0.0	0.0	0.0	0.0
PA–CE–NW	0.0	0.0	0.0	0.0	0.0
PA–TM	0.0	0.0	0.0	0.0	0.0
PA–TM–NW	0.0	0.0	0.0	0.0	0.0
PA–TM–CE	0.0	0.0	0.0	0.0	0.2
PS	0.0	0.6	1.2	2.2	3.1
PH	0.0	0.0	0.6	1.5	2.3
LD	0.0	0.0	0.6	1.3	1.6
LD–TM	0.0	0.0	0.0	0.0	0.1
FE	0.0	0.0	1.0	1.4	2.1
FE–TM	0.0	0.0	0.0	0.0	0.0
FS	0.0	1.0	2.2	3.1	3.7
FS–TM	0.0	0.0	0.0	0.0	0.0
FS–TM–NW	0.0	0.0	0.0	0.0	0.0
CS	0.0	0.0	0.0	0.0	0.0
Q	0.0	0.0	0.0	0.5	0.9

3.4. Modelling Decay Rates of Treated and Modified Wood

A new concept for characterizing the durability of wood-based materials and predicting the service life of wood was recently proposed by Meyer-Veltrup et al. [13,14], based on the material-inherent protective properties, the moisture performance and the climate- and design-induced exposure dose of wooden structures. This approach has been successfully applied to untreated wood [15–17] and was validated on historical objects from WWII made of spruce wood [54].

The main objective of this study was to validate the model approach of Meyer-Veltrup et al. [13], which has been developed and validated for untreated wood of numerous different species. The need to consider both the inherent protective properties and the wetting ability of a wood-based material for service life prediction becomes evident from Figure 6, in which the relationship between the total (moisture- and temperature-induced) exposure dose and the decay rate of the various decking materials is shown. The two parameters were not well correlated, because the effect of the material-inherent properties of the different materials remained unconsidered.

In contrast, the material resistance dose D_{Rd} correlated well with the decay rates of the decking, as shown in Figure 7. By also applying the effect of inherent protective properties, e.g., due to active ingredients of wood preservatives, the material resistance is represented in a more comprehensive manner compared to solely using temperature and MC data for establishing an exposure dose. Furthermore, the positive effect of thermal modification and water-repellent treatments on the outdoor performance of the examined materials is considered, as well as the most likely synergistic effects between moisture performance and inherent durability.

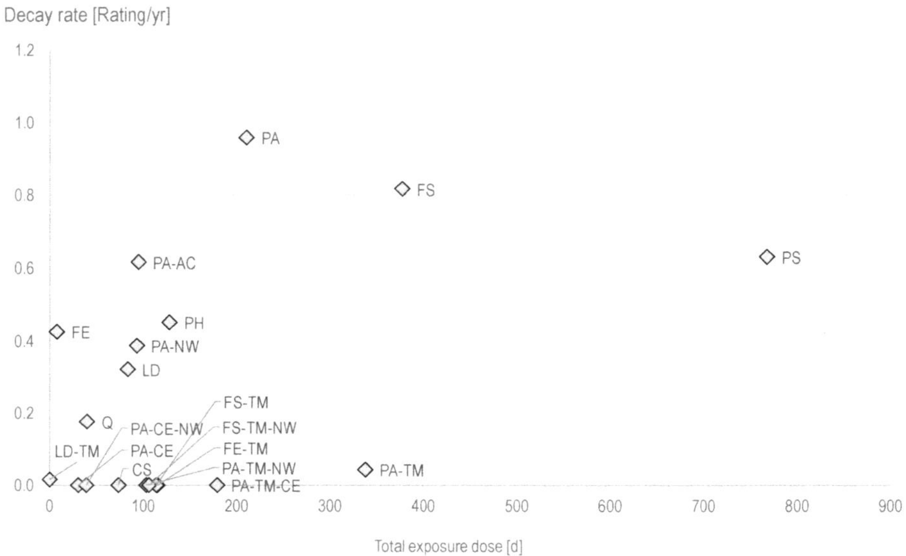

Figure 6. Mean decay rate of decking at the model house in Ljubljana versus the total exposure dose.

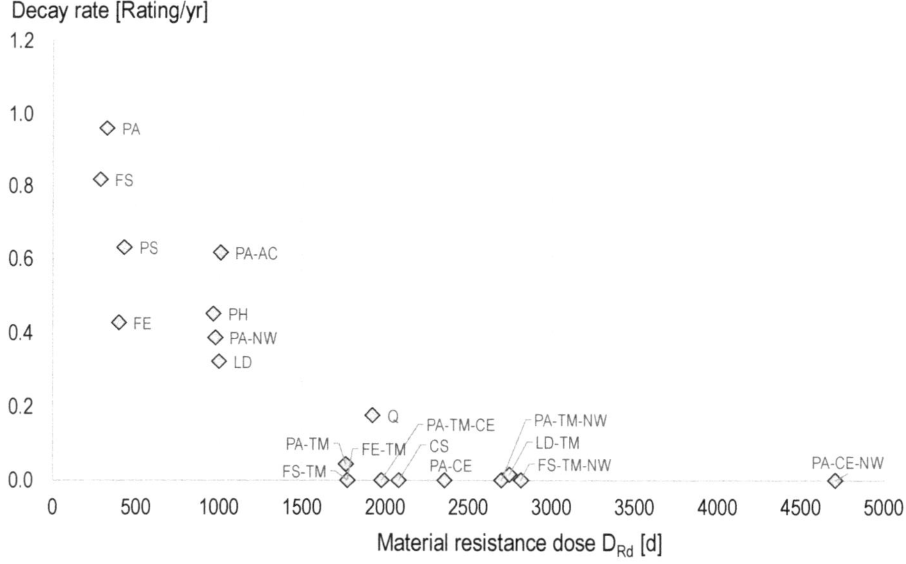

Figure 7. Mean decay rate of the decking at the model house unit in Ljubljana versus the material resistance dose D_{Rd}.

Although the preservative-treated decking, in particular, shows no decay yet, the model fits the decay rates well in general. However, to better distinguish between different highly durable materials, it might be necessary to collect further long-term data (i.e., field test data for an exposure time of several decades) of the latter.

4. Conclusions

The results clearly show that the dose D_{Rd} was well correlated with the decay rates of the decking of the model house. The model approach, taking into account the material-inherent protective properties, the moisture performance and the climate- and design-induced exposure dose of wooden structures, also proved to be accurate for assessing modified and preservative-treated wood. Furthermore, the positive effect of thermal modification and water-repellent treatments on the outdoor performance of the examined materials was evident, as well as a synergistic effect between moisture performance and inherent durability.

Since the number of long-term field tests for which corresponding lab decay and moisture dynamic tests has been performed is scarce, it might be meaningful to sample from longer-running tests for further subsequent validation of the model approach. This might also work for structures in service, with a known service life, that show the first signs of decay.

Author Contributions: Conceptualization, M.H.; methodology, M.H., C.B. and D.K.; validation, C.B.; formal analysis, M.H. and D.K.; investigation, D.K.; resources, M.H.; data curation, M.H., B.L. and D.K.; writing—original draft preparation, M.H.; writing—review and editing, B.L., D.K. and C.B.; visualization, M.H.; supervision, M.H. and B.L.; project administration, B.L. and M.H.; funding acquisition, M.H.

Funding: The authors acknowledge the support of the Slovenian Research Agency within the framework of project L4–7547, program P4–0015 and the infrastructural centre (IC LES PST 0481–09). Part of the research was also supported by the project Sustainable and innovative construction of smart buildings—TIGR4smart (C3330–16–529003) and Wood and Wood Products Over a Lifetime (WOOLF).

Conflicts of Interest: The authors declare no conflict of interest.

References

1. Bulcke, J.V.D.; De Windt, I.; Defoirdt, N.; De Smet, J.; Van Acker, J. Moisture dynamics and fungal susceptibility of plywood. *Int. Biodeterior. Biodegradation* **2011**, *65*, 708–716. [CrossRef]
2. Schmidt, O. *Wood and Tree Fungi: Biology, Damage, Protection, and Use*; Springer: Berlin, Germany, 2006.
3. Reinprecht, L. *Wood Deterioration, Protection and Maintenance*; Wiley: Hoboken, NJ, USA, 2016; p. 366.
4. FTP. The Forest–based Sector Technology Platform Vision 2040 of the European forest–based sector. Brussels. Available online: http://www.forestplatform.org (accessed on 26 August 2019).
5. *Durability of Wood and Wood–based Products—Testing and Classification of the Durability to Biological Agents of Wood and Wood–based Materials*; European Committee for Standardization: Brussels, Belgium, 2016.
6. Isaksson, T.; Thelandersson, S. Experimental investigation on the effect of detail design on wood moisture content in outdoor above ground applications. *Build. Environ.* **2013**, *59*, 239–249. [CrossRef]
7. Brischke, C.; Rapp, A.O. Dose–response relationships between wood moisture content, wood temperature and fungal decay determined for 23 European field test sites. *Wood Sci. Technol.* **2008**, *42*, 507–518. [CrossRef]
8. Van Acker, J.; De Windt, I.; Li, W.; Van den Bulcke, J. Critical parameters on moisture dynamics in relation to time of wetness as factor in service life prediction. In *Proceedings of the 44th Scientific Conference of the International Research Group on Wood Protection*; IRG/WP 14–20555; IRG Secretariat: Stockholm, Sweden, 2014.
9. Marzouk, M.; Azab, S.; Metawie, M. BIM-based approach for optimizing life cycle costs of sustainable buildings. *J. Clean. Prod.* **2018**, *188*, 217–226. [CrossRef]
10. EC (2011) Regulation (EU) No 305/2011 of the European Parliament and of the Council of 9 March 2011 laying down harmonised conditions for the marketing of construction products (CPR) and repealing Council Directive 89/106/EEC. *Off. J. Eur. Union L* **2011**, *88/5*, 5–43.
11. Kutnik, M.; Suttie, E.; Brischke, C. European standards on durability and performance of wood and wood-based products—Trends and challenges. *Wood Mater. Sci. Eng.* **2014**, *9*, 122–133. [CrossRef]
12. *Eurocode—Basis of Structural Design*; European Committee for Standardization: Brussels, Belgium, 2002.
13. Meyer-Veltrup, L.; Brischke, C.; Alfredsen, G.; Humar, M.; Flæte, P.O.; Isaksson, T.; Larsson-Brelid, P.; Westin, M.; Jermer, J. The combined effect of wetting ability and durability on outdoor performance of wood–Development and verification of a new prediction approach. *Wood Sci. Technol.* **2017**, *51*, 615–637. [CrossRef]

14. Meyer-Veltrup, L.; Brischke, C.; Niklewski, J.; Frühwald Hansson, E. Design and performance prediction of timber bridges based on a factorization approach. *Wood Mater. Sci. Eng.* **2018**, *13*, 167–173. [CrossRef]
15. De Angelis, M.; Romagnoli, M.; Vek, V.; Poljanšek, I.; Oven, P.; Thaler, N.; Lesar, B.; Kržišnik, D.; Humar, M. Chemical composition and resistance of Italian stone pine (*Pinus pinea* L.) wood against fungal decay and wetting. *Ind. Crops Prod.* **2018**, *117*, 187–196. [CrossRef]
16. Brischke, C.; Hesse, C.; Meyer-Veltrup, L.; Humar, M. Studies on the material resistance and moisture dynamics of Common juniper, English yew, Black cherry, and Rowan. *Wood Mater. Sci. Eng.* **2018**, *13*, 222–230. [CrossRef]
17. Kržišnik, D.; Lesar, B.; Thaler, N.; Humar, M. Performance of Bark Beetle Damaged Norway Spruce Wood Against Water and Fungal Decay. *Bioresource* **2018**, *13*, 3473–3486. [CrossRef]
18. Rep, G.; Pohleven, F. Wood modification—A promising method for wood preservation. *Wood Ind.* **2001**, *52*, 71–76.
19. Rep, G.; Pohleven, F.; Bučar, B. Characteristics of thermally modified wood in vacuum. In Proceedings of the 34th scientific conference of the International Research Group on Wood Protection, Stockholm, Sweden, 6–10 June 2004.
20. Humar, M.; Pohleven, F. Solution for wood preservation. EP 1791682 (B1). 3 September 2018.
21. *Durability of Wood and Wood-based Products—Use Classes: Definitions, Application to Solid Wood and Wood-based Products*; European Committee for Standardization: Brussels, Belgium, 2013.
22. Humar, M.; Kržišnik, D.; Lesar, B.; Thaler, N.; Ugovšek, A.; Zupančič, K.; Žlahtič, M. Thermal modification of wax-impregnated wood to enhance its physical, mechanical, and biological properties. *Holzforsch* **2017**, *71*, 57–64. [CrossRef]
23. Kržišnik, D.; Lesar, B.; Thaler, N.; Humar, M. Influence of Natural and Artificial Weathering on the Colour Change of Different Wood and Wood-Based Materials. *Forests* **2018**, *9*, 488.
24. *Field Test Method for Determining the Relative Protective Effectiveness of a Wood Preservative in Ground Contact*; European Committee for Standardization: Brussels, Belgium, 2015.
25. Kržišnik, D.; Lesar, B.; Thaler, N.; Planinšič, J.; Humar, M.A. Study of moisture performance of wood determined in laboratory and field trials. *Eur. J. Wood Wood Prod.* Submitted manuscript.
26. Brischke, C.; Lampen, S. Resistance based moisture content measurements on native, modified and preservative treated wood. *Eur. J. Wood Wood Prod.* **2014**, *72*, 289–292. [CrossRef]
27. Otten, K.A.; Brischke, C.; Meyer, C. Material moisture content of wood and cement mortars—Electrical resistance-based measurements in the high ohmic range. *Constr. Build. Mater.* **2017**, *153*, 640–646. [CrossRef]
28. *Durability of Wood and Wood–based Products—Determination of the Natural Durability of Solid Wood against Wood–destroying Fungi, Test Methods—Part 1: Basidiomycetes*; European Committee for Standardization: Brussels, Belgium, 2005.
29. Raspor, P.; Smole Možina, S.; Podjavoršek, J.; Pohleven, F.; Gogala, N.; Nekrep, F.V.; Hacin, J. Collection of industrial microorganisms. In *Catalogue of Cultures*; Biotechnical Faculty, University of Ljubljana: Ljubljana, Slovenia, 1995; p. 98.
30. *Thermal Insulating Products for Building Applications—Determination of Short Term Water Absorption by Partial Immersion*; European Committee for Standardization: Brussels, Belgium, 1997.
31. *Wood Preservatives—Methods for Measuring Losses of Active Ingredients and Other Preservative Ingredients from Treated Timber—Part 2: Laboratory Method for Obtaining Samples for Analysis to Measure Losses by Leaching into Water or Synthetic Sea Water*; European Committee for Standardization: Brussels, Belgium, 1994.
32. Isaksson, T.; Thelandersson, S.; Jermer, J.; Brischke, C. Beständighet för utomhusträ ovan mark. In *Guide för Utformning och Materialval (Rapport TVBK–3066)*; Lund University, Division of Structural Engineering: Lund, Sweden, 2014.
33. Isaksson, T.; Brischke, C.; Thelandersson, S. Development of decay performance models for outdoor timber structures. *Mater. Struct.* **2013**, *46*, 1209–1225. [CrossRef]
34. Brischke, C.; Meyer Veltrup, L. Modelling timber decay caused by brown rot fungi. *Mater. Struct.* **2016**, *49*, 3281–3291. [CrossRef]
35. Freeman, M.H.; McIntyre, C.R. A comprehensive review of copper–based wood preservatives with a focus on new micronized or dispersed copper systems. *For. Prod. J.* **2008**, *58*, 6–27.
36. Lebow, S.T.; Lebow, P.; Woodward, B.; Kirker, G.; Arango, R. Fifty-Year Durability Evaluation of Posts Treated with Industrial Wood Preservatives. *For. Prod. J.* **2015**, *65*, 307–313. [CrossRef]

37. Obanda, D.N.; Shupe, T.F.; Barnes, H.M. Reducing leaching of boron-based wood preservatives—A review of research. *Bioresour. Technol.* **2008**, *99*, 7312–7322. [CrossRef] [PubMed]
38. Lesar, B.; Kralj, P.; Humar, M. Montan wax improves performance of boron-based wood preservatives. *Int. Biodeterior. Biodegradation* **2009**, *63*, 306–310. [CrossRef]
39. Esteves, B.M.; Pereira, H.M. Wood modification by heat treatment: A review. *Bioresources* **2009**, *4*, 370–404.
40. Spear, M.; Binding, T.; Jenkins, D.; Nicholls, J.; Ormondroyd, G. Mild thermal modification to enhance the machinability of larch. In Proceedings of the European Conference on Wood Modification, Lisbon, Portugal, 15–19 May 2014.
41. Höpken, M. Investigations on the Growth and Moisture Transport of Indoor Wood Decay Fungi in Piled Wood Specimens. Master's Thesis, Department Biology, University Hamburg, Hamburg, Germany, June 2015.
42. Meyer, L.; Brischke, C. Fungal decay at different moisture levels of selected European-grown wood species. *Int. Biodeterior. Biodegradation* **2015**, *103*, 23–29. [CrossRef]
43. Popper, R.; Niemz, P. Wasserdampfsorptionsverhalten ausgewählter heimischer und überseeischer Holzarten. *Bauphysik* **2009**, *31*, 117–121. [CrossRef]
44. Žlahtič-Zupanc, M.; Lesar, B.; Humar, M. Changes in moisture performance of wood after weathering. *Constr. Build. Mater.* **2018**, *193*, 529–538. [CrossRef]
45. Humar, M.; Kržišnik, D.; Lesar, B.; Ugovšek, A.; Rep, G.; Šubic, B.; Thaler, N.; Žlahtič, M. Monitoring of window, door, decking and façade elements made of thermally modified spruce wood in use. In Proceedings of the 8th European Conference on Wood Modification (ECWM8), Helsinki, Finland, 26 October 2015; pp. 419–428.
46. Van Acker, J.; Van den Bulcke, J.; De Windt, I.; Colpaert, S.; Li, W. Moisture dynamics of modified Wood and the Relevance Towards Decay Resistance. In Proceedings of the 8th European Conference on Wood Modification (ECWM8), Helsinki, Finland, 26 October 2015; pp. 44–55.
47. Schwarze, F.W.; Landmesser, H.; Zgraggen, B.; Heeb, M. Permeability changes in heartwood of *Picea abies* and *Abies alba* induced by incubation with Physisporinus vitreus. *Holzforschung* **2006**, *60*, 450–454. [CrossRef]
48. Žlahtič, M.; Humar, M. Influence of artificial and natural weathering on the moisture dynamic of wood. *BioResources* **2017**, *12*, 117–142. [CrossRef]
49. Brischke, C.; Stricker, S.; Meyer-Veltrup, L.; Emmerich, L. Changes in sorption and electrical properties of wood caused by fungal decay. *Holzforschung* **2018**, *73*, 445–455. [CrossRef]
50. Meyer, L.; Brischke, C.; Treu, A.; Larsson Brelid, P. Critical moisture conditions for fungal decay of modified wood by basidiomycetes as detected by pile tests. *Holzforschung* **2016**, *70*, 331–339. [CrossRef]
51. Brischke, C.; Meyer, L.; Gry, A.; Humar, M.; Francis, L.; Flæte, P.O.; Larsson Brelid, P. Natural durability of timber exposed above ground: A survey. *Wood Ind.* **2013**, *64*, 113–129. [CrossRef]
52. Augusta, U. Untersuchung der Natürlichen Dauerhaftigkeit Wirtschaftlich Bedeutender Holzarten Bei Verschiedener Beanspruchung im Außenbereich. Ph.D. Thesis, University of Hamburg, Hamburg, Germany, June 2007.
53. Isaksson, T.; Thelandersson, S.; Jermer, J.; Brischke, C. Service life of wood in outdoor above ground applications: Engineering design guideline—Background document. Report TVBK-3067. Div. of Structural Engineering, Lund University, Sweden. 2015. Available online: http://www.kstr.lth.se (accessed on 1 August 2018).
54. Kržišnik, D.; Brischke, C.; Lesar, B.; Thaler, N.; Humar, M. Performance of wood in the Franja partisan hospital. *Wood Mater. Sci. Eng.* **2019**, *14*, 24–32. [CrossRef]

MDPI
St. Alban-Anlage 66
4052 Basel
Switzerland
Tel. +41 61 683 77 34
Fax +41 61 302 89 18
www.mdpi.com

Forests Editorial Office
E-mail: forests@mdpi.com
www.mdpi.com/journal/forests

www.ingramcontent.com/pod-product-compliance
Lightning Source LLC
LaVergne TN
LVHW070128170726
843515LV00007B/1767

* 9 7 8 3 0 3 9 3 6 3 3 2 2 *